# Jewish Eugenics

## John Glad

**Wooden Shore, L.L.C.**

Washington, D.C.　　London　　Tel Aviv

First published 2011
by Wooden Shore, L.L.C.
2601 Woodley Pl. N.W.
Suite 910
Washington, DC 20008-1567

http://www.WoodenShore.org

WoodenShore@gmail.com

Tel.: 202 667-6386

© 2011 John Glad

Library of Congress Control Number: 2010935472

ISBN: 978-0-89703-005-2 (6x9, lithocase binding, 464 pages)

ISBN: 978-0-89703-006-9 (large print, 8x11, perfect binding, 464 pages)

A-14
8/2011

Dedicated to the memory of those Jewish and non-Jewish eugenicists who were defamed and persecuted in the Western world, some of whom had earlier been driven into exile from Hitler's Germany, even as others of their colleagues perished; in homage to both eugenicists and their opponents who were victimized, imprisoned, and murdered under Communist rule; in acknowledgement of today's eugenicists – again both Jewish and non-Jewish – who continue their struggle to defend the genetic patrimony of future generations; and in respect for those scholars and scientists who may disagree with them but who share their selfless concern for the future of humanity.

*When God created the first man, he took him around to all the trees in the Garden of Eden and said to him, "See my handiwork, how beautiful and choice they are.... Be careful not to ruin and destroy my world, for if you do ruin it, there is no one to repair it after you.*

A *midrash*
MR Ecclesiastes 7:13 s.v. reKh.

# Table of Contents

# Illustrations

While the great exodus of Jews from the Russian Empire that lasted from 1880 to 1913 aroused sympathy within the already existing community of Western Jews, the way of life of the former shtetl dwellers was also a source of discomfort among prosperous 'Hebrew' Englishmen, Germans, and Americans, whose aspirations were largely assimilationist and who saw this sudden influx of poor relatives as compromising their own social positions. In 1885 a twenty-five year old Austrian lawyer and atheistic journalist who saw himself as a German and who at one point even proposed a mass baptism of the Jews, published a utopian novel with his own solution for the situation: You Have Only to Want It for It Not to Be a Fairytale. A year later he followed up with The Jewish State: An Attempt at a Modern Solution of the Jewish Question. His name was Theodore Herzl, and he is considered to be 'the father of Zionism.' His proposal was to establish a state – not necessarily Palestine – where hard agricultural work would 'cure' the new arrivals from the Pale of Settlement of their ghetto culture. Among the options advocated were Argentina, Australia, rural Canada, Mesopotamia, Uganda, and Cyrenaica (the eastern coastal region of modern-day Libya). Herzl's partner in Zionism, the popular eugenicist Max Nordau, saw Herzl's proposal as straightforward eugenics: the then popular Lamarckian belief in the heritability of acquired characteristics lent hope that the scrawny, weak, and inferior Jew – an image internalized by many Jews – would become physically strong, sexually potent, and morally fit. Envisaged as replacing both the ghetto Jews and the 'effete' coffee house Jews, this image was subsequently transmogrified into that of Zionist 'fighting Jews' who established the Jewish state by force of arms. Created by Ephraim Moses Lilien, the 'first Zionist artist,' the illustrations were intended to illustrate the ideal of Nordau's 'muscle Jew.' The writer Stefan Zweig recalled that in this "son of a poor orthodox Jewish woodturner from Drohobycz, I encountered for the first time an Eastern Jew and a Judaism which in its strength and stubborn fanaticism, had hitherto been unknown to me."

# About Writing This Book

I recall high school reading assignments in Indiana in the late 1950s on the Jukes and the Kallikaks and later taking a university anthropology course in Bloomington that dwelt on those same brachiocephalic and dolicephalic measurements that were soon to fall out of favor among younger anthropologists. I even attended a lecture by the British eugenicist Julian Huxley (1887-1975) – in the very building that housed the Kinsey Institute, which owed so much to the pioneering research of eugenicist and sexologist Magnus Hirschfeld. In graduate school I was caught up by the great ideological tidal wave that swept over academia in the wake of the Vietnam War and the civil-rights movement. On another level, however, I missed a part of it, having spent 1968 at Moscow University on the US/USSR academic exchange program. When I got back I remember a friend laughing as he described how a student had ripped open his office door and thrown a water-filled balloon at him; still another told of mounted police charging up the steps of the university library.

Like many students of history, I was torn between admiration for human achievement and dismay at seeing so many unable to appreciate the grand unfolding of culture, much less participate in it. Roughly in 1975 I became curious about eugenics. I attended a reception for members of the Genetics Department at the University of Iowa, hoping for guidance in learning more on the subject. To my surprise, the geneticists whom I questioned either lacked any knowledge of the topic or feigned ignorance when speaking in the presence of colleagues. To this day I don't know which explanation is valid. In 1979, co-chairing a department at the University of Maryland in College Park, I was summoned by the Dean and asked in a tone that must have been familiar to victims of the Inquisition about my opinions on race.

Although I had devoted my entire professional life to the defense of human rights and considered my efforts to constitute part of the struggle for the rights of future generations, my efforts had been largely focused on the international scene, and I had never written on the topic of race, nor discussed it any more than the average person. I had participated in a 'Big Brother' program, in which I took Afro-American children to museums on weekends and had supported a Taiwanese orphan, and his hostile tone and angry eyes were as surprising as they were upsetting. An ideological *coup d'état* had taken place both in popular culture and in academia since my undergraduate days, and the new rulers were ferreting out even potential dissenters. In effect, ideology was dictating the resolution of scientific questions.

So I resolved to learn about eugenics on my own – the best way to learn anything, really – and eventually wrote *Future Human Evolution: Eugenics in the Twenty-First Century*, which can be downloaded free in an ever growing number of languages at http://whatwemaybe.org. The site has been visited more than a million times, and the book may well be the most popular book ever written on eugenics.

Recognizing in Washington D.C. (where else?!) the crucial role played by Jews in intellectual life and politics, and by some Jews in the suppression of the eugenics movement, I resolved to go back to basics and create a chronology of the shifting Jewish viewpoints that have led us to where we find ourselves today. And, as the reader will see, the deeper I dug, the more I found. It is now indisputable that much of what might be termed 'accepted eugenics narrative' is in crass discordance with the historical facts.

A timeline is like a skeleton. As the pieces fall into place, the creature gradually forms and seems almost to peer back at us like a living being, gesturing toward our future from the past. And onto this scaffolding we can grow the muscles, organs, and skin of our destiny.

Writing books about Jews used to be a far easier undertaking than it is today, with Jewish anxieties over 'anti-Semitism' having been so elevated as to render dispassionate scholarly discourse nearly impossible.

Another problem is that the very definition of Jewry has become a moving target. After all, to write a book about Jews means to first come up with a working definition of who is a Jew, and that is no easy task. Formerly, Jews were considered to be the descendants of Abraham, and they believed in Judaism. Now both religion and Abrahamic lineage are off the board as generic definers.

Aside from providing a few lists of persons with patently Jewish names, I have in all other cases relied on more weighty evidence of Jewishness. Thus, to cite but one example, even though the name 'Titmuss' is indicated as a Jewish surname in the Family Tree of the Jewish People (184,237 surnames)[1], I omitted Richard Titmuss of the London School of Economics after corresponding with a colleague who had written an article about him. The surname 'Burt' is both a Jewish and a Scottish name, so the famous psychometrician Cyril Burt is discussed in this text, but not as a Jew. In such cases I chose to err on the side of caution with regard to others, leaving out ten Jewish proponents of eugenics for fear of including even one non-Jew as Jewish.

---

[1] Avotaynu, http://www.avotaynu.com/csi/csi-result.html, accessed July 5, 2008.

Second, the web that I cast inevitably had more rips and gaps than netting. Many Jews attach little significance to their Jewishness, or are reluctant to have it known, sometimes even to family members.

Third, there are people not all of whose parents or grandparents were Jewish. Where does one draw the line? Are we going to revert to such lexical monstrosities as *Mischling* and *quadroon*? I confess to being a dyed-in-the-wool universalist, and instinctively recoil from such discussions as invidious, but have taken up the topic only out of sheer necessity. Considering the influence of the Jewish community in America, the mixed attitude of Jews toward eugenics in reshaping the human genome in a number of ways (more about this later) is far too important to be ignored.

Many scholars and scientists would vehemently object to being labeled as 'Jewish' geneticists, anthropologists, historians, etc. on the grounds that the adjective is parochial and even ghettoizing. Given the massive assault on the eugenics movement as a supposedly 'anti-Semitic' ideology of genocide, however, historical veracity requires that the distorted image produced over the last four decades be rectified. The topic is not merely an important fragment in the rich and vibrant mosaic of Jewish intellectual history. Far more important, it will determine the survival of culture itself.

The immense Jewish tragedy during World War II has received its due remembrance, and we are all in debt to those who preserved and preserve a memory of the victims. But when the inevitable distortions forged over the flame of despair pose a new and even greater threat to the future of humanity, and to Jewry in particular, the situation has clearly gotten dangerously out of hand.

The famous geneticist and evolutionary biologist Theodore Dobzhansky commented: "Human evolution has forced man to a crossroad from which there is no escape.... The choice is between a twilight, cultural as well as biological, or a progressive adaptation of man's genes to his culture, and of man's culture to his genes."[1]

If we – in the most fundamental fashion – fail to understand even the recent past, not to mention the present, how can we as a species ever hope to be worthy of the terrible price paid for our genetic patrimony in the form of 'natural selection' (how deceptively banal the phrase now comes across to us) or to cope with our responsibilities to posterity? This is not to say that eugenics – including Jewish eugenics – has not

---

[1] Dobzhansky, T., *Heredity and the Nature of Man*, Harcourt, Brace and World, 1964.

been on occasion abused in the most infamous fashion, and I attempt here to objectively lay out the actual facts, regardless of whose political ox is gored in the process.

The greater part of this book is devoted to the Micro-chronology. The immediate temptation was simply to gather instances of Jewish advocacy of eugenics (some 400 are given here), but such a simplistic approach would have been repetitive and, frankly, tedious. Instead I have attempted to present this fascinating jigsaw puzzle in all its dynamism and with all its relationships as a sort of Jewish Easter egg hunt. Join me and follow a bare-bones narrative that will overturn virtually everything that you previously absorbed about eugenics. These are authentic voices being allowed to speak for themselves as they wrestle with the ultimate questions of existence.

Do not for a moment make the mistake of thinking that this is just one more recondite scholarly book devoted to the obtuse intersection of medicine and Judaism. Even now the ideological struggle documented on its pages determines, to a huge extent, the ideologies underlying contemporary politics and the even greater topic of our ongoing evolution as a biological species. There are many who do not want the facts documented here to come to light. Public discourse has been molded by political realities. In Hans Christian Andersen's tale *The Emperor's New Clothes* a kingdom's naked ruler claims to be wearing clothing that is invisible to those unfit for their positions or incompetent, and everyone is intimidated into silence – until a child exclaims: "But he isn't wearing anything at all!"

You are reading a book on a topic that supposedly not only does not exist, but one that is even inconceivable, a contradiction in terms. This misapprehension stands in gross contradiction to the grand tradition of Jewish culture and is the product of diligent propaganda manufactured by a heavily Jewish group that itself represents a small minority within the Jewish community. Frankly, their task was made easier by the fact that most people have only the vaguest notion of what eugenics is, not to mention realizing its enormous importance. The best way to expose propaganda is *total* honesty with the reader, so read the facts laid out here and judge this political catwalk for yourself.

The bulk of the Micro-Chronology consists of verbatim quotes. This is the way it was and is, without manipulation. For all its excesses, eugenics has been an astounding, indeed an existential success for Jews, molding them into a uniquely resourceful and intelligent people, and the current assault on eugenics by an understandably emotion-driven minori-

ty Jewish faction represents a frontal assault on the very essence of Jewry.

One facet of eugenics is that of cloning. Without any doubt it
will soon be possible to create future da Vincis, Beethovens, Einsteins.
Indeed, human cloning may well already be secretly investigated in Israel
today. Legal specialist Carmel Shalev of Tel Aviv University writes that
the parliamentary debate on cloning went almost unnoticed by the public
and the media, and that the fear of 'playing God' was a virtual non-issue
in Israel.[1]

It is my hope that direct access to a multiplicity of ever evolving,
cross-indexed ideologies will serve future scholars researching the topic.
Obviously, there is more than one Ph.D. dissertation to be mined here.
But even more important are those individuals faced with excruciatingly
painful, highly personal decisions regarding their future families, and
also their religious and/or secular counselors.

I would be happy to hear from readers. I can be reached at Woo
denShore@gmail.com or jglad@umd.edu. But electronic addresses
change; check the website http://whatwemaybe.org.

*

I would like to express my gratitude to Albina Tretiakova-
Birman, Michael Brin, my wife Larisa Glad, Sarah Gorman, Seymour
Itzkoff, Igor Krol, Andrew MacDonald, Gerhard Meisenberg, Oleg
Panczenko, my son Aaron Jon Glad Pearce, Don Peretz, Daniel Vining,
James Woodbury, the Leo Baeck Institute for the Study of the History
and Culture of German-Speaking Jewry, the Center for Jewish History in
New York, the Hebraica and Judaica Collection of the Melvin Gelman
Library of George Washington University, the United States Holocaust
Memorial Museum, and the Interlibrary Loan Department of the Univer-
sity of Maryland in College Park for assistance in preparing this book.

---

[1] Shalev, 2008, 334.

# The Way It Was and Still Is

## Framing the Topic

*Despite, perhaps because of, the growing threats of assimilation, intermarriage, and low birthrates, many Jews are writing about the Jewish future. Many more are thinking and speaking about it. The time is ripe for bringing our disparate ideas together in a collective enterprise devoted to devising a plan – or plans – of action to preserve the Jewish future.... There is no reason why one of the oldest continuing human civilizations cannot turn to the newest of technologies to enhance its prospects for the future. For the first time in our long history, our survival is in our own hands, and not in those of our enemies.*

Alan Dershowitz, *The Vanishing American Jew*[1]

Human ecology transcends political issues, even renders them trivial relative to the long-term survival of our species, which requires four conditions: a supply of natural resources; a clean, biodiverse environment; a population no larger than the planet can comfortably sustain on an indefinite basis, and, at a bare minimum, preservation of our genetic patrimony – what in Yiddish is known as *yichus*, defined by Manhattan Rabbi Simon Jacobson as – "'good blood,' cherished genes."[2]

Human evolution is not confined to the bailiwick of history; it is also present and future. Whereas previous human evolution occurred thanks to genetic selection via differential mortality, current selection operates via differential fertility: a U.S. Census Bureau study of 2006 data revealed that of women 40 to 44 with graduate or professional degrees, 27% were childless, compared with 18% of women who did not continue their education through high school.[3] Thus it should come as no surprise that Diaspora Jews, who constitute an exceptionally high-IQ group, are likewise not having enough children to maintain their population, and this trend is both undermining the quality of the general human gene pool and decimating Jewry in an even more devastating fashion than did the violence of World War II.

---

[1] Dershowitz, 1977, 340-341.
[2] Jacobson, 2004.
[3] Zezima, 2008.

The eugenics gospel has not gone unheeded in Israel. In a 2006 survey conducted by members of the Department of Nursing of Tel Aviv University, 16% of the respondents agreed with the statement "Cloning should be permitted for producing individuals with high IQ," and 35% believed that "cloning should be permitted for avoiding genetic diseases."[1]

Astoundingly, beginning in the late 1960s, a politically active minority within the Jewish community has enjoyed spectacular success in intimidating into abject silence the persons supporting the traditional eugenic values of Jewish society. This is an ideological split within Jewry that coincides with an identical fracture cleaving popular culture from the grand thrust of modern science. The ongoing Jewish demographic implosion is not a 'final solution' imposed by an implacable outside enemy, but one generated from within the Jewish community itself, albeit with the best of intentions. Denounced by egalitarian (anti-hereditarian) thinkers as 'racists,' 'anti-Semites,' and 'self haters,' advocates of eugenics found refuge in testing, demography, genetics, and sociobiology – where popular mythology is barred entrance by so simple a barrier as popular ignorance even of the terminology of these fields.

Taboos change over time. During the Cold War neither the West nor the Soviet Union wanted to undermine the image of their respective German allies, and Jews on either side of the Iron Curtain did not want to be seen as a fifth column undermining the common effort. I was an early participant in the early days of the Holocaust Memorial Movement, having been the chief translator of the *Black Book* (Holocaust Library), compiled by Ilya Ehrenburg and Vasilii Grossman on the horrendous slaughter of Jews in German-occupied Soviet territories. Not surprisingly, the book had been forbidden for publication in the Soviet Union. While the West had no official censorship, Jews there were also reluctant to dwell upon the topic. Moreover there was embarrassment, even contempt among some younger Jews, over the older generation's reported passivity in failing to resist their persecutors. When in the 1950s the future historian Raul Hilberg (1926-2007) insisted on writing his dissertation on the Holocaust, which later became the basis for his *Destruction of the European Jews*, the topic was still – impossible as this may seem today – proscribed by Jewish intellectuals, and his advisor at Columbia University, the Jewish-German social theorist Franz Neumann, warned him that his choice of subject might be his academic funeral. At least five

---

[1] 120 Israelis (68 health professionals and 52 non-health professionals; Barnoy/Ehrenfeld/Sharon/Tabak, 2006, 27.

publishers rejected the book, and it was finally published by a small Chicago house only after a wealthy patron agreed to buy 1,300 copies to go to libraries.[1] Now that the Holocaust topic is no longer taboo, it is eugenics that has taken its place as pariah.

Almost inevitably, whenever the topic of eugenics is raised, it is followed by the puzzled question: "Just what exactly is eugenics?" In a private poll which I conducted in Maryland in May 2009, 76% of the respondents were not even aware that it is not "a method for generating electricity widely employed in Europe," or a "General Motors hybrid car intended to compete with Toyota's Prius."[2]

So let us begin with a definition: *eugenics is a social and scientific movement that seeks to replace natural selection with scientific selection.* No biological population can remain viable without Darwinian selection, and human beings are no exception. Eugenics is all about healthy, intelligent children and parental responsibility to future generations. The basic principle is that which has been successfully applied by animal breeders for millennia: *like breeds like* – at least usually, if not always.[3]

Once the continuity of humankind with the rest of the animal kingdom was established, invigorated attempts to improve the human genome became inevitable. Eugenics is, after all, quite simply, applied human genetics. Five of the first six presidents of the American Society of Human Genetics were also members of the board of directors of the American Eugenics Society. Historically, modern genetics is an offshoot of the eugenics movement, not the reverse.

A frequent criticism of the eugenics movement is that it was a dilettantish salon culture of a privileged but amateurish aristocracy. While it is true that such an element did indeed exist, even a casual perusal of the membership lists of the (British) Eugenics Society and the American Eugenics Society is sufficient to see that their members numbered among the intellectual elite. Both lists indicate a constant stream of Ph.D.s and MDs, and the many Jews on the lists present no exception in this respect.

Although the improvement of health and intelligence is the ultimate goal of the eugenics movement, an even more persistent theme is

---

[1] Martin, 2007.

[2] Unpublished.

[3] I refer readers wishing to learn in greater detail about the eugenics movement, both historical and contemporary, to my book *Future Human Evolution: Eugenics in the Twenty-First Century*, Hermitage Publishers, 2006. Aside from the print edition, it is available in a number of languages free of charge at http://whatwemaybe.org.

how to halt genetic decline. As societies began ensuring greater equality of opportunity, to that very degree they select out young people of ability to pursue career interests over reproduction. At the same time, at the other end of the spectrum, welfare programs provide incentives to young women of low ability to regard reproduction as a greater source of income than employment. The result, eugenicists argue, is a doomed, dysgenic society (i.e., one destructive of genetic patrimony).

At its root, eugenics is an interdisciplinary conceptualization of the genetic consequences of social practices for current human and future. Applied to animals, it would not be controversial. The counterresponse was (and still is) an unspoken denial that human evolution is an ongoing process: hybridization has supposedly eliminated subspecies, so that the fundamental human genotype is now claimed to be virtually immutable, with only trivial intraspecies variation existent. Even while conceding that humankind is indeed the product of evolution, proponents of human particularism assume that human beings are the one species no longer affected by that process. Humanity, they argue, is the issue of a single African woman ('Eve'), and any subsequent or future human evolution is only 'skin deep.' Eugenicists tend to be skeptical of this view, which they regard as rooted more in wishful thinking than in objective science. Their model of human evolution is similar to that of the dog, which was bred independently in different places at different times from various subspecies of wolf. Most of that diversity is between African populations. Even if it could be proved that a human 'Eve' actually existed, 150,000 years of evolution in isolated groups living under the most diverse conditions has produced enormous inter- and intragroup diversity, which is a great resource but also a disability when it takes the form of genetic illness, low intelligence, or lack of altruism.

Human ecology does not limit itself to the present population but defines society as the entire human community over time; we should act as nature's stewards, and simple parental responsibility mandates self-restraint. Thus modern eugenics goes hand in hand with neo-Malthusian thinking, which views the current global population as already exceeding the planet's long-term carrying capacity, and is generally opposed to the view of a Julian Simon (1932-1998), who dismissed concerns regarding overpopulation, resource exhaustion, and global pollution.

*Positive eugenics* refers to approaches intended to raise fertility among the genetically advantaged. These include such genetic techniques as in vitro fertilization, egg transplants, and cloning, and also ways to encourage use of those techniques, for example, targeted demographic analyses and financial and political stimuli. Pronatalist countries (that is, those that wish to stimulate their birth rates) already engage in moderate forms of positive eugenics.

*Negative eugenics*, which is aimed at lowering fertility among the genetically disadvantaged, largely fits under the rubric of family planning and genetic counseling. This includes contraception, abortions, and sterilization. To ensure that such services are available to all on a nondiscriminatory basis, it is advocated that, at a minimum, persons with low income receive such services, free of charge.

*Genetic engineering*, which was unknown to early eugenicists, consists of active intervention in the germ line without necessarily encouraging or discouraging reproduction of advantaged or disadvantaged individuals. It will allow people to have their own biological children without passing on their most problematic genes.

National family policy provides a good illustration of how a eugenics policy might be implemented. A government can opt either to offer subsidized day care to all women, permitting those wishing to work the opportunity to pursue their careers (according to eugenicists, a praiseworthy approach), or it can subsidize only poor women, many of whom are thus encouraged to view childbearing as a source of income (according to eugenicists, a dysgenic approach).

Another example is presented by the starkly different positions of the U.S. and Canadian governments on immigration. The United States imports the underclass of other countries to 'do jobs Americans don't want to do,' while Canada, whose immigrants are easily just as ethnically diverse as are America's, rates immigration applicants according to educational levels and skills, which correlate highly with intelligence.

Simultaneous with and analogous to China's Cultural Revolution in the 1960s, an ideological upheaval arrived in America as a denial of Darwinism, declaring eugenics to be the ideology of Holocaust. As the timeline demonstrates, for Jews, who had practiced eugenics for millennia, it was a repudiation of their own history.

The squandering of a group's genetic patrimony is not by any means an exclusively Jewish affliction. Humankind's elites are generally disappearing. Economists study human fertility in terms of cost-benefit analysis. Children are no longer the economic advantage they once were when the economy centered around agriculture.

As opposed to its two universalist heresies – Christianity and Islam – traditional Judaism is an explicitly tribalist religion, but as Jews left the ghetto and were subsumed by modern secular culture, they attempted to reconcile tribalism with universalism, creating an internal tension which still rives Jewry today. The resolution of this tension was found in America in the 1960s and 1970s in the form of 'multiculturalism.'

The period witnessed a confluence of three major ideology-forming strivings for Jews: the anti-war movement, the civil rights movement, and the Holocaust Memorial Movement. Jews defended blacks in Selma and Little Rock and battled the police at the Chicago Democratic Convention, but it was the visual images of heaped-up corpses discovered in German concentration camps a quarter century earlier that most keenly triggered their protest. The result was anger, 'radicalization,' and the pursuit of poorly compatible goals. The Holocaust had been seared into their collective memory, and they were determined at all cost to avoid the role of outsiders confronted with a unified native and hostile ethnos. Thus they supported open borders, which would make them one minority group of many, even as they fiercely defended the right of the Zionist state to take precisely the opposite tack.

The arrival in Israel, beginning in the late 1960s, of a million immigrants from the Soviet Union, a majority of whom were reportedly not Jewish[1] and whose worldview had been formed by Soviet life, brought equally 'conservative' leaders to the forefront, for example Avigdor [Evet] Lieberman (b. 1958). The new consensus welded firm the inherently contradictory and previously improbable wedding of the 'right,' some of whose Zionist predecessors made no secret of their admiration for Mussolini's fascism, and the 'left' under the banner of unconditional support for the state of Israel. The phrase 'Jewish lobby' became synonymous with 'Israeli lobby,' and all the while the support of the only remaining superpower remained seamless. Not surprisingly, although Barack Obama was elected President in 2008 with the backing of 83% of American Jews, his popularity in Israel was soon in the single-digit range.[2] The decisive role of Russian immigrants in Israeli elections was easily one of the most influential political developments in the post-World War II period, but it was studiously ignored by most political commentators.

Within the scholarly world, deeply mistrustful of biological determinism, members of a 1970s radical Jewish New Left formed the Sociobiology Study Group (SSG). The historian Neil Jumonville commented that the sociobiology debate should be viewed as an inter-

---

[1] Tolts, 2003.

[2] *Washington Post*, 2010; CNN exit poll, http://www.cnn.com/ELECTION/2008/results/polls/#val=USP00p2.

generational conflict, with scholars active before the 'cultural revolution' of the 1960s usually committed to a liberal universalism, as opposed to younger scholars, who were more inclined to owe an allegiance to an ethnos-centered social vision.[1]

Sociobiology is a child of the eugenics movement, and modern cybernetics will produce the next offspring, promising to outstrip the human intellect and reduce man to the status of vehicle rather than end stop: even within the parameters of biology we will soon step beyond simple preservation and venture into improvement. There is no topic more important. Jewish religious tradition makes man a partner with God. How far can we, dare we go?

The shift from a traditional religious worldview to humanism to eugenics follows the classic sequence of Hegelian paradigm shifts: status quo → revolution → counterrevolution. Such fundamental ideological changes create competing and essentially irreconcilable worldviews: divine dictate (for example, Judaism's mandate that Jews abstain from pork, circumcise males, worship God, and observe the Sabbath); and logic-derived systems, as in utilitarian ethics, that proceed from the 'greater good' postulate, which itself is accepted *a priori* and not on the basis of any logical justification. These two systems exist in such separate dimensions that they are often mutually exclusive or, at the very least, irrelevant to each other. Modern thought attempts to find common ground and thus reconcile them, stressing commonalities and glossing over contradictions and irrelevancies. It was a less than harmonious marriage even without the advent of Darwinism, which studies man as just another animal and searches for verifiable cause-and-effect phenomena. If ethics is irrelevant to the lion eating the wildebeest, why should ethics have any relevancy to the human animal? Are we not only Darwin's children, but Nietzsche's as well – 'beyond good and evil'?

I here attempt to demonstrate that both traditionalism Judaism and the modern Jewish reconciliation of divine dictate with secular logical systems *happen* to fall into the domain of Darwinism to an unusual degree, promoting eugenic selection and co-optation of talent from without. One could also make a strong case for polygamy in Islam, whereas monogamous Christianity comes off relatively badly. The priestly celibacy of Zen Buddhism and Catholicism is decidedly dysgenic.

Over the course of the modern period an individualistic ethos has come to dominate that of the *socium*, emphasizing individual rights over duties to society. Such a 'democratic' worldview is based on a skep-

---

[1] Jumonville, 2002, 569.

ticism about the intents of the state, whose goals indeed all too often boil down to a redistribution of wealth to the advantage of those groups which are better organized at the expense of those which are not. As for communism, E. O. Wilson summed up the historical conclusion with charming succinctness: "Wonderful idea, wrong species." Not surprisingly, Wilson's specialty is ants, which are infinitely more altruistic with regard to their own community than are people.

Essentially, we humans are the invasive species *par excellence*, consuming, polluting, and overreproducing, all the while squandering our species' genetic patrimony. But gloomy as the future may appear, thinking, moral individuals (whatever 'moral' means) have no choice other than to do what they can, and while the Jews may not be above criticism, they have in many ways followed a path of social development and, mainly, genetic self-selection that can serve as a model for all of humanity.

So let's get started. We are not dealing here with a narrow, technical area, and the range of topics and disciplines could not be broader. Forget what you think you know about the subject and wait till you've finished reading the book to make your own judgment. The meat is not in my summary remarks, it's in the timeline. As Jack Webb, in the 1950s television detective series *Dragnet*, liked to phrase it: "Just the facts, Ma'am."

## To Be or Not to Be

> *I call heaven and earth to witness against you this day that I have set before you life or death, blessing or curse; choose life therefore that you and your descendants may live.*
>
> Book of Deuteronomy, 30:19

Modern society is in self-destruct mode, but biology-blind models hold sway over biological explanations. Inter-group variance, a qualitative immigration policy, Malthusian overpopulation scenarios, ideology driven policies in education and national achievement, and dysgenic fertility patterns form a soothing dreamworld of taboos forbidding even the mention of genetic differences. After all, no one likes bad news. If only for the sake of consistency, Jewish demographic discussions are shaped by this same ethos, and Jewish *Untergang* is thus treated as a non-event.

Sergio DellaPergola, Director of the Division of Jewish Demography and Statistics at the Hebrew University in Jerusalem, cautiously wondered if "the organized Jewish community was able to withstand objective scrutiny of its own trends," and pointed out the "sometime conflict of interests between researcher-sponsoring organizations and the community of professional investigators."[1]

When in 2002, United Jewish Communities and the Jewish federation system released just some of the doomsday findings of the *National Jewish Population Survey 2000-01*, the Survey's results had to be radically reworked to make them more palatable. The Survey had failed to confirm 1960s optimistic hypotheses about supposedly 'converging' Jewish fertility patterns which would wipe out or at least diminish the negative correlation between educational level and the birth rate.

According to the study, the total average number of children born to Jewish-American women aged 40-44 was 1.86, although a total fertility rate (TFR) of at least 2.1 constitutes the threshold of sustainability in a modern society. But that is not all. If one calculates in the number of children not raised as Jewish, an estimate of only 1.36 remains. This means that every generation the Jewish community effectively loses a third of its population[2] – a new Holocaust every quarter century. But this time the event is not only voluntary, it is even celebrated by Jewish liberals themselves as a triumph of 'multiculturalism.'

Other findings included an ageing Jewish population marrying at later ages with fertility rates below replacement levels. Only after the definition of who is Jewish was broadened so that the intermarriage rate, estimated at 52% in the 1990 NJPS, was lowered to 43%, was the report approved for publication. Despite the manipulation and censorship surrounding the Survey's findings and even its release, its authors optimistically, and perhaps naively, expressed hope that its themes would "serve as the basis of important policy discussions in the American Jewish community."[3]

Intermarriage also has relevance for the qualitative aspects of Jewish demography. DellaPergola has pointed out that "historically outmarriage was strongly related to upward social mobility, and was more frequent among the better-educated, wealthier, and more socially mobile." The obvious conclusion is that the mean Jewish IQ is being lowered by these losses.[4]

---

[1] DellaPergola, 2005, 123.
[2] DellaPergola, 2005, 106; citing F. Mott and J. Abma in "Contemporary Jewish Fertility: Does Region Make a Difference?" *Contemporary Jewry*, 13, 74-94
[3] Updated 2004 version.
[4] DellaPergola, 2003.

## Jewish Intelligence

In 1921 the eminent Jewish-British biologist and eugenicist Redcliffe Salaman (1874-1955) predicted before the Second International Conference of Eugenics that chances were 100 to 1 that the "little bright-eyed Jewish lad hawking newspapers in his ragged clothes" in London's East End would "better himself" if only given the chance thanks to his natural ability.[1] Although Salaman's prediction has proven remarkably accurate, his viewpoint is vehemently attacked nowadays by a veritable eugenics-bashing industry, most of it Jewish. Who are these opponents of eugenics and what motivates them? The very subtitle of Jewish-American historian Sander Gilman's 1996 book *Smart Jews: The Construction of the Image of Jewish Superior Intelligence* rejects the concept of Jewish intelligence as a 'construct,' that is, something invented and not based on reality. And even though the majority of Jews agree with Salaman and disagree with Gilman, it is Gilman's opinion that currently carries the day in the popular media.

In the third volume of *Who Are the Jews?* entitled *Fatal Gift: Jewish Intelligence and Western Civilization*, Jewish-American historian and eugenicist Seymour Itzkoff (*b.* 1928) takes issue with this point of view in general and with Gilman specifically, whom he dismisses as "a representative example of a newer kind of self-hating Jew, a denier of the objective reality of Jewish intelligence, even when it stands before us universally in evidence in the scientific and historical record":

> *Can we ever resolve the dilemmas posed by history as well as our own fragile civilizational existence if we refute fact and truth in favor of momentarily salving mythologies? Here is the essential tragedy of the Holocaust. Had Western civilization been able to proclaim the truth that Jewish accomplishment was not part of a sinister conspiracy to take over the world, here a people apart, tainted with peculiar cultural traditions, could we not have been able to stop the insanity of 'National Socialism' and the other pseudo-egalitarian crusades against human accomplishment?[2]*

The topic is thoroughly covered in Richard Lynn's *The Chosen People: A Study of Jewish Intelligence and Achievements.*[3]

---

[1] Salaman, 1921, 137.
[2] Itzkoff, 2006, 18-19.
[3] Lynn, 2010.

## Silent Holocaust

*This is the way the world ends*
*This is the way the world ends*
*This is the way the world ends*
*Not with a bang but a whimper.*
T. S. Eliot, "The Waste Land," (1922)

While there was a Jewish presence in the United States prior to 1880, the ancestors of the overwhelming majority of American Jews arrived from the Russian Empire between 1880 and 1914. From the very beginning their fertility rates were consistently lower than those of non-Jews. Soon births fell below replacement level.

| | |
|---|---|
| 1851-1923 | In Berlin Jews have lower fertility than do non-Jews.[1] |
| 1851-1962 | Italian Jews have a lower fertility rate than the total population.[2] |
| 1889 | A study of over 10,000 U.S. Jewish families reveals a Jewish birth rate lower than the non-Jewish birth rate.[3] |
| 1896-1934 | In Budapest Jews have lower fertility than do non-Jews.[4] |
| 1900-1936 | In Warsaw Jews have lower fertility than do non-Jews.[5] |
| 1910-1920 | In St. Petersburg, Jews have lower fertility than do non-Jews.[6] |
| 1900-1930 | The Jewish birth rate is lower than the general birth rate in Romania, Hungary, Prussia, Vienna, Amsterdam, and Lenin- |

---

[1] Liebman Hersch, "Jewish Population Trends in Europe," *Jewish People: Past and Present*, II, 11, Table 10, cited in Goldscheider, 1967, 200.
[2] Roberto Bachi, "The Demographic Development of Italian Jewry from the Seventeenth Century," *The Jewish Journal of Sociology*, IV, Dec., 184, Table 13; cited in Goldscheider, 1967, 200.
[3] John S. Billings, "Vital Statistics of the Jews in the United States," *Census Bulletin*, No. 19, Dec. 30, 1889, 49; cited in Goldscheider, 1967, 197.
[4] Liebman Hersch, "Jewish Population Trends in Europe," *Jewish People: Past and Present*, II, 11, Table 10, cited in Goldscheider, 1967, 200.
[5] *Ibid.*
[6] *Ibid.*

| | |
|---|---|
| | grad.[1] |
| 1926 | Canadian census data show a Jewish birth rate only 70% of the total population.[2] |
| 1931-1932 | Not only in Warsaw, but in other Polish towns Jews have lower fertility than do non-Jews.[3] |
| 1938 | In Buffalo New York, the average completed family size of professional Jews is 2.9, in contrast to 3.2 for businessmen, 3.5 for artisans, and 3.7 for peddlers.[4] |
| 1945-1947 | In Great Britain the Jewish fertility rate is 11.6 per 1,000, compared to 16.8 for the total population.[5] |
| 1948 | Jewish families seem to be relatively unaffected by the 'baby boom.'[6] |
| 1948 | In a limited survey of parents of Jewish college students, college-educated Jews are found to have smaller families than do those with only a grammar-school education.[7] |
| 1949 | Canadian data indicate an urban Jewish fertility rate lower than the non-Jewish fertility rate.[8] |
| 1951 | The average size of Jewish families in Canada decreases from 3.6 in 1941 to 3.2, as opposed to a drop of 3.9 to 3.7 for non-Jews during the same period.[9] |
| 1955 | The "Growth of American Families" study indicates an aver- |

[1] Uriah Z. Engelman, "Sources of Jewish Statistics," in Louis Finkelstein (ed.), *The Jews: Their History, Culture, and Religion*; cited in Goldscheider, 1967, 200.

[2] Mortimer Spiegelman, "The Reproduction of Jews in Canada, 1940-42," *Population Studies*, IV, Dec. 1950. 299-313; cited in Goldscheider, 1967, 199.

[3] Liebman Hersch, "Jewish Population Trends in Europe," *Jewish People: Past and Present*, II, 11, Table 10, cited in Goldscheider, 1967, 200.

[4] Uriah Z. Engelman, "A Study of Size of Families in the Jewish Population of Buffalo," *University of Buffalo Series*, XVI, Nov., 195-210; cited in Goldscheider, 1967, 203.

[5] Hannah Neustatter, "Demographic and Other Stastical Aspects of Anglo-Jewry," in Maurice Freedman (ed.), *A Minority in Britain*, 1955, 82; cited in Goldscheider, 1967, 200.

[6] Liebman Hersch, "Jewish Population Trends in Europe," *Jewish People: Past and Present*, II, 11, Table 10, cited in Goldscheider, 1967, 200.

[7] Myer Greenburg, "The Reproductive Rate of the Families of Jewish Students at the University of Maryland," *Jewish Social Studies*, X, July, 230; cited in Goldscheider, 1967, 203.

[8] Nathan Goldberg, "The Jewish Population in Canada," *Jewish People: Past and Present*, II, 35-39; cited in Goldscheider, 1967, 200.

[9] Louis Rosenberg, "The Demography of the Jewish Community in Canada," *The Jewish Journal of Sociology*, I, Dec., 1959, 217-233; cited in Goldscheider, 1967, 199.

| | age size of Jewish families of 1.7, as opposed to 2.1 for Catholics and Protestants. Furthermore, Jews expect significantly fewer children (2.4) than either Catholics (3.4) or Protestants (2.9).[1] |
|---|---|
| 1957 | Swiss Jews are shown to have a lower fertility rate than the total population.[2] |
| 1960 | The "Growth of American Families" study continues to indicate that Jews expect and desire fewer children than do either Catholics or Protestants.[3] |
| 1961 | Dutch Jews are shown to have a lower fertility rate than the total population.[4] |
| 1963 | A sample survey of the Jewish population of the Providence, Rhode Island, metropolitan area shows a clear inverse relationship between socioeconomic status and fertility among first-generation Jews, but other studies seem to indicate greater homogeneity and convergence in the fertility patterns of third-generation Jews.[5] |

So there is the eugenic argument in a nutshell: while natural selection favored intelligence during most of human history, in modern society, intelligent people – including Jews – are not having enough children even to replace their numbers, and society is in genetic decline.

Two poems – separated by millennia – come to mind: Isaiah's reference to the Jews as "a light unto the nations" and Dylan Thomas's famous *villanelle* "Rage, rage against the dying of the light." The Jews are not so much decimated by an external enemy, as being slain by their own hand. The last hour is near. Will they indeed go gentle into that good night?

## Human Particularism

We are the product of the interbreeding of a virtually endless chain of species and subspecies (including Neanderthals) and have lived

---

[1] Freedman/Whelpton/Campbell, "Differential Fertility among Native-White Couples in Indianapolis," XXI, July, 226-271; cited in Goldscheider, 1967, 199.

[2] Kurt B. Mayer, "Recent Demographic Developments in Swtzerland," *Social Research*, XXIV, Summer, 350-351; cited in Goldscheider, 1967, 200.

[3] Campbell/Whelpton/Patterson,1960.

[4] "Dutch Jewry: A Demographic Analysis," *The Jewish Journal of Sociology*, III, Dec., 195-243; cited in Goldscheider, 1967, 200.

[5] Goldscheider, 1967, 202.

in great isolation from each other under the most radically differing conditions over the 500,000 years within which we modern humans share common ancestors.[1] Nevertheless, the International Code of Zoological Nomenclature classifies human beings as *homo sapiens sapiens*, granting special status to humans as a taxonomic rank for which no subspecies exists, nor even can ever exist. According to this view, any physical or mental differences between an Australian aboriginal and an Englishman are too trivial even to be noticed by a respectable taxonomist. (Without waiting for the professionals to reassure them, Englishmen promptly interbred with aboriginals at the first opportunity, demonstrating that if by definition all members of a species can interbreed, the same is also true with regard to relations between subspecies.)

Even as the concept of human particularism rendered the word 'subspecies' unattractive with regard to people, the mighty wave of freedom that swept over the world in the last third of the twentieth century rendered 'race' unacceptable as well. 'Race,' it was declared, was still another 'social construct' that existed only as a fantasy. (Some feminists made the same claim about the differences between men and women, and evolution itself has been referred to as a "social construction."[2]) I was proofreading the manuscript of this book when I received a note from a Jewish intellectual, to whom I had written that it was "silly to have to argue that health and intelligence are better than sickness and stupidity." His response:

> *Health is an unsuccessfully chosen grouping of symptoms. There is no such thing as health, nor can one come to productive conclusions using this concept. Intelligence is not an objective thing. There are more kinds of mental and spiritual activities than your philosophy can dream of. These are hopeless words to use. They distort and lead to obsessive mental circles, they torture you. Even had there been such things as health and intelligence, neither one of them would be genetically determined. With very few exceptions one could teach anyone to be what you would call healthy and smart. We are ideologically too far apart. Too many basic axioms and even definitions are different for us. I never should have opened this discussion.[3]*

---

[1] Green, *et al, 2010.*
[2] Ruse, 1999.
[3] Anonymous at request of author.

Politics generally amounts to the horizontal struggle between organized groups and wealthy individuals in the pursuit of their real or perceived interests. By contrast, eugenics represents a vertical effort – lobbying for the genetic patrimony of future generations. Unaccustomed to protests on behalf of this as yet nonexistent group, those of us who are currently breathing generally find it more comforting to proceed from an assumption of human particularism.

Philosopher David Heyd of the Hebrew University writes of the purported "rift between the human and the natural": while animals are viewed as being instinct driven in a positivist sense, people lay claim to reason and 'free will.'[1] But once the recognition is made of the continuity of humanity with other species, it becomes more and more difficult to "characterize humans in contradistinction to other animals," to use Heyd's phraseology. That continuity in its turn rests on the recognition of causality, as opposed to intervention by deity. Based as it is on a theory of human particularism, the "software heresy" of egalitarianism must be able to stand up to the piercing gaze of scientific observation and thus is doomed to at least partial failure, but at what point exactly does the half-full glass suddenly become half-empty?

The twentieth century can be divided into thirds: the first third being one of eugenic utopian thought, the second one of reassessment, with the last third dominated by an anti-hereditarian utopianism. Eugenicists believe that since we now understand the mechanism of evolution and know that human beings are a biological species, the road to perfection is clearly laid out along the lines of scientific selection. In this sense, eugenicists are entirely accurate in their appraisal of humankind, but unrealistic in their assumption that humans are rational and altruistic enough to implement this knowledge for the good of distant 'future generations.' By contrast, even those egalitarians (anti-hereditarians) who accept Darwinism assume that evolution has produced only insignificant variance within and between human populations, and evolution has come to a grinding halt for human beings. Thus, utopia is to be found at the end of an environmentally determined rainbow. Essentially, anti-hereditarian egalitarianism is secular religion: if we have been created in the image of God, we are divine too. But ultimately the number of frail links in a chain is meaningless. It snaps whether they are one or many.

The current popular assumption is that the normal rules of animal husbandry and population management have little applicability to people. In 1968 Soviet academician Nikolai Dubinin displayed no reticence in

---

[1] Heyd, 2003.

laying out this extraordinary worldview that so appealed to Joseph Stalin and that was soon to become so popular in the West.

> *Research in man's genetics will be based on the fact that man in his development has reached a stage where he is excluded from the evolution of the animal world. Man's evolution is guided by the laws of society, by class struggle, by the development of productive forces in cooperation with superstructures such as culture and science. The process of anthropogenesis and sociogenesis went hand in hand to produce man, and after their completion a very complicated interlacement of primary social and secondary biological factors has taken place in the life of man. This essentially new situation in evolution is known to no creature on earth besides man. Purely biological features of man's development have given way to social ones, which have come to play the leading role. It is time to speak of the initiation of a new science – social biology. It is going to make progress in the future.[1]*

Although Dubinin's version of 'social biology' was diametrically opposed to Darwinism and modern sociobiology, which insist on an unbroken continuity of *homo sapiens* with the biological universe, his prediction proved prescient. Within months the *Eugenics Quarterly*, which reprinted his comments without comment, was renamed *Social Biology*, displacing eugenics with narrowly focused articles on medicine and demography that drew no controversial social conclusions. It was a scenario more than familiar in Dubinin's homeland, where the censor's chief advice to scholars was expressed in the ubiquitous slogan 'Don't generalize!' A bloodless purge had taken place in America. Even as Russia was shaking off the mythology of Lysenkoism, the West was celebrating its betrothal to Lysenko's heirs. An intellectual *coup d'état* had taken place, and many of the purge masters were Jews shoving aside other Jews.

---

[1] Dubinin, 1968, 145.

## Jewish Particularism

The philosopher and rabbi Ludwig Stein (1859-1930) eloquently summed up the dynamics of Jewish universalism and Jewish particularism.

> *Now we understand the true meaning of humanity comprehended by Lessing, Herder, and Schiller as the deepest secret of history. Ourselves an anthropological and philosophical union, we grope backward in longing for that proto-unity lost in the course of our development. Clearly the meaning of history is not human separation, partition, and disjunction, but, rather, the religious unification of the hearts, of the band of peoples in their common language, of science for the spirit, of fantasy for the arts, and lastly the grand unity of the State and the common historical events for the nation. That is why the nation-states consolidated over the course of the nineteenth century, but even these states are only the penultimate, and not the last word of history. The grand longing is for that central unity, for that humanity out of which we have all sprung, become differentiated, and gone our separate ways.[1]*

The nineteenth century had accepted race implicitly and absolutely, and Jewish leaders and thinkers were themselves enthusiastic adherents of racial theory well into the twentieth century. The civil rights movement of the last third of the new century was a great triumph for humanity, and the Jewish community can justly take pride in the indisputable fact that the Jewish used all its impressive political influence to support the rights of Afro-Americans in the 1960s and 1970s, but there was an obvious discrepancy: Jews continued in practice to define themselves in terms of biology. With the majority of Jews either atheists or 'non-religious,' what else remained? Gefilte fish and matzah ball soup? (I know, I'm going to get into big trouble for this.)

As pointed out by the anthropologist Harry L. Shapiro (1902-1990), diaspora life led to a great interbreeding experiment with other peoples, making Jews forerunners in the more and more global game of panmixia. Russia's greatest poet Alexander Pushkin (1799-1837) was part African, as is the president of the United States even as I write these lines. One can with justice regret the loss of diversity, but it is senseless

---

[1] Stein, 1905.

to resist the inevitable – illustrated by 60,000 Ethiopian Jews in Israel. Even without this latest infusion, genetic tests have shown that the present Jewish population would appear to have a total Negro admixture of the order of five to ten percent.[1] The distinctions between races are inexorably being erased. As the popular phrase runs, 'Deal with it.'

## Infiltration Theory (IT)

There have been a number of theories on the origins – both environmental and genetic – of superior Jewish intelligence, none of them entirely convincing, at least to this author.

A 'winnowing out' of persons of lower intelligence by death and/or assimilation would raise the mean IQ, but would probably reduce the absolute number of high IQs. In any case, 'marrying brains' would be a zero sum game if practiced only within one's own community.

My belief that the source of high Jewish IQ may very well be found in selective infiltration into the gene pool has led me to formulate 'infiltration theory.'

Varied as the group unquestionably is, probably a majority of Jews regard their community as an extended family, and most (but not all) animal species practice altruism between family members. Intelligent and energetic outsiders must have observed the advantages to be derived as a member of this particular clan, but Jewish cohesiveness combined with high barriers to exogamy (outbreeding) had to be overcome. Ben Wattenberg (b. 1933), a journalist who has consistently preached a gospel of America as a 'universal nation,' summed up the Jewish attitude toward infiltration: "Unlike Christians, we Jews are not missionaries. If someone wants to join the Jewish people, we're going to make it difficult for them."[2] So infiltration was not easy, but membership provided (and continues to provide) significant advantages within a social system consisting of persons inclined to individualism and, at most, immediate-family relations. A team will regularly win out over a single person. Outsiders seeking to gain membership so as to benefit from such nepotism would have to overcome major obstacles, thus self-selecting for intelligence and persistence.

If the infiltrator had himself achieved social rank and wealth (both of which correlate positively with intelligence) or was himself a 'learned man,' traditional group resistance to infiltration was lessened, and the individual managed to be accepted and 'marry in.' For example,

---

[1] Mourant *et al.*, 1978, 57.
[2] Witte, 2008, A16.

in 1879 Chicago reform rabbi Bernard Felsenthal (1822-1908) commented: "The aversion against entering into family connections is not so
strong any more as it used to be, particularly if the family is in good social and financial position."[1]

Historically, the willingness among both Jews and non-Jews to
enter the Jewish community or flee it has depended on the circumstances
of political and financial advantage/disadvantage. For example, Jews in
the Iberian Peninsula in the late fifteenth century or in Europe during the
period 1933-1945 were understandably eager to conceal their background. By contrast, the collapse of the Soviet Union induced a large but
unknown number of non-Jews to attempt infiltration so as to emigrate
and/or benefit from Jewish financial support. In 2009 the Moscow newspaper *Kommersant* reported that the phenomenon was widespread among
non-Jews in Russia, Belorussia, and Ukraine.[2] My own area of expertise
is the Russian émigré community, and I have personally witnessed a
number of such attempts. Obviously many of these infiltrators emigrated
to Israel or the United States, where there was already a considerable
population of non-Jewish spouses.

An important barrier to outsider infiltration into the genome was
the establishment of the matriline as the determinator of Jewishness.
(While there can always be doubts as to paternity, maternity is tough to
simulate.) At the same time, genetic illnesses engendered by excessive
endogamy (inbreeding) are reduced by the Halachic prohibition of inces-
tuous relationships (as well as of adulterous relationships). Thus, within
Orthodox circles there is a widely held rabbinic opinion that artificial
insemination by third-party donor is permissible as long as the donor is a
non-Jew[3], who is assumed not to be a relative either by blood or by marriage. The barrier to outsider infiltration is also a door that can be intentionally left ajar when male infertility requires it to fulfill the Biblical
imperative to 'be fruitful and multiply.'

When all is said and done, Jews have traditionally been an urban
population, and towns exist only by virtue of constant recruiting. It all
must have begun even before Abraham took his bondwoman Hagar the
Egyptian, with whom he had a son Ishmael. Moses himself married a
Cushite woman and on another occasion took as a wife Zipporah, the
daughter of a Midianite.

---

[1] Goldstein, 1997, 34.
[2] Amlinsky, 2009.
[3] See Kahn, 2006, 470.

Assimilation was always a reality for the Diaspora. In 1920 the Paris-based Russian-language newspaper *Evreiskaya tribuna* (Jewish Tribune) editorialized that the Jews were linked intimately with Russia and that "the Jews represent the sole conquest made in the western provinces, not by Russian arms, but by Russian culture."[1]

That the same was true for Germany can be seen from the following 1916 statement by philosopher Hermann Cohen (1842-1918):

> *We live with a lofty sense of German patriotism, [appreciating] the unity between the German people and the Jews, the entire former history of Jewry that has laid a path from now on as a cultural-historical truth within German politics and in the life of the German people, and which will blaze up in German folk sensibility. As Germans we want to be Jews, and as Jews we want to be Germans. Peering out to the most distant point on the horizon of the historical world, we see the Germans and the Jews as intimately bound up with one another.*[2]

The pathologist and Darwinist Otto Lubarsch (1860-1933) went even further, seeing nothing positive in his Jewish roots and declaring his life's goal had been to marry a Christian woman and become entirely German. So eager was he in pursuing this goal that he called for the German borders to be sealed off from the *Ostjuden*, whom he regarded as an alien body living off a magnificent German host nation, and even welcomed Hitler's coming to power.[3]

Inevitably, assimilation led to greater intermarriage rates with non-Jews. In 1913, physician, demographer, and eugenicist Felix Theilhaber provided statistics on mixed marriages for Berlin Jews[4], whose numbers had increased from 3,373 in 1816 to an estimated 150,000 – largely thanks to the in-migration of Eastern Jews. His data open for us an utterly fascinating window onto the past:

Theilhaber summed up 'Legitimate Jewish Births,' 'Illegitimate Jewish Births,' and half of 'Mixed Marriage Births' to produce 'Total Jewish Biological Offspring.' (Note that there are minor discrepancies for some years.) It is unclear what percentage of parents of illegitimate Jewish children were *both* Jewish, and there is undoubtedly an under-

---

[1]Cited in Lvov-Rogachevsky, *A History of Russian Jewish literature*: including B. Gorev's essay "Russian Literature and the Jews," Ann Arbor, 1979, 50.
[2] " Stimmen zur Entwicklung der deutschen Judenpolitik,"
http://www.vho.org/D/dudj/13.html, accessed Oct. 5, 2008.
[3] Lubarsch, 1931, 529-578.
[4] Theilhaber, 1913b, 71.

count factor resulting from unknown numbers of unregistered children of Jewish fathers and non-Jewish mothers. Bearing these limitations in mind, mixed marriage births can be divided by legitimate Jewish births to calculate to provide a *partial* exogamy (outbreeding) coefficient:

| Year | Legitimate.Jewish Births | Mixed-Marriage Births | Partial Exogamic Coefficient |
|------|------|------|------|
| 1875 | 1,370 | Ca. 125 | 9.1% |
| 1876 | 1,394 | " 130 | 9.3 |
| 1877 | 1,366 | " 135 | 9.8 |
| 1878 | 1,456 | " 140 | 9.6 |
| 1879 | 1,245 | 149 | 9.6 |
| 1880 | 1,345 | 144 | 11.9 |
| 1881 | 1,313 | 159 | 10.7 |
| 1882 | 1,320 | 173 | 12.1 |
| 1883 | 1,294 | 185 | 13.3 |
| 1884 | 1,368 | 177 | 14.2 |
| 1885 | 1,379 | 175 | 12.6 |
| 1886 | 1,288 | 201 | 15.6 |
| 1887 | 1,383 | 198 | 14.3 |
| 1888 | 1,370 | 214 | 15.6 |
| 1889 | 1,490 | 217 | 14.5 |
| 1890 | 1,528 | 218 | 14.2 |
| 1891 | 1,540 | 253 | 16.4 |
| 1892 | 1,587 | 212 | 13.3 |
| 1893 | 1,573 | 200 | 12.7 |
| 1894 | 1,572 | 219 | 13.9 |
| 1895 | 1,519 | 192 | 12.6 |
| 1896 | 1,438 | 204 | 14.1 |
| 1897 | 1,362 | 219 | 16.0 |
| 1898 | 1,476 | 200 | 13.5 |
| 1899 | 1,470 | 202 | 13.7 |
| 1900 | 1,455 | 237 | 16.2 |
| 1901 | 1,452 | 218 | 15.0 |
| 1902 | 1,548 | 232 | 14.9 |
| 1903 | 1,409 | 207 | 14.6 |
| 1904 | 1,458 | 218 | 14.9 |
| 1905 | 1,407 | 246 | 17.4 |
| 1906 | 1,502 | 243 | 16.1 |
| 1907 | 1,341 | 208 | 15.5 |
| 1908 | 1,270 | 230 | 18.1 |
| 1909 | 1,198 | 214 | 17.8 |
| 1910 | 1,108 | 198 | 17.8 |
| 1911 | 1,095 | 205 | 18.7 |

Expressed as a scatter chart, these unique data show a doubling of the known exogamy rate over a span of 36 years, even though it was very significant even in the beginning – approaching 10%:

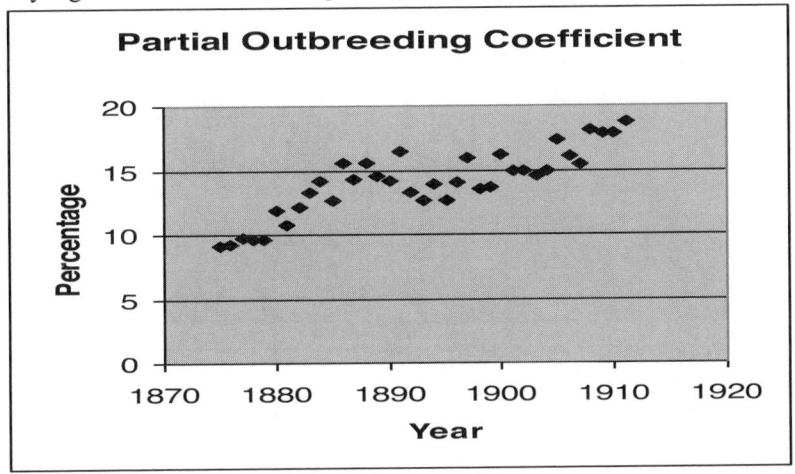

Theilhaber also supplies statistics for Berlin:

## Jewish Population in Berlin

| Year | Legit. Births | Illegit. Births | Mixed-Marriage Births | Jewish Marriages | Mixed Marriages | Total Biological Jewish Offspring | Total Jewish Marriages Persons |
|------|------|------|------|------|------|------|------|
| 1875 | 1,370 | 67 | ca. 125 | 289 | 134 | 1,470 | 712 |
| 1876 | 1,394 | 64 | ca. 130 | 282 | 141 | 1,490 | 705 |
| 1877 | 1,366 | 77 | ca. 135 | 317 | 130 | 1,485 | 764 |
| 1878 | 1,456 | 74 | " 140 | 250 | 152 | 1,566 | 652 |
| 1879 | 1,245 | 78 | 149 | 259 | 162 | 1,401 | 680 |
| 1880 | 1,345 | 75 | 144 | 311 | 158 | 1,492 | 780 |
| 1881 | 1,313 | 83 | 159 | 329 | 168 | 1,495 | 826 |
| 1882 | 1,320 | 85 | 173 | 343 | 212 | 1,481 | 898 |
| 1883 | 1,294 | 93 | 185 | 353 | 159 | 1,479 | 865 |
| 1884 | 1,368 | 77 | 177 | 379 | 180 | 1,535 | 938 |
| 1885 | 1,379 | 70 | 175 | 400 | 185 | 1,530 | 985 |
| 1886 | 1,288 | 71 | 201 | 424 | 146 | 1,459 | 994 |
| 1887 | 1,383 | 82 | 198 | 400 | 167 | 1,565 | 967 |
| 1888 | 1,370 | 68 | 214 | 464 | 166 | 1,545 | 1,094 |
| 1889 | 1,490 | 79 | 217 | 534 | 140 | 1,677 | 1,208 |
| 1890 | 1,528 | 90 | 218 | 544 | 166 | 1,727 | 1,254 |
| 1891 | 1,540 | 74 | 253 | 523 | 144 | 1,740 | 1,190 |
| 1892 | 1,587 | 79 | 212 | 578 | 157 | 1,772 | 1,213 |

| Year | Legit. Births | Ille- git. Births | Mixed- Mar- riage Births | Jewish Mar- riages | Mixed Mar- riages | Total Biological Jewish Offspring | Total Jewish Marriages Persons |
|------|------|------|------|------|------|------|------|
| 1893 | 1,573 | 62 | 200 | 573 | 164 | 1,735 | 1,210 |
| 1894 | 1,572 | 78 | 219 | 525 | 147 | 1,760 | 1,197 |
| 1895 | 1,519 | 79 | 192 | 555 | 169 | 1,694 | 1,279 |
| 1896 | 1,438 | 95 | 204 | 539 | 201 | 1,635 | 1,279 |
| 1897 | 1,362 | 101 | 219 | 644 | 200 | 1,672 | 1,488 |
| 1898 | 1,476 | 90 | 200 | 585 | 204 | 1,666 | 1,374 |
| 1899 | 1,470 | 90 | 202 | 621 | 209 | 1,659 | 1,471 |
| 1900 | 1,455 | 76 | 237 | 594 | 221 | 1,649 | 1,409 |
| 1901 | 1,452 | 97 | 218 | 620 | 201 | 1,658 | 1,441 |
| 1902 | 1,548 | 101 | 232 | 615 | 202 | 1,757 | 1,435 |
| 1903 | 1,409 | 85 | 207 | 597 | 212 | 1,597 | 1,406 |
| 1904 | 1,458 | 85 | 218 | 629 | 246 | 1,662 | 1,504 |
| 1905 | 1,407 | 100 | 246 | 624 | 285 | 1,630 | 1,533 |
| 1906 | 1,502 | 122 | 243 | 627 | 270 | 1,743 | 1,524 |
| 1907 | 1,341 | 111 | 208 | 638 | 286 | 1,556 | 1,562 |
| 1908 | 1,270 | 125 | 230 | 585 | 281 | 1500 | 1,451 |
| 1909 | 1,198 | 104 | 214 | 565 | 303 | 1,409 | 1,433 |
| 1910 | 1,108 | 99 | 198 | 577 | 275 | 1,306 | 1,429 |
| 1911 | 1,095 | 119 | 205 | 570 | 298 | 1,302 | 1,438 |

As can be seen, between 1875 and 1911 the number of mixed marriages rose from 134 to 298, while the number of biologically Jewish children fell from 1470 to 1302.

The U.S. 1957 Current Population Survey conducted by the Bureau of the Census showed a Jewish intermarriage rate of 7.2%,[1] but within a half century that figure had increased arguably to over 50%. The new level inevitably leads to conclusions that proponents of Jewish genetic continuity find utterly dismaying. Even if we hypothesize an obviously impossible population in which the original generation consists exclusively of clones of ancient Hebrews, a 50% intermarriage rate would mean that within just three generations only 12.5% of the uniquely Jewish genes would have survived. We are, after all, not speaking here of ponderous, drawn-out demographic processes, but of grandchildren.

Even within Israel the ongoing infiltration rate is extraordinarily high. Statistician Mark Tolts of the Division of Jewish Demography and Statistics at the Avraham Harman Institute of Contemporary Jewry, Hebrew University of Jerusalem in Israel, provides the following data:

---

[1] Goldstein/Goldscheider, 1966, 386.

**Percentage of Jews among Migrants to Israel
from the Russian Federation and the Entire FSU, 1990-2002**

| Year | Russian Federation | | Entire FSU |
|---|---|---|---|
| | Goskomstat of Russia data[a] | Israel CBS data[b] | Israel CBS data[b] |
| 1990 | -- | -- | 96 |
| 1991 | -- | -- | 91 |
| 1992 | 64[c] | 82 | 84 |
| 1993 | 60 | 82 | 83 |
| 1994 | 58 | 77 | 77 |
| 1995 | 53 | 73 | 72 |
| 1996 | 49 | 67 | 68 |
| 1997 | 36 | 60 | 60 |
| 1998 | 31 | 55 | 54 |
| 1999 | 31 | 51 | 50 |
| 2000 | 27 | 47 | 47 |
| 2001 | 25 | 45 | 44 |
| 2002 | 23.5 | 43 | 43 |

(a) Of all emigrants to Israel whose ethnicity was known.
(b) Of the immigrants who entered to Israel according to the Law of Return whose ethnicity/religion was known.
(c) Second half of the year.[1]

The mirror-image of IT is exfiltration theory (ET). For example, a 2008 study of the population of the current Iberian Peninsula population revealed an estimated 20% mean proportion of ancestry with Sephardic Jews whose ancestors had been expelled in 1492-1496 or who remained but still identified themselves as Jews.[2] Jonathan Ray, a professor of Jewish studies at Georgetown University, comments: "If four generations on I have no knowledge of my genetic past, how does that affect my understanding of my own religious association?"[3] The expulsions were a short-term political event, but is it possible that in less stressful times there was lessened group resistance to exogamous marriage in cases of perceived low social value?

---

[1] Tolts, 2003; citing Goskomstat of Russia data; Israel CBS data.
[2] Adams *et al.*, 2008.
[3] Wade, 2008.

## A Reassessment of Vocabulary

*"When I use a word", Humpty Dumpty said in a rather*
*scornful tone, "it means just what I choose it to mean –*
*neither more nor less." "The question is," said Alice,*
*"whether you can make words mean so many different*
*things." "The question is," said Humpty Dumpty,*
*"which is to be master – that's all."[1]*

Absurd terminology is a reliable barometer of muddled and emotional thinking, and any serious reexamination of Jewish eugenics requires a reconstitution of vocabulary. The word 'anti-Semite,' for example, still hangs on as an anachronistic coinage of late nineteenth-century racial taxonomy, referring not only to Jews, but also to Akkadians, Canaanites, Phoenicians, and Arabs. Semites were supposedly the offspring of Shem, one of Noah's three sons. Historical maps printed at the turn of the nineteenth and twentieth centuries show the southern shores of the Mediterranean as settled by 'Semites' while areas north of the Mediterranean are indicated as inherited by 'Aryans.' The mythology is no longer taken seriously, although in all fairness it must be conceded that the Semitic languages are recognized as a distinct linguistic family.

Thus, even though the concept 'Semite' has been debunked as mythology, 'anti-Semites' continue to populate the writings of even serious scholars. One could with equal success employ the term 'anti-unicornism.' Still more quixotically, the 'Semite' component of 'anti-Semite' does not refer to 'Semites,' but only to Jews. Lastly, at least some Ashkenazi roots reach back, not to Palestine, but to Khazaria, so that calling the Jews 'Semites' is nothing less than nonsense all around.

'Anti-Semitic' is now universally used to mean 'anti-Jewish,' but its lack of etymological coherence becomes especially evident, not in the adjective, but in the noun. If one substitutes 'Semite' with 'Jew,' the result is the hilarious neologism 'Anti-Jew.' And what is the proper terminology when the 'anti-Semite' is, say, an Arab, that is, a 'Semite'? A 'self-hater'?

The continuing usage of the term 'anti-Semite' is all the more paradoxical in that it was made popular and may even have been coined by the German writer Wilhelm Marr (1819-1904) as a racial concept scientifically justifying a hostile attitude toward Jews (a posture that did

---

[1] Lewis Carroll, *Through the Looking Glass and What Alice Found There*, NY, 1993, 124.

not prevent him from marrying four different women, one of whom was Jewish and two half-Jewish. There was even a claim that Marr himself was half-Jewish[1]). In 1879 Marr wrote *Jewry's Victory over the Germanic Peoples.*[2] He also founded the "League of Anti-Semites," which advocated the deportation of Jews from Germany. Marr has been dead for a century, but his nonsensical neologism continues to trade as coin of the realm.

The reason the word has persevered for so long is that its vagueness conflates hostility toward Jews with criticism of Jews, presenting a huge barrier to objective analysis. Thus, vocabulary predetermines not only the categories in which we think, but even the range of permitted topics. Even though the term is a patent absurdity, it has a long history behind it that will allow it to cling to existence for at least a while longer.

The corruption of language has engendered a corruption of thought. One cannot help sympathizing with the historian Yael Hashiloni-Dolev, who as a loyal Israeli finds herself immersed in the intense emotions generated by the Holocaust-from-eugenics claim, but at the same time opts to be the bearer of a classic the-king-is-naked message. Accordingly, she writes that she has opted to largely eschew 'eugenics' as an emotionally laden 'buzzword,' replacing it with 'reproductive genetics,' even though the latter phraseology is far broader than the former: virtually all eugenics does indeed fit under the reproductive-genetics rubric, but reproductive genetics encompasses far more than eugenics. For that matter, 'reproductive genetics' is a redundancy; after all, what is genetics all about? Feebly, she finally concedes that "while writing about the past, the term 'eugenics' is obviously more appropriate."[3]

Actually, the code word for eugenics has now been chosen. A search that troubled many a scholar has produced a startlingly simple result... 'genetics.' Even a brief perusal of *Ha'aretz*, *Forward*, or *Jewish Week* produces glowing recommendations of genetic procedures that are pure eugenics.

When, however, the eugenic thrust of human genetics is pointed out, such phraseology as 'slippery slope' and 'troubling' immediately appears – but only as cautionary notes, not as insurmountable barriers. In this fashion eugenics is actively pursued and decried simultaneously.

---

[1] "War Wilhelm Marr ein Jude?" *Weltkampf*, 1944, No. 2, ff94; cited in Alex Bein and Harry Zohn, *The Jewish Question.*
[2] *Der Sieg des Judenthums über das Germanentum*, Gustav von Linden, Leipzig.
[3] Hashiloni-Dolev, 2006.

As for the historic term 'racial hygiene,' mention is avoided of its original reference to the human race in its entirety, thus creating the impression that it referred to individual races – a confusion that was admittedly abetted by proponents of racial hostility.

In the ears of those people who are at least familiar with the word 'eugenics,' it is frequently associated with 'racism," a term which conflates the legitimate study of biological interracial variance with pathological hatred for other races. How did such an association come about?

First of all, early eugenicists bear part of the blame. Privilege is inevitably drawn to self-justification, like iron to a magnet. To 'racism' in the meaning of prejudice, hostility, and denigration can be added 'classism.' Much of that which was passed off as eugenics was dilettantish, haughty, and frankly inexcusable. Just as kings formerly justified their authority as deriving from God, the upper classes patted themselves on the back at so richly deserving their accumulated wealth and power.

It was the late 1960s that married – in the public mind – eugenics to racial hostility. Outraged over prejudice and hostility, Jews sought out fellow victims for sympathy and support, and Afro-Americans had certainly been victimized, by any standards. 'Negro' civil-rights organizations were Jewish-funded and even Jewish-staffed. Over and over the media repeated the word 'racism' in conjunction with 'anti-Semitism.' Even as their parents fled the cities, young Jewish radicals marched in Alabama and Little Rock. It was really quite idealistic.

But there was an understandable psychological need to understand what motivated the foe. And that slot was filled when the phraseology was expanded to 'racism, anti-Semitism, and... eugenics.' No matter that the majority of eugenicists had publicly decried such hostilities and prejudices.

Ultimately, as we now see, the common front of Afro-Americans and Jews proved to be a *mésalliance*. Incredibly, the Zionist lobby found its true supporters in an altogether different underclass – the religious fundamentalists, who not only reject Darwinism, but also long for Armageddon, when those Jews who do not convert to Christianity will perish during the Second Coming. In the meantime, most Americans not only have forgotten, and probably never knew, the meaning of the word 'eugenics,' but some are aware of negative associations.

But if you think 'anti-Semitism,' 'racism,' and 'eugenics' are problem words, let's look at 'Jew.'

## Are Jews Jews?

The political implications of this, at first blush, perplexing question are enormous: Israel's Declaration of Independence begins with: "Eretz-Israel was the birthplace of the Jewish people." When Israel's newly elected Prime Minister Menachem Begin presented his first cabinet to Parliament in 1977, he reiterated this fundamental position: "We were granted the right to exist by the God of our ancestors.... The Jewish people bears a historic and eternal right to the land of Israel, the heritage of our ancestors. This right is unappealable."[1] And while Jewish atheists are reluctant to accept this celestial decree as the basis of their right to Palestine, most of them take for granted that they are of the only slightly diluted stock of Jews who lived 2 ½ millennia ago. But such an assumption is in blatant contradiction with the findings of science: gynecologists Susan Klugman and Susan J. Gross write that, although "contemporary Jews share several chromosome markers and polymorphisms as well as genetic mutations... there is no such thing as a Jewish genome and Jews are more likely to share sequences with fellow non-Jews than with each other."[2]

Even if one denies the existence of any statute of limitations regarding the migration of Jews to Palestine/Israel, a lone cultural continuity without an underlying genetic continuity contradicts Zionist ideology. I have received several letters from Israeli colleagues complaining of pressure placed on them, some of it by high-placed officials, not to question official ideology. Of all the classic "Yes-but-is-it-good-for-Israel" topics, this one ranks near the top of the censorship hierarchy.

Francis Galton's close colleague Joseph Jacobs, President of the Jewish Historical Society, in 1899 formulated the fundamental question of Jewry – even today – in the title of an article in Popular Science Monthly: "Are Jews Jews?" The mythmaking that is the basis of any 'communal identity' is impossible to comprehend without first defining the meaning of the word 'Jew.' (Not only have I purloined the title of this subchapter from Jacobs, but I now also raid the lexicological pantry of the social historian Sander Gilman).[3]

Considering the centrality of Jewish identity to the group psyche, it is curious that precisely this identity has been repeatedly and radically redefined.

---

[1] June 6; cited in Filc, 2010, 22.
[2] Klugman/Gross, 2010, 37-38.
[3] Gilman, 1986, 2.

The traditional definition of Jewishness assumed implicitly that:

Jews = Judaism = matrilineal descent

That is, all Jews believed in Judaism, and all believers in Judaism were Jews.

Inevitably, the very existence of Jewish converts to other religions necessitated a religious reexamination of this syllogism, so that in the eyes of at least some in the religious community both secular Jews and persons who could not claim matrilineal descent were not Jews, thus replacing the second equal sign in the formula with a plus sign:

Jews = Judaism + matrilineal descent

But the loss of Jews to agnosticism, atheism, and the Christian heresy was not the sole taxonomic issue. There remained the thorny problem of converts to Judaism. Even though their reception by the group often ranged from straight-forward hostility to polite but tight-lipped restraint, their very existence required a redefinition of Judaism itself that was not to the liking of many:

Jews = Judaism

The Union of Orthodox Rabbis of the United States and Canada (Agudath Harabonim) has issued a statement declaring that "Reform and Conservative are not Judaism at all. Their adherents are Jews, according to the Jewish Law, but their religion is not Judaism."[1] Thus, at least in the opinion of this group, which does not represent all Orthodox Jews, religion does not define Jewishness:

Jews ≠ Judaism

The National Jewish Population Survey goes even further, defining as Jewish "a person... whose religion is Jewish and something else":

Jews = Judaism + "something else"

Science, on the other hand, was advancing its own view of nature and, thus, of Jews as a part of that nature. When Carl Linnaeus

---

[1] http://truejews.org/Igud_Historic_Declaration.htm, accessed Dec. 22, 2008.

(1707-1778) created the binomial system of taxonomy and Darwin demonstrated the mechanism by which species and subspecies evolved, human subspecies were defined as races, and Jews were classified as members of the "Semitic race" – a linguistic category. Quickly, however, the vigor of Jewish particularism imposed a definition of Jewry as a separate subspecies or race:

Jews = matrilineal descent = race

But this definition too proved unacceptable. If intermarriage had always existed, it became rampant in many countries of Western Europe in the nineteenth century. In America today a majority of Jews marry non-Jews, and many Jewish parents do not want their sons and daughters to be dismissed as mamsers. If the children of Jewish fathers and gentile mothers are ejected from the fold, the decline becomes even more catastrophic. In order to accommodate those persons who wish to be considered Jewish, American Reform Jews resolved in 1983 to accept patrilineal descent, necessitating a new formula:

Jews = matrilineal descent = patrilineal descent

Since this formulation obviously failed to reconcile the religious and biological definitions, a still broader definition was proposed – that a Jew is anyone who either "adheres to Judaism or is of Jewish lineage."[1] Thus

Jews = matrilineal descent=patrilineal descent=Judaism[2]

Here too the definition remains inadequate – both with regard to biology and religion.

On the biological front we see a huge range of physical types among persons who consider themselves Jewish. One minor example of many: in 1886 the German pathologist Rudolf Virchow surveyed over

---

[1] Jewish Virtual Library, http://www.jewishvirtuallibrary.org/jsource/glossJ.html, accessed June 26, 2008.

[2] For those of a more mathematical bent:

If B $\vee$ R, then J, or B $\vee$ R => J, or

If B(A) $\vee$ R(A), then J(A)=true (A is a person, B(A)=true if the person is Jewish by birth, etc.).

B(A) $\vee$ R(A) => J(A)

(" $\vee$ " is the "disjunction sign," and "=>" is the "implication arrow.")

ten thousand German children and determined that eleven percent of the Jewish children were blond, as opposed to 31 percent of the German.[1] In 1891 the Polish historian J. Krzywicki commented: "Despite their fanatical self-isolation the Jews have much non-Semitic blood in them. In India, this side of the Ganges, they are black; in England they are blue-eyed and blond; in Russia's western guberniyas they have broad faces and Slavic noses."[2] Jewish anthropologists supported this conclusion – over and over – and it is as disconcerting to have to refute this belief as it would be to be forced to once more disprove the theory of Martian canals constructed by intelligent beings. The claim that today's Jews have substantially retained the genetic heritage of ancient Jews is myth, seductive and powerful as myth often is, but nonetheless myth. Jews have not retained, at least not to any significant degree, the Abrahamic genotype (whatever it may have been). Ancient Jews were themselves a disparate amalgam of Sumerians, Hurrians, Hittites, "Sea Peoples," Hamites, Akkadians, and Amorites, among others. As genetic entitities, all those peoples long ago interbred themselves out of existence, and the Jews are no exception; only their cultural tradition has been preserved, albeit not, of course, in its entirety. Harry L. Shapiro, Chairman of the Department of Anthropology in New York's American Museum of Natural History and President of the American Eugenics Society, pointed out that the ancient Jews were but one tribe of many in the Mesopotamian region:

> *Since there is no reason to believe that the founders of the Hebrew people were a distinctive group in their homeland – linguistically, religiously, or culturally – it would place too great a strain on probability to assume they were, in any significant way, genetically or racially differentiated from the general population to be found there. Any such assumption would, indeed, demand circumstances not known to have existed there. Experience among better known people quite clearly demonstrates that where barriers of culture, religion, or language do not exist, even genetically distinct groups living in one area tend to interbreed, thus maintaining a gene flow that would lead to eventual amalgamation.[3]*

Even were we to make the utterly impossible assumption that ancient Jews were all each other's clones and that their descendants ex-

---

[1] Virchow, 1885.

[2] Krzywicki, J. 1891. *Lyudi.* Warsaw. Quoted in Judt, 1903, 14.

[3] Shapiro, 1960, 26.

perienced a mean rate of endogamy (inbreeding) of 97% a generation over the course of 2½ millennia of diaspora (e.g., 100 generations), more than 95% of that patrimony has still been lost irretrievably.[1]

## Modern Jews ≠ Ancient Jews

The agonizing over a loss of genetic commonality clearly comes too late for the barn door to be closed: the horse escaped two millennia ago.

Much understandable publicity has been attracted by attempts to trace genealogy via genetic testing, including the identification of certain haplotypes found with heightened frequency (but not exclusivity) among bearers of the various permutations of the name Cohen-Kogan.[2] For that matter, every human being carries genetic sequences that were inherited from viruses whose sole function in the human genome appears to be self-replication. This search for statistical associations amounts to fishing in genetic waters that are far murkier than the layman has been led to believe.

Female slaves obviously have only limited freedom or none whatsoever in choosing sexual partners – a prerogative no doubt exercised by the Israelites when they began to conquer the Canaanites in the thirteenth century BCE. In turn Assyria conquered Israel in 722 BCE, and Babylon repeated the conquest in 587 BCE, the two empires making off with the majority of the population as slaves. Only in 538 BCE did Cyrus permit them to rejoin those who had managed to remain behind. During the following centuries (the period of the Second Temple), the Jews were ruled successively by the Persians, the Greeks, and the Egyptians. In 70 CE the Romans put down a revolt and took many Jews as slaves to Rome. The same happened after the revolt of Bar Kochba in 132-135 CE. In the seventh or possibly eighth century the Khazars, a semi-nomadic Turkic people, converted to Judaism, providing perhaps most of the ancestors of Ashkenazic Jews, who suffer from different genetic illnesses and have different genetic markers than do Sephardic Jews. Throughout the Diaspora, sometimes openly, sometimes under

---

[1] The mathematics is straightforward: PC = previous genetic commonality, CC = current genetic carryover, 2,594 years ≈ 104 generations:

$$CC = PC(1-.01)^{104} = .3516$$
$$CC = PC(1-.02)^{104} = .1223$$
$$CC = PC(1-.03)^{104} = .0421$$

[2] See: "Jewish Genetics: Abstracts and Summaries,"
http://www.khazaria.com/genetics/abstracts-jews.html.

conditions of stealth, Jews and non-Jews produced children, the above-mentioned dark Jews of India being just one example of many. As a result, Jewry is like the ax whose head has been replaced three times and the handle eight times. In an interview granted to *Ha'aretz* the eminent Israeli geneticist Raphael Falk wrote that it is impossible to define who is a Jew and that "apart from the Jews who seek to do so, those who wish to define Jews biologically are the anti-Semites and the Nazis."[1]

To put the matter in perspective, even the claims of modern Greeks and Egyptians to be the direct, undiluted descendants of ancient Greeks and Egyptians are of dubious validity, although we are dealing here with peoples who inhabited the same geographical space over millennia. Other periods also introduced massive exogamy among the Jews, for example, the great forced conversions of Jews to Christianity in Spain are reflected in the genes of 20% of the population of the Iberian peninsula.[2]

The prominent anthropologist Maurice Fishberg was hostile to the idea – popular among Zionists – that the Jews were a distinct, non-European race which had preserved itself in its original purity in spite of the Jews' wanderings all over the globe,[3] and physical anthropologists had indeed already disproven this theory in the late nineteenth century. Within-group physical similarities exist, not so much because they have been inherited from ancient Jews but, rather, thanks to strong (but not total) endogamy. Thus:

$$\text{Jews} \neq \text{race}$$

Jacobs may have perceived "no evidence of any... large admixture of alien elements in the race since its dispersion from Palestine," but the answer to his 1899 question "Are Jews Jews?" was already as clear to scientists then as it remains today:

$$\text{modern Jews} \neq \text{ancient Jews}$$

By 1911 leery Jewish leaders had asked Fishberg to debunk the heretofore popular racial definition of Jewishness, leaving the Jewish community wondering just who they were. During the interwar period

---

[1] Karpel, 2006.
[2] Wade 2008.
[3] Hart, 1999, 295; citing *The Jews: A Study of Race and Environment*, New York, 470.

the term 'ethnicity' came into usage. In the words of historian Eric L. Goldstein, the term found acceptance as "emotionally satisfying and politically safe."[1] In equating the difference between Jew and gentile as comparable to, say, that between Germans and Englishmen, the new usage tacitly cast overboard religion as well as genes as defining elements.

$$Jews \neq race \neq Judaism = ?$$

In addition to scientific objections, the broad definition leaves unbridgeable differences of religious opinion, with not only one end of the spectrum denying the legitimacy and sometimes even the Jewishness of the other, but even the middle denouncing both ends. Indeed it could not be otherwise; while Jews generally attempt to present a unified face to the gentile world, they are ideologically a very heterogenious group, causing Ben-Gurion himself to produce the absolutely broadest definition of Jewishness:

$$Jews = \text{"anyone who says he is"}$$

Some African tribes desiring to emigrate took him at his word and declared themselves to be Jewish.

The waters become even more muddied. The astonishingly prolific Gilman promotes his book *Jewish Self-Hatred* by placing on the back cover a quote from the London Review of Books: "Jews are not the invention of the anti-semite – or vice versa. It is a reciprocal relationship: each invents the other,"[2] thus producing two mutually exclusive definitions:

$$Jews \neq \text{the invention of the anti-Semite}$$

$$Jews = \text{the invention of the anti-Semite}$$

Or perhaps:

$$Jews \leftrightarrow \text{the invention of the anti-Semite}$$

---

[1] Goldstein, 1997, 54-55; private correspondence from Goldstein to John Glad.
[2] Gilman, 1986, back jacket of paper back edition.

So we return to the eternal question: 'Who then are the Jews?' The problem with all these definitions is that they are barking up the wrong tree. The essence of Jewry lies not in any purported genetic preservationism, but rather in its eugenic dynamism. By maintaining high barriers to genetic interlopers, but not excluding them entirely, a constant influx of high-quality genes was achieved, producing superior intelligence. The Jewish-British geneticist and eugenicist Redcliffe Salaman formulated the most accurate definition: an endogamous family[1]:

Jews = members of a cultural and breeding alliance

Still another added twist is the concept of Jewishness as a cancelable condition, as when the London newspaper Jewish Chronicle in a 1930 editorial referred to Leon Trotsky as a "former Jew" on the grounds that he had abandoned all the religious and cultural accoutrements of Jewish culture. (Trotsky himself referred to himself as a Russian.) Much the same concept was presented by the philosopher Jean-Paul Sartre (1905-1980) in dismissing Jewish disparagers of Jewry as "inauthentic Jews,"[2] and more recently by the Israeli demographer Sergio DellaPergola in referring to persons who are "currently Jewish."[3] One American Zionist (a self-contradictory phrase, some would argue) reader of the manuscript of this book commented about Karl Marx that it was insufficient to be born Jewish, one had also to "qualify."

All this begs the question of *who were the ancient Jews?* Israeli 'post-Zionism' rejects such fundamental categories of the national narrative as 'the Jewish people,' 'the ancestral land,' 'exile,' 'diaspora,' 'aliyah,' and 'Eretz Israel.' The 'minimalist' school of archeologists promotes the view that the Bible is mythology and that no self-respecting scientist or scholar will accept anything that is not proven by actual archeological digs. According to Rabbi David Wexler, President of the American Jewish University, it is likely that the Genesis story may well have originated in Mesopotamia with its story of the flood that grew out of the periodic overflowing of the Tigris and Euphrates Rivers.[4] Lee I. Levine, a professor at the Hebrew University of Jerusalem, writes, "There is no reference in Egyptian sources to Israel's sojourn in that country, and the evidence that does exist is negligible and indirect." Le-

---

[1] Endelman, 2004, 82.
[2] Sartre, 1948, 109 (original French text in 1946).
[3] Feb. 19, cited in Gilbert, 2007, 42; DellaPergola, 2003.
[4] Lieber/Harlow, 2001.

vine also wrote that excavations showed there had been no walls at Jericho.[1]

## Darwinism

The controversial aspect of eugenics is that its proponents advocate the application of Darwinian theory to humanity today, and not just to its pre-historic roots. Scientists acknowledge the continuity of *homo sapiens* with the rest of the animal and plant kingdom, and bioethicist Peter Singer and philosopher Paolo Cavalieri go so far as to view apes as part of our "community of equals."[2] Nor is the effectiveness of artificial selection in doubt. Since the beginning of human history, people have not hesitated to selectively breed non-human species. Thus Galton launched the eugenics movement quite straightforwardly as applied Darwinism.

In point of fact, the hostility toward eugenics is tantamount to a rejection of Darwinism, which posits selection – natural or artificial – as its core teaching. Indeed, were it not for influence of social elites, popular sentiment would mandate the teaching of creationism in U.S. schools even today. The dilemma for egalitarians is how to discriminate between Darwinism and eugenics, accepting the former and rejecting the latter. It is an impossible task, leaving egalitarians with the sole option of glossing over the inherent contradictions in their own worldview. The egalitarian position was still tenable up into the 1930s, when Lamarckian tradition held that an improved environment would improve the genes (Darwin himself had doubts), but science has now totally disproven this view.

The German biologist August Weismann (1834-1924), who argued for a continuity of 'germ plasm,' dealt Lamarckism a deadly blow as early as 1889, when he cut off the tails of mice for generation after generation without affecting subsequent births, leading the writer Isaac Asimov to quip that Weismann could have saved himself the trouble by considering that after many generations of circumcision, Jewish males display no reduction in their foreskin at birth. Nevertheless, although Lamarckian tradition has been disproven, its egalitarian tail is still very much alive and twitching.

A popular misconception is a conflation of eugenics with Social Darwinism that, in the words of economist Thomas C. Leonard, "has become canonical."[3] In point of fact, the two stand in fundamental opposition to each other; Social Darwinism favored natural selection, meaning

---

[1] Hevesi, 2008, 34.
[2] McNeil, 2008.
[3] Leonard, 2005, 230.

that the weak should be left to die, while eugenics explicitly proposes replacing natural selection with artificial selection. For example, eugenicist and president of both Indiana and Stanford Universities David Starr Jordan (1851-1931) argued in 1915 that war impoverishes the 'breed,'[1] as opposed to Herbert Spencer's (1820-1903) 'survival of the fittest.'

Another Social Darwinist, Ludwig Gumplowicz (1838-1909), was the son of a Kraków rabbi and a later convert to Christianity.[2] In his book *Racial Struggle*, this professor of law at the University of Graz and influential Zionist theoretician wrote that a stronger people enslaves a weaker people and takes over its land.[3] The book was widely reprinted and translated into other languages. Sadly, Gumplowicz's analysis of the cause of war is hard to overturn. How else can the overwhelming propensity of our species for armed conflict be explained? What is, at the very least, debatable is that he considered attempts to structure society along principles contrary to human nature to be delusional, and thus the very existence of such a militaristic propensity constitutes its own justification.

Fundamentally, Social Darwinism posited that human beings were animals like any other, and in this regard its heritage has been retained by modern sociobiology. But if this is the case, in the words of Dostoevsky, "all is permitted" and might makes right. Herein lies the cruel dilemma of modern secular thinkers: the inherent lack of morality on the part of science. Three decades after Gumplowicz's death the German government was headed by another Social Darwinist who promptly began to convert theory into practice.

## Ethical Implications of Darwinism

How to reconcile social and individual interests and their often intractable conflicts of interest? The phrase 'human rights' is now understood as the rights of the individual over the collective – as long as these rights do not impinge on the perceived interests of *contemporary* group members capable of resistance. By contrast, in cases of intergenerational conflict of interest, people of the future by definition cannot be our contemporaries and thus are incapable even of protest, not to mention resistance. When a person suffering from a heritable genetic illness or low intelligence wishes to have children, the currently accepted view is

---

[1] Jordan, 1915.
[2] Doron, 1980, 398.
[3] Gumplowicz, 1883.

that he/she has a 'right' to pass on these traits to untold millions of his or her potential posterity.

Environmental experts and wildlife population managers accept as axiomatic the priority of the species over its individual members. Both the individual wildebeest and its eternal foe the lion perish if they are slow runners, and this cruelty is accepted as the price necessarily paid for the process that created us all – evolution. We have managed to extract ourselves from this horror, raising the question as to whether humankind should now be classified as a self-bred, artificial species comparable, say, to the Chihuahua.

The Judaic view of creation is pointedly anthropocentric:

> *So God created man in his own image, in the image of God he created him; male and female he created them. God blessed them and said to them, "Be fruitful and increase in number; fill the earth and subdue it. Rule over the fish of the sea and the birds of the air and over every living creature that moves on the ground." Then God said, "I give you every seed-bearing plant on the face of the whole earth and every tree that has fruit with seed in it. They will be yours for food."*

Genesis 1, 1 – 31

There are two assertions here: a) that our species is so perfect – even divine – that no further improvement or evolution is possible and b) that the world exists as an object of human consumption. Contrast this worldview with that of other religions:

**Buddhism:**
*He who experiences the unity of life sees his own Self in all beings, and all beings in his own Self, and looks on everything with an impartial eye.... To live a pure unselfish life, one must count nothing as one's own in the midst of abundance.*

Buddha

**Hinduism:**
*As the air is everywhere, Flowing around a pot And filling it, So God is everywhere, filling all things And flowing through them forever.*

Ashtavakra Gita 1: 18-20

**Zoroastrian Tradition:**
*That nature alone is good which refrains from doing unto another whatsoever is not good for itself.*

*Avesta, Dadistan-i-dinik 94:5*

But all these religions have in common a view of behavior as a 'moral' choice determined by 'free will,' as opposed to modern science, which proposes an entirely different conception, summed up by biochemist and evolutionary geneticist Gerhard Meisenberg, who sees humanity as "lumbering robots and digital (or possibly analog) computers, and poorly constructed ones to boot."[1]

It is obviously difficult to maintain a continuity of ethical tradition when a former religious ethos is rejected by many in favor of humanism, which in its turn is displaced by value-neutral science. Both Jewry and eugenics are based on communal spirit as a core value, and the current dualistic Jewish attitude toward eugenics demonstrates just how difficult and perhaps even impossible such a shifting process really is.

## The Anti-Darwinian Rebellion

Revolted by Thomas Hobbes's (1588-1679) analysis of human history as a 'war of all against all,' Jewish idealists reached out for a common humanity, proclaiming a social ideal that was more than idealistic; it was even romantic. Three Jewish intellectuals – Karl Marx, Sigmund Freud, and the anthropologist Franz Boas – posited a model of human behavior shaped by environment rather than genes. It was a software rebellion against the dictatorship foisted upon us by our aggressive genetic hardware. We could, it was hoped, be reprogrammed if only the environment in which we found ourselves could be properly restructured, and these new characteristics could somehow be passed on to future generations by yet to be discovered biological mechanisms. Influential as this software heresy was, it eventually crashed against an even more massive iceberg than Jewish self-definition. That iceberg was science.

It cannot be overemphasized that the political assault on a 'biological paradigm' of human nature was not an exclusively Jewish enter-

---

[1] Meisenberg, 2007, vii-ix.

prise, but Jews still play a disproportionate role in the action, as can be seen from the signatories of a 1979 collective letter, published in the *New York Review of Books,* attacking E. O. Wilson's book *On Human Nature.*[1]

The authors of the complaint maintained that crucial flaws undermine the entire structure of sociobiology; they termed it 'pseudo-scientific,' claimed it is intended to justify the *status quo,* and dismissed it as sexist, spurious, and outdated.[2] If the attackers were at pains to make their complaints comprehensible to the man in the street, Wilson's professional terminology is replete with such opaque terminology as 'inclusive fitness theory' and 'eusociality.' And he confines his writings largely to insects, and sometimes mole-rats, venturing very cautiously and rarely into discussions of the human animal. The result is that the general public really does not even suspect the political implications of sociobiology. In point of fact, today's sociobiology or 'evolutionary psychology' is more replete with political significance than even the protesters suspected back in 1979.

In a 2007 article on the explosive topic of 'group selection,' E. O. Wilson collaborated with the biologist and anthropologist David Sloan Wilson (b. 1949) to author an article that appeared in the *Quarterly Review of Biology.* The authors based their conclusions on studies of 12,000 known ant species, but toward the end they took up a timid gauntlet and touched upon human evolution as well.

Sociobiology is the study of social behavior from a biological perspective. The two Wilsons proceeded from this point of view to lay out the basics of 'multilevel selection theory,' in which 'group-advantageous' behaviors based on within-group altruism permit the altruistic group to out-compete groups consisting of selfish individuals, each pursuing his own goals. Nowhere in the article did they touch upon

---

[1] Joseph Alper, professor of chemistry, University of Massachusetts, Boston; Jonathan Beckwith, professor of microbiology and molecular genetics, Harvard Medical School; Bertram Bruce, scientist, artificial intelligence, Bolt, Beranek, and Newman, Inc.; Robin Crompton, graduate student, bioanthropology, Harvard University; Val Dusek, professor of philosophy, University of New Hampshire; Edward Egelman, graduate student, biophysics, Brandeis University; Stephen Jay Gould, professor in the Museum of Comparative Zoology, Harvard University; Ruth Hubbard, professor of biology, Harvard University; Hiroshi Inouye, research fellow, Harvard Medical School; Robert Lange, professor of physics, Brandeis University; Lila Leibowitz, professor of anthropology, Northeastern University; Richard Lewontin, professor of biology, Harvard University; Freda Salzman, professor of physics, University of Massachusetts, Boston

[2] Alper *et al.,* 1979.

Jewish topics, but the potential applicability to Jewry is clear. The authors concluded that while selfishness beats altruism within single groups, altruistic groups beat selfish groups. In their analysis population structures can be spatial or based on kin recognition, but kinship is supposedly the consequence and not the cause of success.[1]

The 1960s' infancy of sociobiology coincided with the civil rights movement, so that it is no surprise that group differences and also between-group competition and selection with their overtones of nineteenth century Social Darwinism were largely taboo topics at the time, as was the conceptualization of human societies as a 'super-organisms.' Even today, the two Wilsons point out, professional researchers bold enough to pursue the venue of 'evolutionary psychology' can easily see their careers ruined.

## Judaism and Science

Science is all about causality; even the nature/nurture debate amounts to no more than estimating the relative contributions to human nature of two interacting determinisms – environmental and genetic. Free will and 'morality,' whose very existence is posited on a rejection of causality, are not part of either paradigm. Protest over this model of the universe lies at the core of Existentialism, which insists that individuals create the meaning and essence of their own lives, rather than having them assigned to them by either science or religion. An old French saying runs: "tout comprendre, c'est tout pardonner." Forgive the Holocaust? This was the essence of Hannah Arendt's intellectual rebellion in writing in *The Banality of Evil* that "causality... is an altogether alien and falsifying category in the historical sciences." And it goes a long way toward explaining Richard Lewontin's and Steven Jay Gould's frontal assault on Wilson's *Sociobiology*. But when Leon Kass searched for an argument against the paradigm of genetic determinism, he came up empty-handed and was forced to fall back on name calling: for him the paradigm elicited only "repugnance." Kass's reaction has been referred to as the "yuk factor."[2]

Throughout history, man has been torn between the desire to achieve immortality via a deity and the realization that if that deity really is omniscient and omnipotent, how is it that this truly good entity permits evil? The Church's response is that man is incapable of comprehending Divine Wisdom. St. Augustine (354-430) told a parable of a man seeing

---

[1] Wilson/Wilson, 2007.

[2] Midgley, 2000.

a child on the seashore attempting to pour the entire ocean into a hole he had scooped in the sand. When the man observed that this was impossible, the boy replied that neither could the man understand the mystery of the Trinity... and disappeared.[1] Such a parable would be alien to the Judaic tradition, in which the deity is not almighty, but is more like a parent who sends his children off into the world and then relinquishes control. This rabbinic image of humans as partners with God is far more common in the world's religions than is the monotheism of Christianity, possibly inherited in part from Akhenaton, who even in Egypt did not manage to make the one-God model stick. In the majority of the world's religions gods and men interact, and often can even reverse roles.

Science is the usurper of God. The human brain uses written speech to create a collective brain extending backward in time to benefit from the entire experience of the species and extending horizontally throughout all areas of knowledge. The God of Judaism may be wholly good and a sole deity to boot, but his authority is limited... and thus, by implication, contestable; Judaism is inherently more compatible with science than is Christianity.

Not all Judaic authorities are comfortable with this worldview. Rabbi Lawrence Troster in the journal *Conservative Judaism* warns that genetic interventionism could go awry, committing "cross-generational retribution" – a power to be exercised by God alone. He cautions against "the sin of despotism over our descendants" and "the sin of complacency" in automatically accepting new technologies, and that man can exercise his freedom "in true partnership with God."[2] Having established his 'partnership,' Troster then concedes that "there are genetic illnesses that no one would argue should be cured if possible... substituting human manipulation for natural selection." It is an ambiguous position that once more illustrates the illusoriness of the phrase 'Judeo-Christian.'

The paradigm of 'genetic determinism' or 'reductionism' sets the stage for still another conflict with religion: if man was created from the interbreeding of non-human species, science goes still further, maintaining that evolution has not ceased. Man can advance or regress within the limits of his species or flow into a new species, or perhaps even a mechanical one created by man, which event would leave biology abandoned in the evolution of thought and in the pursuit of a usurpial "God-building."

---

[1] Arroyo, 2007.
[2] Troster, 2002, 40.

## 'Jewish Studies'

In the nineteenth century, scientific study of and scholarly speculation about Jewry were largely a gentile undertaking revolving around physical anthropology. By the turn of the century, however, cultural anthropologists, led by Franz Boas, many of whose pupils were Jewish (albeit by no means all, Margaret Mead [1901-1978] and Ruth Benedict [1887-1948], for example), had launched a coup that took over a half-century to truly come into its own. A parallel situation developed in psychiatry under the Freudians.

A large proportion of these anthropologists and psychiatrists not only spoke German as their mother tongue, but were assimilationists who, in the words of anthropologist Gelya Frank, "deemphasized the Jewishness of Jews who contributed to mainstream institutions as the price of social inclusion under the universal values of secular humanism."[1] Startling as it may be in today's climate, German Jews in America were generally eager to present themselves as German and in the days of the Weimar Republic were decidedly pro-German in their general orientation.

Israeli historian Joachim Doron (b. 1923) writes that "the juxtaposition of 'Aryan' racial ideologues and Jewish intellectuals stirs emotional resistance that renders a free discussion of this problem virtually impossible."[2] In a different essay Doron is even more explicit: "The Zionist 'self-criticism' that necessarily attended the longing for a 'new Jew' has been forgotten or suppressed over the last generation.... It cannot be denied that the Jewish self-criticism so widespread among the German Zionist intelligentsia often seemed dangerously similar to the plaints of the German anti-Semites. The Zionists were keenly aware of this problem but they were not deterred by it."[3]

Today Jewish studies have become an almost exclusively Jewish intellectual ghetto, thus rendering irrelevant the ambiguity of the phrase 'Jewish Studies.' In practice the discipline is enforced in the universities as 'studies by Jews about Jews.' There is an obvious conflict of interests when *all* the researchers in any field themselves constitute the field of study. Hence the glossing over Jewish eugenics on the part of Jewish researchers whose worldview, honest as many of them may have tried to

---

[1] Frank, 1997, 731.
[2] Doron, 1980, 389.
[3] Doron, 1983, 170, 171.

be, has nevertheless been shaped to various degrees by emotional and political considerations.

## Zionism

Discrepancies between popular opinion and reality will always be with us. The French thinker Jacques Ellul (1912-1994) pointed out that, paradoxically, intellectuals are the social group most vulnerable to propaganda because they possess more information and thus are not only accustomed to dealing with unverified claims but also feel themselves capable of forming opinions on virtually everything.[1]

Jewish attitudes toward Zionism had originally been overwhelmingly negative. There was a consensus that it was a utopian fantasy that would, moreover, lead to ill will toward Jews. Up to the founding of Israel in 1948, many Jews still regarded Zionism with a jaundiced eye. Even when the Jewish state was established, support was still weak and ambivalent.

On January 19, 1902, the *New York Times* reprinted without comment an article under the title "The Evil of Zionism," taken from the Cincinnati newspaper, *The American Israelite*, which in turn quoted an article penned by the editor of London's *Jewish Chronicle*:

> *As Mr. J[oseph] H[iam] Levy* [British author and economist, 1838 – 1913] *has well said in this connection: "That Great Britain would long tolerate the unlimited inflow of a population proclaiming their intention to remain aliens to the furthest generation, and sneering at anglicization as a wretched shibboleth," is not to be thought of by sane politicians.... The Zionists... must understand that their ostentatious proclamation of a Jewish nationality that cannot be content with anything else but a Jewish state is merely playing into the hands of the enemies of their race. It is a confirmation of the contention that English citizenship has been conferred on a number of people who can never be Englishmen, and Jews may wake up one day to find that while Zionism has failed to hew out a separate Jewish nationality, it has destroyed that which years of laborious work have achieved in free countries like England.... This is the position which The Israelite has assumed from the very beginning of this pernicious agitation. Motives should always be considered, but wise men have ever held a fool to be more dangerous than a*

---

[1] Ellul, 1967.

*deliberate evildoer, especially to those whom he seeks to serve.*
*We therefore firmly believe that Jewish Zionism has in the few*
*years of its existence done more harm to Israel than has Chris-*
*tian anti-Semitism.... Zionism and anti-Semitism are the twin*
*enemies of the Jews, and the former is the potentially more dan-*
*gerous.*[1]

Two world wars failed to wipe out such attitudes. On November
16, 1945, New York Rabbi Arthur Lelyfeld complained that an unrepre-
sentative clique of anti-Zionist Jews had been spreading "the falsehood
that Jews are divided on the question of Palestine." Lelyfeld charged that
such persons were "the unwitting stooges of the old-line British imperial-
ists" and that "they have played into the hands of America's leading anti-
Semites, who have embarrassingly praised them as 'good Jews.'"[2]

In a spectacularly successful public-relations campaign, the Holocaust Memorial Movement crushed anti-Zionist moods within the traditionally liberal humanist Jewish community, causing it to embrace such
radical rightist Israeli political leaders as Menachem Begin (1913-1992),
Ariel Sharon (b. 1928), and Binyamin Netanyahu (b. 1948). In the words
of the American historian Eran Kaplan, today's Likud party is "the modern political incarnation" of Jabotinsky's (1880-1940) Revisionism,
which presented Italian fascism as a model for the Jewish state: "Revisionism was, first and foremost, an attack on modernity; it was an attempt to revise the course of Jewish history and release it from the hands
of the champions of such ideals as progress, rationality, and universal
rights."[3] The support of Western Jewish liberals for a government of the
extreme right represents a striking instance of the primacy of family ties
over ideology. The impetus for this corporate model was provided by the
influx into Israel of Eastern European Jews, themselves the products of
the Soviet totalitarian state. Herzl conceived Zionism as a conveniently
distant place to cure the *shtetl* culture that he so despised, but now those
Eastern Europeans have received *carte blanche* from their doting Western cousins.

Inevitably, the consequences of World War II for European Jewry created an intellectual need to understand the motivation of the per-
petrators of Jewish deaths during the war. Largely ignored prior to the
Arab-Israel war of 1967, the Holocaust is now cited as justification for
the creation of Israel. When eugenics was proposed as the worldview

---

[1] "The Evil of Zionism," *New York Times*, pg. 28.
[2] Gordon, 1945.
[3] Kaplan, 2005, 159, 177.

underlying Hitler's motivation, and eugenicists were described as *Schreibtischtäter* (bureaucrat criminals), everything seemed to fall in place. As for emotions, rather than taper off over time, they actually intensified, driven by the massive efforts of the Holocaust Memorial Movement, which in its turn is driven by a concern for the survival of the Jewish state. American scholar of Judaism Jacob Neusner (b. 1932) wrote: "you have to remember that the State of Israel came into being because of the Holocaust...."[1] The Jewish eugenicist Arthur Ruppin was even more frank in 1923, when he commented: "Were it not for the Jews' racial affinity with the peoples of the Near East, it would not be possible to justify Zionism."[2] Geneticist Raphael Falk of the Hebrew University in Jerusalem was equally blunt, pointing out that from the late 19th century on, the Zionists defined Jews in a biological sense with no connection necessarily to religion or culture.[3]

But the task of history is to portray the past accurately, not to hold up a distorted mirror in the hope of influencing the future.

The bulk of Jewry consists of Ashkenazis who arrived in North America and Europe from the tsarist pale of settlement and quickly overwhelmed their Western brethren by sheer virtue of numbers. The United States had accepted millions of them from the Russian Empire, actually starting slightly before the assassination of Alexander II in 1881 up to the beginning of submarine warfare in 1914.

Burdened by centuries of discrimination, they were generally perceived as backward, sickly, bereft of culture, and prone to crime. Quickly the new arrivals made the dismaying discovery that even German Jews, including German Jews in America, had been pursuing a far more assimilationist strategy than they, and that these 'Germans' not only were uncomfortable with their 'Russian' cousins, but were even inclined to consider the 'uncouth' habits of the new arrivals in terms of deterministic explanations that presented 'inherited' characteristics as immutable.

In England the *Jewish Chronicle* commented in 1880 that the new arrivals "have the Russian habit of living in dirt, and of not being offended at unsavoury smells and a general appearance of squalor.... [T]hey must be taught some elementary lessons before they can be brought up to the level of their poor English brethren." In America the *Jewish Messenger* sniffed that same year that "there is a lack of refine-

---

[1] Neusner, 2003, 113.
[2] Karpel, 2006.
[3] Karpel, 2006.

ment and true spirituality despite the exactness with which they adhere to their traditional habits."[1] In Germany in 1897 the Zionist leader Theodor Herzl in the infamous article "Mauschel" went even further, describing the Yiddish-speaking Eastern Jew as "something unspeakably low and repugnant" (*etwas unsagbar Niedriges und Widerwärtiges*).[2] It was precisely this disdain on the part of many German-speaking Jews for their Yiddish-speaking cousins that stimulated the new exiles' attachment to Zionism and its promise of a restorative national home. They had until recently hoped to find refuge in the Russian Empire,[3] then emigrated in despair, and now saw that they must once again move on – to their 'historic homeland.'

Salazar, Franco, Hitler, Mussolini, and even such great 'democrats' as Roosevelt and Churchill were 'strong men' intent on pursuing specific goals without being excessively preoccupied by such fictions as *vox populi*. Zionism was a creation precisely of such a mentality. It is no accident that the eugenicist Zeev (Vladimir) Jabotinsky idolized Mussolini. Zion was a place of refuge not encumbered by the insidious threat of intermarriage, a place where a *bodenständig*[4] people could be cured by invigorating agricultural labor, a place of "ingathering of the exiles" (kibbutz galuyot), a place where the original racial integrity could be reestablished and the alien elements introduced by intermarriage could be sloughed off.

Zionism was, in Falk's formulation, a eugenic experiment, but eugenics with a Lamarckian twist. Alexander Schüler in 1912 wrote a booklet entitled *Jewish Racial Nobility* (*Der Rassenadel der Juden*) that was actually an enthusiastic and extensive retelling of *The Racial Problem* (*Das Rassenproblem*) by the Jewish eugenicist Ignaz Zollschan (1877-1948). Schüler takes up the historic theme of Jewry's mission as a light unto the nations, but stresses that to achieve its goal, it has to preserve its "racial purity,"[5] already threatened by assimilation.

Now this Lamarckian eugenics has been rejected by scientists, but its conclusion – the State of Israel – remains as a 'fact on the ground.'

---

[1] *Jewish Chronicle*, Oct. 1, 1980 ; JM, Sept. 17, 1980, 17, cited in Tananbaum, 2001, 941, 956.

[2] Herzl, 1897, 1.

[3] See John Glad, *Russia Abroad*, Hermitage Publishers, 1999 for a broader discussion.

[4] Schüler, 1912.

[5] Schüler, 1912, 56-58; also Joseph Jacobs (see: Efron, 1994, 81-82); Weiss, Meira, 2002, 2.

## Lamarckian Egalitarianism

Jean-Baptiste Lamarck (1744-1829) had posited that environmental factors created heritable characteristics. Applied to the Jews, habits viewed as negative were explained as having their origins in ghetto life but were considered curable within an altered environment, with the new characteristics transmitted to future generations. Thus, while not all Lamarckians were Jewish and not all anti-Lamarckians were gentiles, the divide separating these two camps clearly lay along these fault lines. When Lamarckism was decisively overturned by science, the Jewish fallback position was to assert that biological variance in humans was trivial and that 'nurture' trumped 'nature' – hands down.

The 'nature/nurture' controversy (an 1874 coinage of Galton's[1]) had been presented to the public as a strawman intended to discredit the hereditarians. In point of fact, while there is indeed still a good deal of uncertainty as to the relative importance of the one factor relative to the other, absolute denial of any role to environment is a position that was never espoused by anyone. By contrast, the absolute nurture model was and is still popularly presented to the general public as the only correct paradigm, denouncing even moderately hereditarian views as 'racist,' 'classist,' 'sexist,' or 'misogynist.'

The Soviet Union was the great bastion of Lamarckian ideology. The Jewish-American geneticist Herman J. Muller, who was doing research in the U.S.S.R., thought he could reverse this line of thought and wrote Stalin a letter suggesting the creation of a eugenic state. Muller barely escaped the U.S.S.R. with his life.[2]

The geneticist and eugenicist Solomon Grigorievich Levit (1894-1938) perished in the purges, as did other Jewish-Russian Lysenko opponents, including Israèl I. Agol (1886-1936) and Max Levin (1885-1938?). Agol was shot the very day that Muller fled, and even the translator of Muller's book was reportedly shot.[3] The Jewish professor of law and co-editor of the Russian Eugenics Journal Pavel Isaakovich (Isaevich) Liublinsky (1938-1982) supposedly died as the result of a 'fall' from the platform of the Leningrad commuter train that he had taken for 25 years.[4] The geneticist Aleksandr Sergeevich Serebrovsky (1892-1948) managed to repent and survived.

---

[1] Galton, Francis. 1874. *English Men of Science: Their Nature and Nurture*. London, Frank Cass Publishers.
[2] Muller, 1936.
[3] Adams, 1990, 197.
[4] Baranovsky, 2005.

Stalin, the great egalitarian, favored the theories of the Lamarck-ist Trofim Denisovich Lysenko (1898-1976), whose chief advisor was Isai Izrailovich Prezent (1902-1969). At least four other senior Jewish officials supported Lysenko: Yakov Arkadievich Yakovlev (born Jacob Epstein, 1896-1938), Central Committee Chairman of Agriculture; Mikhail Aleksandrovich Chernov (1891-1938), People's Commissar for Agriculture; Aleksandr Ivanovich Muralov (1886-1937), President of the All-Union Academy of Russian Sciences; and Pavel Petrovich Postyshev (1887-1939), Secretary of the Ukrainian Committee of the Central Committee of the Communist Party.[1] Evidently none of them had any previous agricultural expertise.

But Stalin's purges were indiscriminate. If the fate of Soviet eugenicists was series of tragedies, their Lamarckian opponents fared no better.

Yakovlev had been active during the collectivization campaign that had doomed millions of people to death from starvation. Arrested on September 12, 1937, he died the next year in prison.[2]

Chernov was a former member of the (largely Jewish) Menshevik Party who had gone over to the Bolsheviks and directed grain confiscations from the peasantry when the collectivization went awry. Accused of being a Trotskyite, at his trial he begged to be spared the death penalty: "So great are my crimes that I cannot make up for them. By working in the future I hoped to compensate for at least a minuscule portion of my grievous crimes." Chernov's pious petition to end his days at hard labor was rejected in favor of execution by shooting.[3]

Like Chernov, Muralov was active in the physical confiscations, but he was also a theoretician of collectivization. Accused of bad management rather than political opposition, he was executed a year earlier than Chernov.

Postyshev originally played the good-hearted traditionalist by proposing that the 'New Year's' tree tradition, forbidden after the overthrow of tsarism, be reestablished for the sake of the children. Later, terrified that he would be blamed for the failure of collectivization, he declared "100%" of members of the Ukrainian government to be saboteurs. This was too much even for Stalin, who did not get around to executing him until a year after Chernov.

---

[1] Zhuravsky, 1993.

[2] Vronskaya/Chuguev, 1994, 639.

[3] Zalesskii, K.A. 2000. *Imperiia Stalina*, Veche, Moscow. Cited by KHRONOS, http://www.hrono.ru/biograf/chernov_ma.html, accessed Dec. 27, 2007.

By contrast, Prezent continued to thrive and was even appointed a university dean (I refrain from comment here), but when it proved politically impossible to criticize Lysenko, Prezent became the chief object of attack. Being largely ignorant of biology, he made an easy target. The heritage of Lamarckism was to find an echo in the fierce egalitarian climate of America in the last third of the twentieth century.

## Anglo-American Eugenicists

It is a simple fact that the major Anglo-American eugenicists came out forthrightly against racial hatred and that eugenicists were arrested, exiled, and murdered under both Hitler and Stalin, not to mention facing fierce hostility in the United States.

In September 1939, the journal *Nature* published a joint statement issued by America's and Britain's most prominent biologists, some of them Jewish.[1] The document is widely referred to as the 'Eugenics Manifesto.' The authors explicitly decried antagonism between races and theories according to which some good or bad genes are the monopoly of certain peoples.[2]

But it is also true that antipathy toward Jews was evident among an undetermined minority of eugenicists. In 1916 the lawyer, environmentalist, and eugenicist Madison Grant (1865-1937) wrote the popular book *The Passing of the Great Race or the Racial Basis of European History*, which went through numerous printings. Aside from its revealing vituperations, it is interesting to note Grant's comment that the belief in the preeminence of nurture over nature was already popular:

> *There exists to-day a widespread and fatuous belief in the power of environment, as well as of education and opportunity to alter heredity, which arises from the dogma of the brotherhood of man, derived in turn from the loose thinkers of the French Revolution and their American mimics. Such beliefs have done much damage in the past, and if allowed to go uncontradicted, may do much more serious damage in the future. Thus the view that the negro slave was an unfortunate cousin of the white man, deeply tanned by the tropic sun, and denied the blessings of Christianity and civilization, played no small part with the sentimentalists of the Civil War period, and it has taken us fifty years to learn that speaking English, wearing good clothes, and going to school*

---

[1] Jenkins, 2007, 1011.

[2] "Social Biology and Population Improvement,"http://whatwemaybe.org.

*and to church, does not transform a negro into a white man. Nor was a Syrian or Egyptian freedman transformed into a Roman by wearing a toga, and applauding his favorite gladiator in the amphitheatre. We shall have a similar experience with the Polish Jew, whose dwarf stature, peculiar mentality, and ruthless concentration on self-interest are being engrafted upon the stock of the nation.[1]*

Grant goes on:

*Whether we like to admit it or not, the result of the mixture of two races, in the long run, gives us a race reverting to the more ancient, generalized and lower type. The cross between a white man and an Indian is an Indian; the cross between a white man and a negro is a negro; the cross between a white man and a Hindu is a Hindu; and the cross between any of the three European races and a Jew is a Jew.[2]*

Grant was not alone in his views; the second edition of his book contained a preface by the prominent eugenicist Henry Fairfield Osborne (1857-1935).

## The Popular Image of Eugenics

In the United States Holocaust Memorial Museum's exhibit 'Deadly Medicine: Creating the Master Race,' curator Susan Bachrach maintains that eugenics "culminated" in the Holocaust.[3] And even though most people would be unable to provide a definition of eugenics, they nevertheless generally concur with Bachrach, who represents an institution that, according to its official Web site, "teaches millions of people each year,"[4] and there are at least sixty-two other Holocaust museums in addition.[5]

But even this outreach is dwarfed by mass media that in their totality reach audiences numbering in the billions – and this on a daily basis. A Yahoo search for 'eugenics' and 'anti-Semitism' produced 311,000 items.[6] A Google search for 'holocaust+eugenics' produced a

---

[1] Grant, 1916, 14.
[2] Grant, 1916, 15-16.
[3] http://www.ushmm.org/museum/exhibit/online/deadlymedicine/overview/, accessed May 7, 2008.
[4] http://www.ushmm.org/museum/mission/, accessed Dec. 30, 2007.
[5] http://www.science.co.il/holocaust-museums.asp.
[6] July 3, 2008.

lengthy series of Web sites, the first of which (the most widely read) began with the sentence "Probably one of the most hideous aspects of the Third Reich was their notorious fascination and experimentation with Eugenics."[1] The first sentence of the second item reads that "The eugenics movement... was the forerunner of the Holocaust."[2] The third item was a review of the above-mentioned eugenics exhibit at the United States Holocaust Memorial Museum in Washington, D.C., making little distinction between German racial hygiene laws and eugenics and characterizing them as "diabolical."[3]

The political alliance of Jewish and Christian Zionism strengthened Jewish Creationism and a negative view of theories of human selection. For example, Jerry Bergman, who was raised as a Jehovah's witness, claims that Darwinism was a defining element in Hitler's ideology, and repeats Bachrach's assertion that this philosophy "culminated in the final solution."[4]

That hereditarian thought was on occasion viciously distorted is not open to dispute, but to present it solely as an instrument of the Holocaust is a grotesque falsification. The following passage from Rabbi George Benedict's 1926 Mother's Day Sermon at Temple Emanu-El, Roanoke, Virginia, presents a far different image of the eugenics movement than one might expect on the basis of the above:

> *And what but eugenic development of the race of Israel, whom, according to the conception of Moses, God had chosen to be a pattern to mankind as a nation consecrated to holiness, is the prime intent beneath every one of the laws of Moses? Whether regarded hygienically, morally, or religiously, the whole purport of the Torah, the Law of Moses, is to separate Israel from the rest of mankind as a Chosen People, in order to be a noble people, a well-born race of men for their own superior happiness, as well as, by way of example, to be a blessing to the world.*[5]

A current version of Jewish eugenics is to be found on the Web site of the Chicago Center for Jewish Genetic Disorders:

---

[1] "Eugenics," http://www.shoaheducation.com/pNEW.html, accessed Dec. 30, 2007.
[2] "Eugenics and the German Medical Establishment," http://www.humanitas-international.org/holocaust/eugenics.htm, accessed Dec. 30, 2007.
[3] Curran, 2004.
[4] Bergman, 1999.
[5] Benedict, 1926.

> *Dor Yeshorim* ['Upright Generations' in Hebrew] *is an interna-
> tional, confidential genetic screening system used mainly by Or-
> thodox Jews, which attempts to prevent the transmission of ge-
> netic disorders that have an increased frequency among mem-
> bers of the Ashkenazi Jewish community. The system was estab-
> lished to follow Jewish law, under which abortion is not al-
> lowed, while acknowledging that testing might prevent the birth
> of an affected child. Designed in the early 1980s by an Orthodox
> rabbi, the system tests young adults before they begin to con-
> template marriage. Participants can then use the system to learn
> their genetic compatibility with potential marital partners.*[1]

Accepted Holocaust narrative has painted the unfortunate Dor
Yeshorim into a corner. Its founder, Rabbi Joseph Eckstein, told me in
December 2007 that he still received only "a little help" in his work,[2] but
this is not the first time in history that theory radically diverges from
practice.

This David-and-Goliath discrepancy of forces explains the need
for a timeline of specifically Jewish eugenics, so obviously and so radically is accepted Holocaust narrative out of whack with reality. Eugenics
– inadvertent and explicit, historical and current – lies at the very core of
Jewish identity. Jewish eugenics cannot be understood as standing apart
from human eugenics.[3]

In addition to the accusations of "anti-Semitism" and "racism,"
the standard follow-up claim is that the eugenics movement was dismissed as bad science in the 1920s and 1930s and thus withered away on
its own. To check the veracity of these claims, I looked through the entire
run of the *Eugenics Quarterly*, published by the American Eugenics Society from 1954, when it took over from the *Eugenical News* (1916-
1953) until 1985, when members of the Society ran for collective cover,
replacing the word "eugenics" with "Social Biology."

As I read through the back issues of the *Eugenics Quarterly*, it
became almost immediately evident that the situation was entirely differ-

---

[1] "Dor Yeshorim," 2003.

[2] Telephone conversation. Israeli sociologists Aviad Raz and Yava Vizner made the
interesting observation that Dor Yeshorim "has been selectively incorporated into the
traditional match-making process." Formally, this is correct, but traditional Jewish
match-making laid emphasis on quantity at the expense of quality, encouraging
people with very serious illness to have children. (Raz/Vizner, 2008, 1361)

[3] Readers who wish to acquaint themselves with a more integrated approach are referred to my *Future Human Evolution: Eugenics in the Twenty-First Century* (Hermitage 2006), also available free online at: http://whatwemaybe.org.

ent from that now being popularly asserted. The contributors were among the leading scientists in their field and they were working on the cutting edge of technology at the time. Their tone was restrained, proper, and entirely professional. As for the 'anti-Semitism' charge, I could find not the slightest trace of it. Some of the persons whose work was published there or whose works were reviewed or advertised (please forgive the length of the list; I could easily have made it longer) were:

K. Z. Altshuler, Baruch S. Blumberg, Lauretta Bender, Bernard Berelson, Marianne E. Bernsteif, Jack B. Bresler, B. Catz, B. Cohen, Leon Jacob Cole, Melvin Embep, Arthur Falek, William E. Feinberg, Joseph Felsenstein, J. D. Finkelstein, I. Lester Firschein, Morris Fisbein, B. Fish, Bertram Fleshler, Ronald Freedman, Benson E. Ginzburg, A. M. Gittelsohn, Paul C. Glick, Jacob A. Goldberg, Calvin Goldscheider, Sidney Goldstein, H. O. Goodman, H. Green, Bernard Greenburg, Alan F. Guttmacher, Melville Herskovits, J. Hirsch, P. A. Jacobs, A. J. Jaffe, Kurt Hirschhorn, John F. Kantner, Franz Kallmann, Arnold A. Kaplan, A. Katz, Aviva B. Kesselman, P. Kunstadter, Samuel M. Levin, Louis Levine, Philip Levine, Richard Levins, Max Levitan, Sarah Lewit, Richard Lewontin, B. Malzberg, B. M. Mandelbrote, N. Mantel, Emmanuel Margolis, Gitta Meier, S. Milham, Jr., Ashley Montagu (né Israel Ehrenberg), Melvin Moss, H. V. Muhsam, H. J. Muller, Edward Pohlman, Stefan Possony, Erich Rosenthal, Ina Samuels, J. Samuelson, Lee E. Schacht, Sam Shapiro, Erwin S. Solomon, Amram Scheinfeld, William Schull, Sheldon J. Segal, Harry L. Shapiro, Hirsch Lazar Silverman, S. E. Snyderman, Mortimer Spiegelman, J. N. Spuhler, Robert M. Stecher, Medora Steedman-Bass, Arthur G. Steinberg, Gary A. Steiner, Curt Stern, Abraham Stone, W. F. Wertheim, Irving B. Wexler, Nathaniel Weyl, Melvin Zelnik (Melvin Zelnick), Anthony Zimmerman.

The journal's persistently upbeat tone startles, making it evident that editors and authors alike had little inkling of the abrupt and massive take-no-prisoners assault that would be launched against them in the late 1960s. Article after article eagerly calls for further research in anticipation of scientific breakthroughs and a qualitatively different understanding of the human species, anticipating ways to improve both its nature and its lot in the framework of Darwinism, and the coming of a new, scientifically grounded worldview. If someone had produced a crystal ball and shown this prestigious international team of scientists how their

efforts on behalf of future generations would soon be portrayed, they –
Jews and non-Jews alike – would have been totally incredulous.

As the historian Mark B. Adams commented, "producing
healthy babies is about as uncontroversial a goal as can be imagined,"[1] so
how is it that the very word 'eugenics' has been so thoroughly demo-
nized?

In 1936 the biologist Julian Huxley expressed the majority view
that "Eugenics falls within the province of the Social Sciences, not of the
Natural Sciences," and two years later the sociologist David Victor Glass
declared 'differential fertility' to be a key aspect of the then coalescing
discipline of population studies.[2]

By the 1930s a global consensus of support had been achieved
for eugenics, especially among physicians, including German physicians.
(If Jews made up 1.2% of the Reich population, in 1925 they constituted
0.9%, and by 1933 the figure had been reduced to a mere 0.76%. But in
large cities such as Berlin, Frankfurt, and Hamburg as many as 30% to
40% of the doctors were Jewish.[3])

But, beginning in the late 1960s, eugenics with its focus on the
selectionary consequences of differential fertility was declared beyond
the pale, replaced by its own offspring – the natural science of genetics,
and also demography with its modern reluctance to probe further than
'total fertility.' Indeed, why differentiate if genetic variance in human
populations is preordained to be inconsequential? Physical anthropology
came to be viewed as a poor second to cultural anthropology, with sociology, criminology, and pedagogy following in lock step, and the arts –
as always – bringing up the rear.

Noting that Darwinism is more influential now than ever and also that eugenics had strongly influenced economic theory in the early
twentieth century, the economist Thomas C. Leonard dryly notes "the
amnesia about the influence of eugenics upon the nascent social sciences
of a century ago," and asks how the movement came to be erased from
the history of American economics:

---

[1] Adams, 1990, 72.

[2] J. Huxley "Eugenics and Society," Galton Lecture, *Eugenics Review* 27(1), 1936,
12; D. V. Glass and C. P. Becker, *Populatiion and Fertility*, London, Population Investigation Committee, 1938, pg 50; both cited in Oakley, 1992, 165.

[3] Kater, 1987, 35; citing Esra Bennathan, "Die demographische und wirtschaftliche
Struktur der Juden," in *Entscheidungsjahr 1932: Zur Judenfrage in der Endphase der
Weimarer Republik* (Tübingen, 1966, 111-112.

*If the relationship between American labor reform and the biology of human inheritance seems to the modern reader unexpected, it is, in part, because eugenics, new scholarship notwithstanding, is still widely misunderstood.... Eugenics was, in actual fact, the broadest of churches. Eugenics was not aberrant; it was not seen as a pseudoscience; it was not laissez-faire; it rejected social Darwinism; and it was not abandoned after Nazi atrocities. Eugenics was mainstream; it was popular to the point of faddishness; it was supported by leading figures in the still-emerging science of genetics; it appealed to an extraordinary range of political ideologies, not least to the progressives; it was – as state control of human breeding – a program that no proponent of laissez-faire could consistently endorse; and it survived the Nazis.* [1]

## Eugenics and Hitler

Perhaps the most thorny question, but one that has to be at least mentioned, given the widespread nature of the rumor and the topic of this book, is a claim made by Hans Frank (1900-1946), Hitler's Regent in Poland. Over the years there has been much speculation over Hitler's purported partly Jewish ancestry, almost all of it leading back to Frank's memoirs *In Sight of the Gallows* (*Im Angesicht des Galgens*), written during Frank's Nuremberg imprisonment prior to being hanged. Frank, who converted to Catholicism while in prison, upon hearing the death sentence, responded that his execution was his penance for collaborating with Hitler, although he claimed to have learned of Hitler's "enormous mass crimes of the most terrible nature" only during the trial.

Frank writes that in late 1930 Hitler directed him to investigate a "disgusting blackmail affair" regarding rumors about Hitler's paternal grandmother Maria Schickelgruber (1795-1847), who supposedly had worked as a cook for a Jewish family by the name of Frankenberger and had given birth to Hitler's father out of wedlock. Frank seems to recall that the blackmail came from Alois Hitler, a stepbrother of Hitler's of a different marriage of Hitler's father, but different sources claim that it was Alois's son William Patrick Hitler (1911-1987). [2] Frank claims that this Frankenberger (Leopold according to some sources [3]) paid child support until the child reached the age of fourteen, and that there had been a

---

[1] Leonard, 2005, 212, 205-206.
[2] http://en.wikipedia.org/wiki/William_Patrick_Hitler.
[3] http://en.wikipedia.org/wiki/Alois_Hitler#Leopold_Frankenberger.

lengthy correspondence. On the one hand, Frank gives the traditional this-cannot-be-because-it-cannot-be argument:

> *That Adolf Hitler had no Jewish blood in his veins strikes me as so blatantly obvious on the basis of his entire nature that it merits no further discussion.*

But then he goes on to discuss precisely this topic:

> *I must therefore say that it is not totally out of the question that Hitler's father was a half-Jew who had sprung from the out-of-wedlock relationship of Schicklgruber to the Graz Jews. If this is true, then Hitler would have been a quarter-Jew and his hatred for Jews could have been partially engendered by a blood rage over a familial hatred psychosis. Who can possibly interpret all this!* [1]

It is impossible to believe that Hitler's Regent in Poland was unaware of the murders of Jews in Eastern Europe during the war, and it stretches credulity that Frank, having been entrusted with such a top-secret mission, would be uncertain as to the source of the blackmail. Furthermore, a subsequent analysis of Frank's statement by Simon Wiesenthal disclosed that there was no evidence of any Jewish family named Frankenberger ever living in Graz. My personal opinion is that Frank must have assumed he would be found guilty but calculated he had a slim chance of escaping execution by feigning religious fervor. And in launching a rumor that cannot be definitively refuted, he obtained his posthumous revenge on his executioners.

*

When Hitler put forward his list of 25 points in 1920, none of them dealt with eugenics, but by the time the second volume of Mein Kampf appeared in 1926, most of which was evidently written in prison in 1924, he had clearly become a believer in 'racial hygiene,' much in the spirit of French Nordic-supremacy theoretician Joseph Arthur Gobineau (1816-1882), whose name is never mentioned in *Mein Kampf*:

> *The* folkish *state must make up for what everyone else today has neglected in this field. It must set race in the center of all life. It must take care to keep it pure. It must declare the child to be the most precious treasure of the people. It must see to it that only the healthy beget children; that there is only one disgrace: despite one's own sickness and deficiencies to bring children into*

---

[1] Frank, 1953, 330-331.

*the world, and one highest honor: to renounce doing so. And conversely it must be considered reprehensible: to withhold healthy children from the nation. Here the state must act as the guardian of a millennial future in the face of which the wishes and the selfishness of the individual must appear as nothing and submit. It must put the most modern medical means in the service of this knowledge. It must declare unfit for propagation all who are in any way visibly sick or who have inherited a disease and can therefore pass it on, and put this into actual practice. Conversely, it must take care that the fertility of the healthy woman is not limited by the financial irresponsibility of a state regime which turns the blessing of children into a curse for the parents. It must put an end to that lazy, nay criminal, indifference with which the social premises for a fecund family are treated today, and must instead feel itself to be the highest guardian of this most precious blessing of a people. Its concern belongs more to the child than to the adult. Those who are physically and mentally unhealthy and unworthy must not perpetuate their suffering in the body of their children. In this the folkish state must perform the most gigantic educational task. And some day this will seem to be a greater deed than the most victorious wars of our present bourgeois era. By education it must teach the individual that it is no disgrace, but only a misfortune deserving of pity, to be sick and weakly, but that it is a crime and hence at the same time a disgrace to dishonor one's misfortune by one's own egotism in burdening innocent creatures with it; that by comparison it bespeaks a nobility of highest idealism and the most admirable humanity if the innocently sick, renouncing a child of his own, bestows his love and tenderness upon a poor, unknown young scion of his own nationality, who with his health promises to become some day a powerful member of a powerful community.[1]*

Hitler's position proved to be a near-fatal embrace, leading the famous philosopher and Zionist member of the Jewish Academy Leo Strauss (1899-1973) to coin the maxim *reductio ad Hitlerum*: "Hitler believed in eugenics. X believes in eugenics. Therefore X is a Nazi."[2] Contrary to claims advanced by eugenics foes, Hitler's position on such

---

[1] Volume Two (published in 1926): The National Socialist Movement, Chapter II: The State, http://www.crusader.net/texts/mk/mkv2ch02.html, accessed Oct. 5, 2008.

[2] Drouard, 1999, 7.

topics was not a bone of contention vis-à-vis his American and British World War II foes. His arch-nemisis Winston Churchill, Churchill, was an ardent eugenicist.

In 1904, the Conservative government of Arthur Balfour had established a Royal Commission "On the Care and Control of the Feebleminded," which reported to the Liberal government and recommended compulsory detention and sterilization of the unfit. In 1912 Balfour had personally addressed the First Eugenics Conference in London, which was attended by then Home Secretary Winston Churchill, who called for "a simple surgical operation (sterilization) so the inferior could be permitted freely in the world without causing much inconvenience to others." In 1910, Churchill had asked the civil service to investigate implementation of the Indiana law on sterilization.[1] Even earlier, in 1899, he had written to his cousin Ivor Guest: "The improvement of the British breed is my aim in life." The poet Wilfrid Scawen Blunt (1840-1922) wrote in his diary:

> *Winston is also a strong eugenist. He told us he had himself drafted the [Mental Deficiency] Bill which is to give power of shutting up people of weak intellect and so prevent their breeding. He thought it might be arranged to sterilise them.[2]*

In 1910 Churchill had written to Prime Minister H. H. Asquith (1852-1928):

> *"The unnatural and increasingly rapid growth of the Feeble-Minded and Insane classes, coupled as it is with a steady restriction among all the thrifty, energetic and superior stocks, constitutes a national and race danger which it is impossible to exaggerate. I am convinced that the multiplication of the Feeble-Minded, which is proceeding now at an artificial rate, unchecked by any of the old restraints of nature, and actually fostered by civilised conditions, is a terrible danger to the race."[3]*

In 1937, pressed by William Peel (1867-1937), head of the Palestine Royal Commission, that Britain "might have some compunction if she felt she was downing the Arabs year after year when they wanted to

---

[1] Sparkes, 1999.
[2] Gilbert, 2009, Jan. 19.
[3] Gilbert, 2009, Jan. 19.

remain in their own country," Churchill showed himself to be not only a supporter of eugenics, but of the rankest 'racial hygiene' as well:

> *I do not admit that the dog in the manger has the final right to the manger, even though he may have lain there for a very long time. I do not admit that right. I do not admit, for instance, that a great wrong has been done to the Red Indians of America, or the black people of Australia. I do not admit that a wrong has been done to those people by the fact that a stronger race, a higher grade race, or, at any rate, a more worldly-wise race, to put it that way, has come in and taken their place."[1]*

Hans Fenske, Dieter Mertens, Wolfgang Reinhard, and Klaus Rosen in their *History of Political Ideas* argue that the rise to power of Germany's National Socialist party was a fundamentally "non-ideological phenomenon" best understood as a national longing to establish community and unity in the chaos of the Weimar Republic years,[2] and the historian Richard Weikart rejects as "absurd" the claim that Darwinism inevitably leads to Nazism, but sees Darwinism as a "necessary but insufficient" cause for National Socialist ideology.[3]

To determine whether biological theories were truly crucial to the rise of the National Socialist German Workers' Party, I selected one hundred books dealing with the Weimar and National-Socialist periods. All contained indexes covering not only proper names but topics as well. I made no attempt to pre-select other than choosing volumes that deal with the period.

The authors of these books range from National Socialist ideologues to recognized Western scholars. Ninety-six of these indexes did not contain the word 'eugenics,' and even four that did contained only a handful of mentions. Even the indexes to *Mein Kampf* and Hitler's speeches do not list eugenics as a topic, although they contain numerous references to race.[4] Eugenics was not the ideological motor it is made out to be. This should not surprise: intergroup hostilities are entirely possible without a belief in a conflict's ideological underpinnings.

The timeline of Jewish eugenics provided here demonstrates beyond doubt that Jews were welcome, active participants in the eugen-

---

[1] Peel Commission Report, proof copy of Churchill's evidence: Churchill papers 2/317; cited in Gilbert, 2007, 120.
[2] Fenske *et al.*, 1987, 531.
[3] Weikart, 2004, 9.
[4] Glad, *Future Human Evolution*, 2006.

ics movement and that Jews even today are still in the vanguard of a eugenic worldview, a fact entirely unknown to most people.

Eugenics is now viewed by many in the United States largely through the lens of the Holocaust and is to such a degree awash with understandably raw emotion as to quash any cries of protest. The upshot of the situation is that a group of largely Jewish activists have so successfully undermined the very eugenic mechanism that made Jewry what it is as to pose an existential threat to Jewry. But Jewish common sense, plain and old-fashioned as chicken soup, has not only continued to hold sway in the practice of eugenics, it has even managed to surf the scientific tide of newly found genetic knowledge – all the while paying lip service to the Holocaust-from-eugenics gospel!

Unlike the U.S. situation, this anti-eugenics view never even got off the ground in Israel. Behavioral scientist Aviad Raz (b. 1968) of Ben Gurion University is quite open in pointing out that both the word 'eugenics' and the actual practice of eugenics enjoy broad approval in that country, and objections to eugenics – at least as far as genetic screening combined with eugenic abortions – are a 'non-issue' in Israel:

> *Eugenic ideologies and practices have persisted in Israel, in a thinly disguised mode, even after the holocaust, because they were an inherent and formative part of Zionism....*[1] *For many of the above-mentioned respondents, prenatal genetic testing was eugenic and was indeed supported precisely for that reason, since "eugenic" for them meant the improvement of the health of progeny and carried positive rather than negative connotations.*[2]

## The Holocaust

Jewish population statistics are so beset with gaps and uncertainties that even today the Israeli demographer Sergio DellaPergola has noted the "permanently provisional" character of Jewish population studies.[3] Obviously these waters were infinitely more muddied at the end of World War II.

Huge Jewish death losses due to violence targeted specifically at Jews toward the end of World War II are a simple fact. Any dispassionate investigator has only to read the testimony of survivors who list their perished family members. As for the effort to produce at least an approximate estimate of the number of victims, it was not merely legiti-

---

[1] Raz, 2005, 184-186.
[2] Raz, 2005, 185.
[3] DellaPergola, 2007, 90.

mate, it was inevitable, but its success is limited by the opaque nature of the violence it attempts to measure. Even before the war, Jews emigrated in large numbers from Germany and Austria, and when the war began, they fled the occupied areas.

During this incredibly chaotic period Jews attempted to pass as non-Jews. When the war was over, Jews from the Soviet empire were terrified they would be forcibly returned home and desperately continued to conceal their Jewishness. Jews in Poland and the U.S.S.R. were hardly eager to announce their roots. As for the German armies, they were intent on destroying evidence of their atrocities as they retreated. And since it is impossible to determine how many managed to escape and how many concealed their ethnicity, we will never know how many actually perished.

Hitler viewed the Jews as culturally and genetically different, but far from dismissing them as primitive in their evolutionary development, he regarded them as powerful competitors to the 'Nordics' whom he championed. On January 30, 1939, in a speech before the Reichstag, he was explicit in explaining both his threat against the Jews and his motivation in making it:

> *If international finance Jewry within Europe and abroad should succeed once more in plunging the peoples into a world war, then the consequence will be not the Bolshevization of the world and therewith a victory of Jewry, but on the contrary, the destruction of the Jewish race in Europe.*[1]

The Holocaust was not about eugenics, but about revenge. And the result was decidedly dysgenic. A double tragedy occurred: the tragic fate of the individual victims, and the severance of a brilliant genetic lineage. Demographer Sergio DellaPergola estimates that, if not for the Holocaust, there would be as many as 32 million Jews worldwide, instead of the current 13 million.[2] Still, such calculations are fraught with uncertainty. As recently as 2004 Berl Lazar, the Director of the Outreach Department of the Federation of Russian Jewish Communities (FEOR), noted that estimates of the number of Jews in Russia ranged from 230,000 to 10,000,000.[3] Obviously, it is not possible to derive reliable figures from such conflicting data. Whatever the losses, the world will

---

[1] http://www.stevenlehrer.com/Hitler_threat.htm
[2] Ilani, 2009.
[3] *New York Times, 1920.*

never know, and will never benefit from, the unborn children of the slain, from the children of the children....

## Deconstructing the Eugenics Bashers

How is it that informed, sincere people have drawn such radically different conclusions regarding eugenics? And how did a small group of secular Jewish intellectuals come to launch a massive attack on the eugenic core of Jewish religious and secular thought? And how were they so successful in depicting eugenics in such a baneful light?

While Stalin was opposed to eugenics, both Hitler and Churchill were ardent proponents even of 'racial hygiene,' and in the United States eugenicists were writing laws on sterilization, isolation, and immigration. In all these countries Jews were eager participants in the eugenics movement. At the very least, the argument was one-sided and distorted, even though individual figures in the movement provided fodder for such an interpretation.

Generally speaking, eugenicists view human genetic variance as too great to permit the majority of the people to figure significantly in the process of civilization, while their opponents view such variance as a desirable source of 'diversity,' even if that diversity consists of illness and low intelligence. The Biblical tradition teaches that people are created in the image of their Creator; if we are thus god-like, why would anyone want any other fellow human being to refrain from having children? Such abstinence is akin to deicide.

The attitude toward the role of society in human affairs can be either communal or individualistic. The communitarian defines society diachronically (over time) while the individualist is the zealous champion of synchronicity – the status quo – either distrusting the state, or perhaps simply not caring about future generations. After all, we as a species have been bred by natural selection to defend first and foremost our own interests and those of our *immediate* offspring, not distant future generations.

In the 'hard sciences' researchers generally hold to a neutral, non-value approach: observed data establish that thus and such an atomic particle flies either left or right in thus and such a magnetic field, and that is that. This detached search for theory confirmed by replicable fact also carries the day in the study of biological species – with one exception: our own. Assigned fellow human beings as the object of examination, the naturalist who formerly dispassionately studied bees, lions, or chimpanzees without preconceptions now begins to furtively glance at the mental notebook of his own hidden agenda. He pretends to be pursuing only the truth, and more often than not he even sincerely believes in this claim. But it is rarely true. Truth be told, we are – all of us – both blessed and burdened with ideologies, which both determine and are determined by our experience and value systems.

Darwinian social concepts involve judgments that are inevitably found to be unacceptable, even outrageous, by large segments of society, forcing even the most sincere and well-meaning intellectuals into defensive postures whose opaqueness is perceived by their opponents as hypocrisy. After all, no one is eager to fall upon his sword. Thus, evidence is selectively mustered and adumbrated that seems to lead to a specific conclusion, but that conclusion is rarely specified. The 'investigator' presents himself, not as an advocate, but as a witness, and if his audience draws certain conclusions, he is simply presenting the facts.

If eugenics survived World War II in America and Western Europe, Stalin had driven it underground during the Great Purges of the late 1930s, in which he eliminated most of his former comrades in arms, very many of whom were Jews. Even earlier, in 1930, he had created the Jewish Autonomous Region in a desolate region of Siberia where the mean January temperature fell to -30 degrees Celcius (-22 Fahrenheit). Not surprisingly, it failed to attract Soviet Jewish immigrants in any significant numbers. In 1948 he simultaneously took two steps whose common motivation no one seems to want to ponder: the campaign against the "rootless cosmopolites" and the UN vote to recognize the creation of the State of Israel. Hitler too had cooperated in the Jewish exodus to Palestine.

1949 saw the arrest of a number of Russia's Jewish cultural figures, and January 1952 marked the formal initiation of the "Doctors' Plot" – a purported conspiracy to murder ailing Soviet leaders. Soviet publications pointed out the predominance of Jews in the ranks of both the 'cosmopolites' and the accused physicians. There were numerous unconfirmed rumors of Siberian camps to which the Jews were supposedly to be deported.

In 1952, volume 15 of the *Great Soviet Encyclopedia* was approved by the censor for publication; it contained the article on eugenics, laying out Stalin's official position on the movement:

> **Eugenics:** *a false science current in the capitalist countries on improving 'human nature.' Eugenics has its source in racist fantasies of the supposed physical and mental 'superiority' of the ruling classes and 'higher' races over the working masses and the 'lower' races. The basic principles of eugenics were ad-*

*vanced by reactionary bourgeois scientists and scholars to mask the real social-economic causes of inequality under capitalism.*[1]

Whatever Stalin's plans for the Jews may have been, they came to an end with his death in March 1953, but his position on eugenics was adopted virtually word for word by a group of American Jewish intellectuals in the late 1960s.

The reality that few academics will concede is that logic is often secondary to fundamental ideological values, so that discussions of eugenics have tended to be polemical rather than objective. Prior to World War II, the tone was far more dispassionate, even collegial. The Jewish Zionist and eugenicist Arthur Ruppin and the German racial theoretician Hans F. K. Günther could even enjoy a pleasant lunch in each other's company and find areas of agreement.

The driving force behind the Holocaust Memorial Movement is not just grief over the tragedy of the past but also the desire to legitimize the Zionist state. In 1975, by a vote of 72 to 35, with 32 abstentions, the UN General Assembly declared that Zionism was "a form of racism and racial discrimination," essentially declaring the State of Israel to be illegitimate. As a counterbalance, Jewish groups massively funded the Holocaust Memorial Movement. In its turn, the Holocaust Memorial Movement attacked the eugenics movement with ever increasing fury.

As the historian Peter Weingart has observed, cultural and political contexts select from the pool of scientific ideas rather than determine them.[2] In the case of eugenics, scientific findings are constantly corrupted by emotional attitudes and political calculation. If the topic at hand were, say, an unresolved problem in physics, even though its resolution might be as yet uncertain, one could at least be confident that researchers were doing their best to come up with an answer and that they were doing so on the basis of accepted methodology. For their part the physicists would not have to fear for their careers and even for their personal safety. In contrast, despite the due diligence and sincerity exercised by a number of researchers in this most sensitive area of biopolitics, any student of biological determinants of human behavior is keenly aware that he is picking his way through a very dangerous minefield. When the Berkeley psychologist Arthur Jensen (b. 1923) maintained that IQ was largely heritable and that there thus were limits to compensatory education, demonstrators disrupted his classes, his car tires were slashed, and swastikas were painted on his office door. There were threats against him

---

[1] *Bol'shaia sovetskaia entsiklopediia*, vol. 15, 372.
[2] Weingart, 2005, 163.

and his family, the university had to hire bodyguards to protect him, and a bomb squad screened his mail.

The preeminence of replicable scientific observation over ideology was a lengthy hard-fought battle – a battle that now rages over the last hurdle: the human brain. That is why the appearance of E. O. Wilson's 1975 *Sociobiology: The New Synthesis* encountered such resistance. When Wilson began applying Darwinian theory to ants and termites, it was inevitable and even obvious that the study of mammalian brains would immediately follow, leading to conclusions that are simply unacceptable to many, albeit not all for the same reasons.

Wilson's attackers were not Bible-belt fundamentalist preachers with eighth-grade educations, but his sophisticated secular Jewish colleagues – evolutionary biologist Stephen Jay Gould and geneticist Richard Lewontin – precisely those who logically could have been expected to be his most enthusiastic supporters. Wilson was exposed as a "counterrevolutionary" and he was attacked in teach-ins, student demonstrations, and articles. By 1982 emotions had become so intense that he had to have a police escort to deliver a lecture on "the coevolution of biology and culture."[1]

This was more than a run-of-the-mill departmental feud, bloodthirsty and pitiless as those can be. Wilson's opponents were soon emulated by a largely Jewish cottage industry of anti-Darwinian scholars and activists who reviewed each other's books and appointed each other to academic positions.

Classical Marxism posits a fundamental human egalitarianism, with economic relationships serving as the 'foundation' of the social order and predetermining social position within that order. It is opposed by Darwinian thinkers, who attach relatively greater significance to biology than do their opponents. It is Boas versus Galton in a different hypostasis.

Boas himself was torn between assimilation and what is nowadays known as 'ethnic identification,' as can be seen from the following absolutely remarkable comment in his address at the 1908 annual meeting of the American Association for the Advancement of Science:

> *With the economic development of Germany, German immigration has dwindled down; while at the same time Italians, the various Slavic people of Austria, Russia, and the Balkan Peninsula, Hungarians, Roumanians, **east European Hebrews**, not to mention the numerous other nationalities, have arrived in ever-*

---

[1] Wilson, 1995.

*increasing numbers. There is no doubt that these people of east-*
*ern and southern Europe represent a physical type distinct from*
*the physical type of northwestern Europe; and it is clear, even to*
*the most casual observer, that their present social standards dif-*
*fer fundamentally from our own. Since the number of new arriv-*
*als may be counted in normal years by hundreds of thousands,*
*the question may well be asked, What will be the result of this in-*
*flux of **types distinct from our own**?* [emphasis added][1]

The coincidence of the civil-rights movement and the Vietnam
War radicalized not only young Americans in general, but Jewish Boa-
sians in particular, whose political leanings had traditionally been to the
left of the political spectrum. The New Left fused with the counter-
culture to produce a 'revolutionary consciousness' with overwhelming
Jewish participation and leadership. As documented by the historians
Stanley Rothman and Robert S. Lichter in their fine study *Roots of Radicalism* (Oxford University Press, 1982), campus leftist movements were,
to a very significant extent, Jewish affairs, be they in Madison or Berkeley. One study indicated that fully 90% of the radical subjects in an Ann
Arbor study had Jewish backgrounds.[2] Susan Stern described her experience in the 'Weather Underground' (where the greetings were four fingers slightly spread, symbolizing the carving fork which Charles Man-
son's gang had driven into the belly of the 8½ months pregnant Sharon
Tate[3]):

*Our aim was to make ourselves equal, man and women, practi-
cally interchangeable. We had no guidelines, no scruples; we
simply started.... No amount of anguish was intolerable when
one considered the end result: a revolutionary warrior, worthy
to fight in the world-wide struggle for liberation.... [We] were
committed to the notion of transforming ourselves into Ameri-
cong.[4]*

The 'Weather Underground,' which had taken its name from
Bob Dylan's Subterranean Homesick Blues ("You don't need a wea-
therman to know which way the wind blows"), was not a typical youth

---

[1] "Race Problems in America," *Science*, No. 1909, No. 29, 840; cited by Leonard B.
Glick, "Types Distinct from Our Own: Franz Boas on Jewish Identity and Assimilation," *American Anthropologist*, No. 84, 545; and also Frank, 1997, 739.
[2] Rothman/Lichter, 1982, 81.
[3] Rothman/Lichter, 1982, 137.
[4] Rothman/Lichter, 1982, 40.

organization; young persons of Jewish background generally had abruptly been politically activated, and the message they heard was a straightforward one: the rightist establishment seeks to justify its exploitation of the people by claiming that 'negroes' and Jews are 'inferior.'

In describing Marxist organizations at that time, the *Encyclopedia Judaica* notes that "Jewish prominence in the New Left was not noted in the mass media, either because it was truly not remarked or because in the aftermath of Hitler's murderous anti-Semitism the media were reluctant to make this observation."[1]

In 1967 Israel won the six-day war, creating a crisis for many Jewish members of the American and European 'New Left' when the Holocaust Memorial Movement was launched to support the Zionist state, and international Marxist organizations, including black militants, denounced Zionism as "kosher imperialism." Jewish participation in leftist activities fell off, and the belief that eugenics had been the driving ideological motor triggering genocide of the Jews became accepted Holocaust narrative.

During the first 47 years following the end of World War II – nearly a half century – only one book associating 'eugenics' with 'Holocaust' is shown by a 'Worldcat' search.

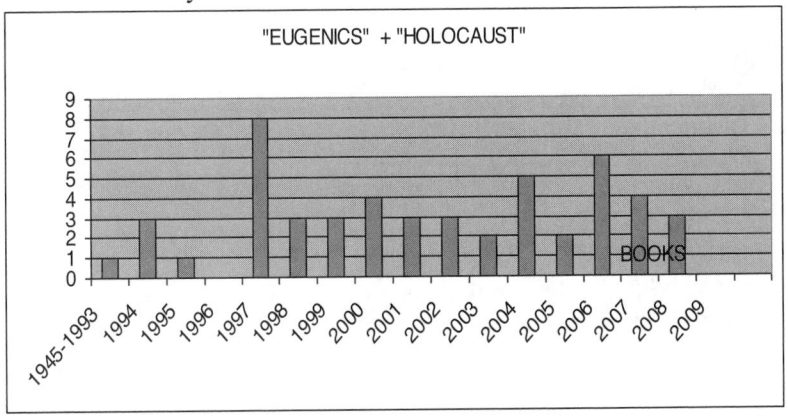

The supposedly causal relationship was accepted with no mention of Jewish participation in the eugenics movement, of the fact that eugenics was popular among the left and the right, of the condemnation by Anglo-American eugenicists of Germany's National Socialist regime, or of Jewish eugenicists who had perished in Hitler's Germany and Sta-

---

[1] N. Gt., 1971-1972, 1031-1032.

lin's Russia. The eugenics-is-evil message was imprinted, almost indelibly.

In the meantime the number of publications on the Holocaust continues to snowball. A search for the word 'Holocaust' on the Online Computer Library Center (OCLC, or 'Worldcat') showed that 94% of books (59,895 vs. 3,708) were produced during the period Jan. 1, 1968-April 11, 2010.

What do publication statistics tell us about the ups and downs of the eugenics movement?

**Average Number of Books Appearing Annually
with the Word "Eugenics" as Keyword
(OCLC Search)**

| PERIOD | TOTAL NUMBER OF BOOKS | AVERAGE NUMBER ANNUALLY |
|---|---|---|
| 1901-1913 | 595 | 45.8 |
| 1914-1918 (World War I) | 281 | 56.2 |
| 1919-1929 | 645 | 58.6 |
| 1930-1938 (Great Depression) | 567 | 63.0 |
| 1939-1945 (World War II) | 196 | 28.0 |
| 1946-1967 (up to Arab-Israeli war) | 343 | 15.6 |
| 1968-1975 (up to 1975 UN Resolution condemning Zionism as racism) | 164 | 20.5 |
| 1976-2010 | 2,850 | 57.4 |

As can be seen, eugenics was not only not abandoned as bad science in the 1930s, as is so often falsely claimed, but was continuing to strengthen, even within the Jewish community.

The number of books published between 1945 and 1967 declined significantly but their content was stilil generally pro-eugenics in tone, with such titles as *Preface to Eugenics; Genetic and Environmental Factors in Human Ability;* or *How Heredity Builds Our Lives: An Introduction to Human Genetics and Eugenics.*

After 1967 works had titles like *A Corrupt Tree Bringeth Forth Evil Fruit: Religion and the American Eugenics Movement; The Logic of Eugenics: The Path from Social Darwinism to the Holocaust;* or '*Hideous Progeny': Eugenics, Disability, and Classic Horror Cinema.* And, of course there are dozens of volumes identifying eugenics as 'racist.'

Since then four decades have passed, and all the while a cohort of Jewish writers continues to attack the eugenics movement, which supposedly threatens at any moment to rip out the stake driven into its vampire heart and once more stalk the planet in search of new victims. In the meantime, precisely as feared by Jewish eugenicists for over a century, the Jews are decimating their own ranks by low fertility and high intermarriage rates. Soon there will be no need for a Shabbat goy to turn out the lights on Shabbat; there won't be any Jews left.

Journalist Ben Wattenberg and Congressional staffer Jeremy Kadden fret that American Jewish women in their thirties are nearly twice as likely to be childless than their non-Jewish counterparts, and that the Jewish community must begin to face this problem.[1]

In 1974 the geneticist Richard Lewontin made the following statement: "For Muller, human progress meant enriching the species for a few superior genotypes while for Dobzhansky it means increasing, or at least maintaining, genetic diversity. Neither view admits the possibility that genetic variation is irrelevant to the present and future structure of human institutions, that the unique feature of man's biological nature is that he is not constrained by it."[2] The model thus proposed is one driven by the human brain's software, with its hardware being essentially identical for everyone and thus irrelevant.

The immense power of the anti-eugenics juggernaut is illustrated by the case of the book *In the Name of Eugenics: Genetics and the Uses of Human Heredity* by Daniel Kevles, who characterizes the eugenics movement as "insidious." The book first appeared in serial form in the *New Yorker*, which is owned by Advance Publications, which also owns dozens of other magazines, newspapers, television and internet operations. Its chairman and CEO is Samuel Irving Newhouse (b. 1926), who was ranked the 38th richest American by *Forbes* magazine in 2008.[3] The work was brought out in book form by Alfred A. Knopf, which is owned by Random House, which is owned by Bertelsmann Media Worldwide,

---

[1] Wattenberg/Kadden, 2005.
[2] Lewontin, 1974, 31.
[3] http://www.forbes.com/lists/2006/54/biz_06rich400_The-400-Richest-Americans_Rank_2.html, accessed May 14, 2008.

which operates in 63 countries and employed 102,397 persons as of December 31, 2007.

On top of all of this comes the popular association of eugenics with Holocaust – even when no such claim is explicitly made. NBC's nine-and-a-half-hour television miniseries *Holocaust* was watched by more than 120 million Americans over four consecutive evenings. Steven Spielberg's 1993 movie *Schindler's List* netted over $221 million at foreign box offices and seven Academy Awards and was watched by 25 million Americans at movie theaters and 65 million on television. It was one of hundreds of films and books on the Holocaust. Or there is Ira Levin's 1976 novel, *The Boys from Brazil*, which was made into a screenplay by Heywood Gould, starring James Mason, Gregory Peck, and Laurence Olivier (production budget $12,000,000). Going up against such a juggernaut is not a task to be taken lightly.

Just who are the heedless Hebrew knights so quixotically assaulting the evil eugenic windmill? The Jewish-Hungarian anthropologist Raphael Patai (1910-1996) took a decidedly uncharitable view of them, describing them as "an extreme manifestation of Jewish self-hate":

> *The New Left was a left-wing radical trend which, in the 1960s, attracted many students and other young people in the United States and Western Europe, and in which Jews played a prominent role. The Jewish participation in the New Left was explained by some analysts as a result of the rationalistic, child-centered, and psychologically understanding home environment of the Jewish middle-class family, which produced children intolerant of rules and restrictions and insistent on the rapid achievement of an ideal society. Another explanation emphasized the historical background of American Jews, which led them, more than the Gentiles, to embrace liberalism, socialism, and communism. The Six Day War of 1967 brought about a crisis and a split in the ranks of the Jewish New Leftists; some became most virulent enemies of 'Zionist imperialism,' while others began to organize distinctly Jewish and pro-Jewish New Left splinter groups and to claim support from the Jewish establishment for their ideas.[1]*

---

[1] Patai, 1977, 479.

Patai maintained that by the mid-1970s this group was "a thing of the past,"[1] but some of their ideas continue to thrive – particularly with regard to human biology.

In 1984 Pantheon Books brought out *Not in Our Genes* by geneticist R. C. Lewontin, biologist Steven Rose (b.1938), and psychologist Leon J. Kamin (b. 1928). The inner front leaf of the book jacket announces that this is a book that "dismantles... the entire myth of scientific neutrality." Not surprisingly, the rear flap contains an evaluation by Ashley Montagu, who calls it "a riproaring dismantling of the recent rise of biologistic interpretations of why we behave as we do." The authors themselves write that they were motivated to write the book out of concern over "the rising tide of biological determinist writing" and that they "share a commitment to the prospect of the creation of a... socialist society."[2] They identify themselves as members of the 'radical science movement' of the 1970s and 1980s:[3]

> *Black radical intellectuals like Malcom X changed the interpretation of crime and imprisonment from individual social pathology into a form of political struggle. If "all property is theft," then theft is just a form of redistribution of property.... The welfare rights movement transformed support payments to women and dependent children from a dole to be received silently into a right to be demanded loudly. The 1960s were marked, in general, by an extraordinary breakdown of a previously accepted consensus and an increase in social struggle.... In the end the owners of capital must control the process of production; the state must control the police and the courts; the schools and universities must control the curricula and students. The growth of biological determinist thought and argument in the early 1970s was precisely a response to the militant demands that increasingly could not be met. It was an attempt to deflect the force of their pressure by denying their legitimacy....*[4]

The authors then proceed to attack the 'bourgeois scientific ideology' in general, declaring even the most abstract pronouncements of physics, such as Newton's laws of motion to have arisen out of the social needs of an emergent class.[5] Having put Newton in his place, the crimson

---

[1] Patai, 1977, 480.
[2] ix.
[3] x.
[4] 21-22.
[5] 41.

troika turns specifically to biology, declaring that only "social organization" stands in the way of social amelioration, and not biology[1]:

> *In one sense, evolutionary theory itself represents the apotheosis*
> *of a bourgeois world view, just as its subsequent development*
> *reflects the contradictions with that world view. The breakdown*
> *of the old static feudal order and its replacement with a conti-*
> *nually changing and developing capitalism helped introduce the*
> *concept of mutability into biology.*[2]

*

> *...high IQ scores... [are] a consequence of... childhood advan-*
> *tages.*[3]

*

> *Twin studies as a whole, then, cannot be taken as evidence for*
> *the heritability of IQ.... In fact, despite the massive devotion of*
> *research effort to studying it, the question of heritability of IQ is*
> *irrelevant to the matters at issue. The great importance attached*
> *by determinists to the demonstration of heritability is a conse-*
> *quence of their erroneous belief that heritability means unchan-*
> *geability.*[4]

Leon Kamin begins his widely read 1974 book *The Science and
Politics of I.Q.* by stating his two "major conclusions" up front: a) "There
exist no data which should lead a prudent man to accept the hypothesis
that I.Q. test scores are in any degree heritable"; and b) "The I.Q. test has
served as an instrument of oppression against the poor." He then goes on
to state that he wrote the book with an explicitly political goal in mind –
influencing policy makers.[5]

Within a few pages the reader is informed that Henry Goddard,
who in 1912 had tested a sampling of Ellis-Island immigrants, had
claimed that "83% of the Jews, 80% of the Hungarians, 79% of the Italians, and 87% of the Italians were feeble-minded."[6] In point of fact, contrary to Kamin's claim, Goddard had specifically stated that "This is a
study not of immigrants in general, but of six small highly selected

---

[1] 129.
[2] 49.
[3] 111.
[4] 116.
[5] 1, 2, 3.
[6] 16.

groups... The study makes no determination of the actual percentage, even of those groups, who are feeble-minded." But Goddard is not consistent: he goes on to write that "One can hardly escape the conviction that the intelligence of the average 'third class' immigrant is low, perhaps of moron, grade." (Goddard leaves open the question as to whether this surmised low IQ is due to bad genes or lack of education.[1]) Thus, even if incorrect, Kamin's now notorious interpretation is at least partially understandable.

Kamin then proceeds to attack the prominent English psychologist Cyril Burt, who had located a number of identical twins who had been raised separately. In 1966 Burt reported an IQ correlation of 0.77 among 53 pairs of identical twins whom he had studied. When Burt, who died in 1971, was posthumously accused of having falsified his data, the purported scandal made for major news. Kamin writes that "the numbers left behind by Professor Burt are simply not worthy of our scientific attention."[2] Kamin even goes so far as to claim that the IQ correlation of identical twins might well be zero.[3] Now, however, a great deal more research has been done on the topic, and Burt's findings have been replicated repeatedly, including Thomas Bouchard's study of 8,000 twin pairs, which came up with a correlation of 0.76 for identical twins reared separately and 0.87 for those reared together.[4] As for broader based kinship correlations, Kamin views them as "in part the product of systematic bias and in part wholly imaginary."[5] Kamin's view is flatly contradicted by a later study of adopted children, conducted by Sandra Scarr and Richard A. Weinberg, also at the University of Minnesota; the adoptee's IQ scores correlated significantly more positively with those of their biological than with those of their adoptive parents.[6]

Kamin sums up his conclusion: "To assert that there is *no* genetic determination of IQ would be a... scientifically meaningless statement. We cannot prove the null hypothesis, nor should we be asked to do so."[7] Kamin's position is thus that of extreme egalitarianism, denying virtually any role to genetic factors and not simply attacking testers, but even questioning their motivation on a moral plane. What is surprising is that such a book, which flies in the face not only of science, but even of

---

[1] Goddard, 1917, 243.
[2] 47.
[3] 52.
[4] Wright, 1997, 61.
[5] 105.
[6] Wright, 1997, 63.
[7] 175.

common sense, achieved its stated political goals. In all fairness to Kamin, it must be conceded that he was justified in pillorying incidents of outrageous insensitivity and sloppy methodology of early testers.

In 1981, W. W. Norton published *The Mismeasure of Man*, by Stephen Jay Gould (1941-2002). Although Gould presented himself as an "evolutionary biologist," the fundamental thrust of his publications was the rejection of evolutionary processes in modern humans, and thus he can be categorized as being entirely in the Boasian tradition. Fully conceding that humans evolved from other species, Gould at the same time preached that Darwinian selection had been petrified in time for our species. Gould himself did not hesitate to point out the radical differences between his views and "the brouhaha over sociobiology"[1] – already then the dominant biological paradigm for studying human behavior. For him the term 'sociobiology' was bereft of meaning and "might as well be dropped."[2]

> *...my dispute with human sociobiology is not just a quantitative debate.... It will not be settled amicably at some golden midpoint.... [These are] two qualitatively different theories about the biological nature of human behavior.... [S]pecific behaviors are... not objects of Darwinian attention in their own right.*[3]

Gould's assertions included the following:

- There are no significant inherited, inborn distinctions in human society.[4]
- The view of science as an "objective enterprise" is a "myth."
- "Science must be understood as a social phenomenon."[5]
- "Facts are not pure and unsullied bits of information."[6]
- Eugenics is no longer a valid worldview in a post-Hitler world.[7]
- General intelligence does not exist.[8]
- Not only does intelligence not exist, it cannot be measured.[1]

---

[1] Pg. 326.
[2] 327.
[3] 329.
[4] 20.
[5] 21.
[6] 22.
[7] 22.
[8] 24.

- Neither is intelligence heritable.[2]
- Scientists searching for causality in human behavior (the eternal goal of all science) are in actuality motivated by their own "social prejudice."
- Even "a few hundred thousand years" of modern human evolution in radically differing environments have been insufficient to have produced more than tiny differences between human races.[3]
- The view that a human being is "nothing but an animal is... fallacious."[4]
- Personality traits such as aggression "cannot be coded in our genes."[5]
- Even if one concedes that the fertility patterns of modern society are dysgenic, evolution does not always follow Darwin's gradualist model, in which minor alterations lead over time to major evolutionary changes. Rather a "punctuated equilibrium" governs lengthy periods of genetic stasis. This seemingly scientific argument, applied, for example, to crustaceans, is a true Trojan horse really intended to be dragged into the gates of the human city. (The idea was not Gould's, but that of Jewish-German geneticist and eugenicist Richard Goldschmidt [1878-1958], who postulated that large sudden macromutations, which he christened "hopeful monsters," were more important than small and gradual changes.)

For Gould, human evolution had come to a grinding halt. He concludes his *Mismeasure of Man* with a quote from T. H. White's fantasy novel *The Once and Future King*, in which the human embryo asks just one favor of God:

*Please God,... if I am to have my choice, I will stay as I am.*[6]

---

[1] 25.
[2] 25.
[3] 323.
[4] 324.
[5] 330.
[6] 334.

The scientific (or, rather, anti-scientific) paradigm proclaimed by Gould is nothing less than breathtaking. How was his message received by society? The back cover of *Mismeasure of Man* boasts a quote from Kamin praising it as "a major contribution toward deflating pseudo-biological 'explanations' of our present social woes" and states that "Mr. Gould has won the National Book Award, the National Book Critics' Circle Award, and the Phi Beta Kappa Science Award (twice) and was in the first group of MacArthur Award winners," and that he had been awarded a teaching position at so prestigious a university as Harvard itself. To say that Gould was 'lionized' is an understatement.

Sander Gilman (b. 1944), referred to earlier, has written extensively on eugenics, but his training is as a Germanist literary scholar. He has opened for us the private notebook of his personal scholarly agenda, and, although it is devoted to his approach to Germanics, he frankly and sincerely reveals in it the core of his negative attitude toward eugenics. His article is entitled "Why and How I study the German," referring not to the study of the German language or the Germans as a people, but to 'the German' as an ethnic category or type:

> ... *I will no longer hear the libel of anti-Semites within the field; I reject their claim for a 'fair hearing' within the profession because their fair hearing will be used, as it always has been, to vilify me, to dehumanize me and my pain. The Holocaust remains and must remain for me and, I hope, for my students the central event of modern German culture, the event toward which every text, every moment in German history and, yes, culture moved inexorably. I am not neutral, I am not distanced, for serving as an outsider does not mean to be cool and clinical, it must mean to burn with those fires that define you as an outsider. My stereotypes of the German (and my awareness of them) lead me to examine the stereotypes that the German has of me. It is from the centrality of the Holocaust in the study of German culture that I must move. For me this is not the age of 'post-modernism,' it is the post Holocaust age. That is the salient marker for our present world, and our work is to understand the world of the German in the light of that moment in history.[1]*

Thus Gilman, who rejects "value-neutral scholarship" and who writes that his own work is "of a piece," reduces all of German civiliza-

---

[1] Gilman, 1989, 200-201.

tion – Kant and Hegel, Mozart, Bach, and Beethoven – to genocide, and he sees his life's goal as inculcating this view in his students.

Gilman describes himself, and thus I would argue also his like thinkers, as a generational phenomenon, rejecting the view of at least some older German Jews that Hitler's Germany was an aberration from German culture.

But this is not all about Holocaust. Gilman names as his 'basic sources' Freud and Freud's followers Karen Horney (1885-1952), Margaret Mahler (1897-1985), Herbert Marcuse (1898-1979), and Wilhelm Reich (1897-1957). Of his contemporaries Gilman names the psychiatrist Otto Kernberg (b. 1928), the cultural anthropologist James Boon (b. 1945), the Afro-American cultural critic Henry Gates (b. 1950), and the feminist critic Elaine Showalter (b. 1941). All of these are proponents of nurture over nature.

Thus the egalitarian heritage of Marx-Freud-Boas achieved enormous influence on the youth culture of the 1960s, and those young people are now persons of influence – editors, writers, university professors. Even more important, others of them are persons of considerable wealth and influence.

The rewards reaped by writers adopting similar views can be illustrated by Edwin Black and his book War against the Weak: Eugenics and America's Campaign to Create a Master Race, which appeared in 2003.

*War against the Weak* was reviewed by Dan Vergano, who discusses eugenics in terms of "historical bigotry," "malevolent scientists," "racists," "fanaticism," "sexism," "elitism," "mistreatment of the weak and poor," and "evil movements."[1]

Vergano's review was published by *USA Today* (1991 year-end circulation 2,274,621[2]), which is owned by Gannett Co., Inc., whose 2008 Web site states that it "employs approximately 49,675 full-time and part-time employees worldwide. Gannett publishes 85 daily newspapers in the USA, including *USA TODAY*, and 18 dailies in the United Kingdom. In addition, the company owns in excess of 1,000 non-daily publications around the world and *USA WEEKEND*, a weekly newspaper magazine. Gannett owns and operates 23 television stations in the United States. The company also has a national group of commercial printing

---

[1] Vergano, 2003.

[2] *USA Today Timeline*,
http://www.usatoday.com/media_kit/pressroom/pr_Timeline.htm, accessed May 14, 2008.

facilities and subsidiaries involved in survey research, direct marketing and new media development."[1]

Black writes that he was supported by his own "fact and foot-note verification team" and "numerous translators."[2] Expressing grati-tude to his "mostly volunteer" assistants in four countries,[3] he writes that more than fifty researchers in fifteen cities in four countries, assisted by scores of archivists and librarians at more than one hundred institutions, helped him to work through 50,000 documents. His Washington, D.C., research staff alone consisted of a dozen assistants. Enigmatically he states that "Many more state officials worked with me on a confidential basis to reveal closed records. Their names cannot be revealed, but they know who they are."

Mr. Black's home page reads that his books have been published in 14 languages in 61 countries, he has been interviewed on hundreds of network broadcasts, and his speaking tours include hundreds of events in dozens of cities each year.[4] He recommends airlines, and of the many restaurants that he recommends he writes that: "The driving factor is quality of cuisine first and foremost, and then style, decor, and service." While Mr. Black has every right to expound his views, which are no doubt sincere, the grotesquely slanted field on which he jousts with op-posing researchers is straight out of *Alice in Wonderland.*[5]

I recall going to a lecture in approximately 1980 by Ashley Montagu at the University of Maryland, where I was teaching at the time. It took place in what may well have been the largest auditorium in the University. Montagu gave the expected environment-is-everything-genes-are-nothing presentation, which he must have delivered hundreds of times. When he finished, he brushed off several skeptical questions with humor until finally a graduate student stood up and said that Monta-gu's views were scientific nonsense and left the hall. But for the most part the students were still in their teens and had little or no inkling of the politics underlying Montagu's message. For that part, neither did the University president, a mathematician, who had been induced to attend to impress the audience. What was remarkable about the lecture was the list of departments and programs that had contributed to Montagu's obvious-ly sizeable honorarium. There must have been fifteen. And there were

---

[1] Gannett: A Brief Company History, http://www.gannett.com/about/history.htm, accessed May 14, 2008.
[2] Black, 2003, ix-xii.
[3] Black, 2003, ix-xiv.
[4] Black, 2008.
[5] Black, 2007.

several hundred students whose attendance was largely mandatory as a condition of this or that course. It was an impressive demonstration of indoctrination, and anyone attending who held to different views could not have helped but sense the futility of opposition.

What are the rewards for presenting the opposite point of view? Dissidents are subjected to academic shunning. Their books and articles are not recommended for publication or are ignored if published, and are certainly not assigned to students. Many librarians not only will not order them, but will refuse to accept them as gifts. Such authors are not invited to participate in conferences or deliver guest lectures, are not awarded grants or academic appointments, and even their correspondence goes unanswered. Since academic specialty fields are normally quite small, such censorship is devastating and unquestionably far more effective than criticism, which might draw journalistic attention to dissident views. This *de facto* blacklisting easily carries the day in the newspapers and on television-radio talk shows, scooping out an ever widening chasm between popular opinion and science. It is a scenario that has been repeatedly played out in academia in the past. Galileo ultimately wins out over the Inquisition, but that can be a *very* lengthy process.

The biased presentation of eugenics has been made possible by the hothouse nature of much of the social sciences in general, and Jewish studies in particular. Already in the nineteenth century, to quote the contemporary American historian John Efron, "the intellectual agenda of Jewish race scientists was to wrest control of the anthropological discourse on Jews from gentiles."[1] Still another contemporary anthropologist Gelya Frank has noted the extraordinary success of this effort: "THERE HAS ALWAYS BEEN [capital letters in original] a lively, if sometimes hushed, in-house discourse about American anthropology's Jewish origins and their meaning. The preponderance of Jewish intellectuals in the early years of Boasian anthropology and the Jewish identities of anthropologists in subsequent generations have been downplayed in standard histories of the discipline."[2]

Outsiders invading foreign scholarly turf have often enjoyed enormous popularity by their insights. For example, the generally positive tone and broad perspective of the two-volume *Democracy in America* (1835, 1840) by Alexis de Tocqville (1805-1859) gained it enormous prestige in America and France. By contrast, when his contemporary, the Marquis de Custine (1790-1857), launched a major assault on Russian

---

[1] Efron, 1994, 29.
[2] Frank, 1997, 731.

culture in his 1839 *Empire of the Czar: A Journey Through Eternal Russia*, the book was banned in that country. History would be vastly impoverished without such outsiders as Herodotus (c. 484 BC-c. 425 BC) or Marco Polo (c. 1254-1324). This seems so obvious that it is strange to have to make the argument.

Even though eugenics is as cross-disciplinary a topic as is possible, encompassing anthropology, bioethics, ecology, genetics, history, philosophy, political science, religion, etc., for many it has been reduced to a branch of Jewish studies – 'Holocaust.'

A multiplicity of views is obviously preferable to the straitjacket of an artificially imposed group cohesiveness, but history teaches that disproportionate political power inevitably seeks to transform scholarly discourse into a propaganda tool.

When we study our own species, we all shoulder a heavy burden of preconceived emotional baggage, but we have no Martians to enrich our range of perspectives. A similar inevitability does not exist with regard to Jewish studies, which have been vitiated by the virtual lockout of non-Jewish scholars from the field. The conduct of dispassionate research is, at best, poorly compatible with political agendas, de facto censorship, an implicit assumption of outsider untrustworthiness, and shrill emotions, understandable as the latter may be. On the brighter side, even though Jewish studies in America and Israel are obviously connecting vessels, the intellectual climate in Israel is vastly more open in this regard than that prevalent in the United States.

## Eugenics in Camouflage

The assault on eugenics has painted bioethicists into a corner. On the one hand they are obliged to condemn eugenics, but on the other they do not wish to plead for sickness. One ruse is the coinage of code words that really mean eugenics – such as 'the new genetics.' Still another solution is to roundly condemn eugenics as a social movement or government program while at the same time advocating it on an individual basis. Note the following two statements by bioethicists Arthur L. Caplan, Glen McGee, and David Magnus, all three employed by the University of Pennsylvania Health System's Center for Bioethics. Both statements are contained in the same three-page article:

"It is a 'given' in discussions of genetic engineering that no sensible person can be in favour of eugenics. The main reason for this presumption is that so much horror, misery, and mayhem have been carried out in the name of eugenics that no person with any moral sense could think otherwise. In fact, the abysmal history of murder and sterilization undertaken in the name of race hygiene and the 'improvement' of the human species again and again in this century is so overpowering that the risk of reoccurrence, sliding down what has proved time and time again to be an extremely slick, slippery slope, does seem enough to bring all ethical argument in favour of eugenics to an end."

"Given the power and authority granted to parents to seek to improve or better their children by environmental interventions, at least some forms of genetic selection or alteration seem equally ethically defensible if they are undertaken freely and do not disempower or disadvantage children.... No moral principle seems to provide sufficient reason to condemn individual eugenic goals."[1]

Conservative Israeli rabbi David Golinkin attempts to make a similar distinction between "gene therapy," which he sees not only permitted but even *encouraged* by Biblical and Talmudic precedent, and "eugenic engineering for non-therapeutic reasons," which supposedly is not: "May we engineer children with blond hair and blue eyes, or children who will grow to be seven feet tall and play basketball? Or piano virtuosi or people with an IQ of 220?"

Scientifically speaking, genetic intervention is genetic intervention, and in a strictly technical sense the distinction is specious. Gene therapy = eugenic engineering = eugenics. Golinkin's intent is actually to approve of eugenics in cases of unambiguous genetic illness and disapprove of it in cases of non-pathological normal distribution (hair and eye color), but even height and intelligence can be problematic. Should not extreme dwarfism and feeble-mindedness be considered pathologies? Thus Golinkin attempts to defend the grand eugenic tradition of Jewish culture while simultaneously paying lip service to condemning the bug-

---

[1] Caplan/McGee/Magnus, 1999, 1,3.

bear of 'eugenics': "As Jews we must be doubly sensitive to eugenics, which was practiced by Nazi doctors in their quest for a master race."[1]

Openly complaining abut the pressure exercised by the constant, often fictional, references in the popular media to Hitler's Germany, Israeli obstetricians Vered H. Eisenberg and Joseph G. Schenker of the Hadassah University Medical Center in Jerusalem define 'germline alteration' to be essentially different from 'eugenic genetic engineering,' the former referring to the insertion of a single gene, while the latter applies to traits determined by more than one gene in interaction with another. One is reminded of a scene in the television series 'I, Claudius' in which Augustus confronts a line of men accused of having slept with his daughter, and one of them defends himself by saying "Only once, Caesar." Not cracking a smile, Eisenberg and Schenker then go on, in effect, to advocate 'germline therapy' for everyone; otherwise society will be even more unequal than it is now.[2] The ruse was about as transparent as is possible, but within a year this specious distinction was repeated by Yossi Segal of the Israeli Academy of Sciences and Humanities, writing in the journal *Jewish Medical Ethics*.[3]

In 2004, writing in the *Israel Medical Association Journal,* Dr. Frida Simonstein of the Ben-Gurion University of the Negev decided to dispense with the verbal equilibristics and openly called for "germ-line engineering," which "may be considered as 'eugenics' but if pursued freely is a noble goal." Using a term coined by molecular biologist Lee M. Silver, she stated categorically that "'self-evolution'… is not only inevitable, but also morally justified." Like countless eugenicists before her, she points out that modern medicine has virtually eliminated selection in human populations, creating a true dilemma with regard to the human genome, which is part of a continuum shared by non-human genomes. We are under a parental obligation, she writes, to protect the genetic patrimony of future generations.[4]

## Euthanasia

A bill was drafted in 1932 by the Prussian Governmental Council – before Hitler's accession to power – to lay the groundwork for selective sterilization in cases of heritable diseases. Although sterilization had been discussed for twenty years, the legislation took the leading

---

[1] Golinkin, 1994, 29.
[2] Eisenberg/Schenker, 1998.
[3] Segal, 1998.
[4] Simonstein, 2004.

German eugenicists by surprise, who were critical of it as counterproductive and inefficient with regard to genetic improvement.[1] On July 14, 1933, the legislation was passed by the German parliament, entering into force in 1934, but now it permitted sterilization against the wishes of the individual concerned, specifically for the surgical sterilization of persons whose offspring would have a high probability of suffering from physical or mental illness, of hereditary feeble-mindedness, schizophrenia, manic-depressive syndrome, hereditary epilepsy, Huntington's disease, hereditary blindness, deafness, or severe physical defects, as well as severe alcoholism.[2] No mention was made of race. From 1934 to 1939 an estimated 300,000 to 350,000 persons were sterilized.[3] Most sterilizations were for feeble-mindedness, followed by schizophrenia.[4] At the time, sterilizations were also being practiced in a number of European countries and the United States. Eugenic considerations did not play a significant role in the German debate. Rather, legislators misguidedly saw sterilization as a cheap alternative to welfare.[5] The Catholic Church was opposed to sterilization, but the Evangelical Church supported it.[6]

The debate over euthanasia was launched by Karl Binding and Alfred Hoche's 1920 book *Legalizing the Destruction of Life Not Worth Living*. The authors, a lawyer and a physician, put forward a strictly economic argument. While there may have been some eugenic relevance in the case of the sterilization legislation, the euthanasia question was, at most, peripheral to eugenics, since its targets were persons who were already institutionally segregated and in many cases sterilized, with limited opportunities for procreation. To their credit, German eugenicists vehemently attacked euthanasia proposals. In 1926, the eugenicist Karl H. Bauer, for example, stated that if selection were used as a principle for killing people, "then we all have to die"; the eugenicist Hans Luxenburger, in 1931, called for "the unconditional respect of the life of a human individual"; in 1933, the eugenicist Lothar Loeffler argued not only against euthanasia, but also against eugenically indicated pregnancy terminations: "we justifiably reject euthanasia and the destruction of *life not*

---

[1] Weingart/Kroll/Bayertz, 1992, 298.
[2] Das "Gesetz zur Verhütung erbkranken Nachwuchses" vom 14. Juli 1933; quoted in Kaiser *et al.*, 1992, 126.
[3] Missa/Susanne, 1999, 18-19 ;Weingart/Kroll/Bayertz, 1992, 470.
[4] Weingart/Kroll/Bayertz, 1992, 469.
[5] Weingart/Kroll/Bayertz, 1992, 22, 174, 263-265, 283, 294.
[6] Weingart/Kroll/Bayertz, 1992, 300.

*worth living.*"[1] Hitler, however, regarded the institutionalized as "useless eaters" who were taking up the time of hospital personnel and occupying bed space to no worthwhile purpose.[2] When, in September 1939, he issued a secret order initiating a national euthanasia program, he did so strictly to free up as many as 800,000 hospital beds for expected war casualties.[3]

The right-to-die movement keeps the topic of euthanasia alive even today. Christian bioethicist Nicholas Capaldi of the University of Tulsa equates Jewish bioethics with Christian ethics:

> *Christianity or the Judeo-Christian (or perhaps we should now characterize it as the 'Abrahamic') inheritance is an intrinsic element of Western civilization. Putting it in as strong a way as I can, there is no ethics other than Judeo-Christian ethics. There can be, as a consequence, no bioethics other than Judeo-Christian bioethics.*"[4]

That Capaldi is misinformed is evident from the following passage by Rabbi Susan Bulba Carvutto of Temple Beth El, in Augusta, Maine:

> *Jewish law does not regard a fetus as a child. A child is a baby that is born. Abortion, even late term abortion, is not murder.... According to Jewish law, an infant does not actually reach full personhood until it is 30 days old. In the days when children commonly died soon after birth, the rabbis decided that parents of infants under 30 days old would be exempt from the requirements of mourning. The customs of sitting shiva, avoiding celebration for 30 days, and reciting Kaddish for eleven months are not traditionally observed for infants under 30 days old. Wishing to spare expense to families, the rabbis even decreed that a funeral is not conducted for such an infant.... Jewish law regard-*

---

[1] Karl H. Bauer, *Rassenhygiene: Ihre biologischen Grundlagen*, Leipzig, 1926, 207; Hans Luxenburger, "Möglichkeiten und Notwendigkeiten für die psychiatrischeugenische Praxis," *Münchener Medizinische Wochenschrift*, 1931, 78: 753-758, 753; Lothar Loeffler, "Ist die gesetzliche Freigabe der eugenischen Indikation zur Schwangerschaftsunterbrechung rassenhygienisch notwendig?" *Deutsches Ärzteblatt*, 1933, 63: 368-369, 369. All quoted in Weingart/Kroll/Bayertz, 1992, 524, 526.
[2] Aktion "T4"/"Wilde Euthanasie" (1939-1945); Aussage des "T4"-Leiters Viktor Brack: "Nutzlose Esser" 1946); Aus: DOC-NO426, in GSTA, Rep. 335, Fall 1, Nr. 202, Bl. 11; quoted in Kaiser *et al.*, 1992, 250.
[3] David Irving, *Hitler's War*, Viking Press, 1977; quoted in Saetz, 1985.
[4] Capaldi, 1999, 247.

*ing abortion is very different from Catholic law, which considers life to begin at conception.*[1]

But Carvutto, who claims that "eugenics is not a Jewish value," is herself not aware of all the facts. At least some interpretations of Jewish law permit not only late abortion, but also neonaticide. During the early 1950s the chief rabbi of Israel, Ben Zion Uziel (1880-1954), maintained that the killer of an infant within thirty days of birth could not be executed because the infant's life is still in doubt.[2] The timeline contains still more items, some of them frankly shocking, on this painful topic.

The very phrase "Judeo-Christian" is, to a significant degree, rooted in a twentieth-century political striving to gloss over differences and suggest more commonalities than actually exist. Christianity has far greater affinities with Islam, both being universalist religions, whereas Judaism is explicitly tribalist. If the Christian-Islamic tradition is based upon total submission to a supreme being, Judaism advocates partnership. Central to both Islam and Christianity is the belief in eternal life, whereas, in the words of Rabbi and bioethicist Elliot N. Dorff, Judaism merely "holds out hope that in some way we continue to live after death."[3] And, of course, there is the incredible but enormously powerful political alliance between Christian fundamentalist eschatology and Zionism, with the former preaching that those Jews who refuse to convert to Christianity will be cast into Gehenna (the lake of fire and brimstone) as a result of the battle of Armageddon. Note the following 2008 incident, as reported by CNN News:

> [Presidential candidate John] *McCain* [b. 1936] *told CNN's Brian Todd that he rejected* [Reverend John] *Hagee's* [b. 1940] *endorsement after Todd brought to his attention Hagee's comments that Adolf Hitler had been fulfilling God's will by hastening the desire of Jews to return to Israel in accordance with biblical prophecy. 'God says in Jeremiah 16: 'Behold, I will bring them the Jewish people again unto their land that I gave to their fathers. ... Behold, I will send for many fishers, and after will I send for many hunters. And they the hunters shall hunt them.' That would be the Jews. ... Then God sent a hunter. A hunter is someone who comes with a gun and he forces you. Hitler was a*

---

[1] Carvutto, 2004.

[2] "Infanticide," *Encyclopedia of Death and Dying*, http://www.deathreference.com/Ho-Ka/Infanticide.html. See also http://faqs.org/faqs/Judaism/FAQ/12-Kids/section-23.html.

[3] Zohar, 2006, 70.

*hunter," Hagee said, according to a transcript of his sermon. In a statement to CNN on Thursday, McCain said "Obviously, I find these remarks and others deeply offensive and indefensible, and I repudiate them. I did not know of them before Rev. Hagee's endorsement, and I feel I must reject his endorsement as well."[1]*

## Abortion

*On Rosh Hashanah it is written, on Yom Kippur it is sealed. How many shall pass on, how many shall come to be; who shall live and who shall die.*

The prayer "Unetane Tokef"

According to the Israel Human Center for People with Disabilities, persons with disabilities made up over 10% of the Israeli population in 2001. This is roughly 600,000 individuals.[2] Dr. Mark Levin of the Brookdale University Hospital in Brooklyn writes that it is estimated that 20-25% of Ashkenazi Jews carry a mutation for one of the so-called 'Jewish' genetic diseases.[3] Such a figure would pose a dilemma for any group, but it is particularly thorny for religious Jews in light of the unambiguously eugenic thrust of Jewish law, according to which families that appear to transmit a heredity illness should be avoided.

In contrast to a 2006 Harris Poll that showed that only 11% of Americans deny the existence of God,[4] an Israeli government poll shows that 69% of Israeli Jews identify themselves as "non-religious,"[5] so that rabbinical opinion is decidedly a minority view in Israel – despite its political influence as a 'swing-vote' lobby. Thus it should come as no surprise that genetic screening followed up by selective abortion is not controversial in Israel – even among those associations of the handicapped whose U.S. analogs are deeply offended by the practice (as opposed to U.S. *parental* organizations, which are supportive). In one Israeli study of the frequency of genetic screening, an astounding 94% of secular women performed amniocentesis, leading behaviorial scientist Aviad Raz of Ben Gurion University to comment that the Israeli view of disa-

---

[1] CNN, 2008.
[2] Raz, 2005, 185.
[3] Levin, 1999, 208.
[4] Harris Poll, 2006.
[5] Israel: Like thisF, as if," 2007.

bility is a secular construction which is also utilitarian and collectivist in nature. He surveyed the major Israeli associations of the handicapped and found that

> *prenatal testing was eugenic and was indeed supported precisely for that, since 'eugenic' for them meant the improvement of the health of progeny and carried positive rather than negative connotations. The two-fold view of disability, comprising both a eugenic prenatal policy and the support of people with disabilities, was captured in the motto of the Israeli Organization for Rare Disorder: Important to Know, Important to Test. Important to Support. Other respondents, who repeated the basic argument in support of genetic testing, nevertheless also acknowledged that it constituted a dilemma regarding eugenics.... It may be surprising to many that Israeli culture, with Judaism as its state religion and the holocaust as its antecedent, is in favor of eugenics.[1]*

Among the three chief branches of Judaism, Reform Jews do not consider themselves bound by Halacha (Jewish religious law) while Conservatives interpret it quite flexibly, both groups placing reliance on individual responsibility. That leaves really only the Orthodox for whom abortion is a particular issue, and even their understanding of Halachic law has evolved over the ages. As pointed out by Dr. Mark Levin of Brookdale University Hospital in Brooklyn, the Orthodox religious mandate is not unified – a circumstance well know in Rabbinic circles, but largely unfamiliar to the majority of Orthodox laymen, who tend to liken abortion to murder (except when essential to preserve the mother's life). In reality there is a second Orthodox view that extends back to medieval scholars other than Maimonides, according to which abortion is merely undesirable rather than criminal.[2] Israel's abortion law, which permits abortion for *any* defect in the embryo, is a classic example of unadulterated negative eugenics.[3]

---

[1] Raz, 2004, 186.

[2] Levin, 1999, 209.

[3] This is now changing for late term abortion. the Israeli Ministry of Health (memorandum #23/07, issued 19.12.07) decided that requests for late term abortions (beyond week 24) will be discussed only by special committees (which include more members), and to justify termination of pregnancy the embryopathy must be medically considered severe as well as probable (probability of more than 30%). (Raz, 2009b)

## Eugenics Recovers

Political scientist Diane Paul in 1984:

*Virtually all of the left geneticists whose views were formed in the first three decades of the century died believing in a link between biological and social progress. Their students, coming to intellectual maturity in a radically different social climate, either did not agree or, in a social climate inhospitable to determinism, were not willing to defend that position. The appearance of socio-biology probably signifies a fading of the bitter memories surrounding the events of the 1940s. As those memories recede, it would not be surprising to witness the re-emergence of a doctrine that was never defeated in the scientific arena but rather submerged by political and social events.*[1]

Just fourteen years later she recalled her 1984 article in which she had characterized as

*"hereditarian" or "biological determinist" the view that differences in mentality and temperament were substantially influenced by genes – employing these terms as though their meanings were unproblematic. That usage today would surely be contested. For the view implicitly disparaged by these labels is once again widely accepted by scientists and the public alike.*[2]

But the word *eugenics* has been detached in the public consciousness from its meaning and remains for many an object of opprobrium.

## Israel and Reproductive Cloning

*And HaShem G-d caused a deep sleep to fall upon the man, and he slept; and He took one of his ribs, and closed up the place with flesh instead thereof. And the rib, which HaShem G-d had taken from the man, made He a woman...*

Book of Genesis, Chapter 2, Bereshit 1

---

[1] Paul, 1984, 590.
[2] Paul, 1998, 29.

Reproductive cloning is a technology used to create animals that are genetically identical, just as are identical twins (who, for the most part, do not agree that their own existence is a monstrous event). The process entails the transfer of a nucleus from a (somatic) donor cell to an egg which has no nucleus. The reconstructed egg containing the DNA from the donor cell must be treated with chemicals or electric current in order to stimulate cell division. If the egg begins to divide normally it is transferred into the uterus of a surrogate mother.

The process has already been employed to reproduce tadpoles, camels, cats, carp, cattle, horses, mice, mules, sheep, and water buffalos. A different technique has been used to clone rhesus monkeys – by splitting the embryo at the eight-cell stage into four genetically identical two-cell embryos.

The first claim of a successfully cloned mammal dates back to 1979, when a professor of biology at the University of Geneva, Karl Illmensee, asserted that he had extracted the nucleus of a four-day-old mouse cell by sucking it into a pipette smaller than the diameter of the cell and then inserting the same pipette into a fertilized mouse egg – much the same technique used successfully today. When a colleague attempted to replicate the experiment and failed, Illmensee was accused of having falsified his research and forced to resign.[1] The affair illustrates why so much of cloning research is conducted surreptitiously.

The Scottish embryologist Ian Wilmut (b. 1944), who cloned the sheep "Dolly" in 1996, went through 277 eggs before succeeding and some of the sheep were born dead or deformed.[2] Since animals reproduced in this fashion can experience significant health problems, most reproductive scientists consider cloning unsafe for use on humans before it is perfected on animals, but there is no doubt that human babies could be produced today with existing technology.

Despite enormous misgivings over potential genetic defects, in 2001, the Italian gynecologist Severino Antinori was quoted by the Italian news agency ANSA as saying human reproductive cloning would "very probably" occur first in Israel, and the German news magazine *Der Spiegel* identified the venue as Caesarea, an Israeli coastal resort. Antinori was quoted by the Italian News Agency ANSA as saying he would seek "political and scientific asylum" in Israel if hostility to his project continued in Italy,[3] but the Israeli Health Ministry said that cloning hu-

---

[1] *Nature*, 1985.
[2] Demick, 2001.
[3] http://www.pakistaneconomist.com/issue2001/issue13/etc3.htm.

man beings was illegal in Israel and dismissed reports that a reproduction team planned to begin the first cloning of a person in Israel within a year.[1] According to United Press International, Antinori told Italian television reporters he had helped three women become pregnant with clones, a claim his office confirmed: "The [pregnancies] are progressing nicely."[2] There is no evidence now that these pregnancies ever resulted in births, if they ever occurred in the first place.

London's *Sunday Times* quoted the Israeli-American biotechnologist Avi Ben-Abraham, a participant in the project: "People claim we are moving too fast. They are right. We are stretching God-given intelligence as far as we can. We are breaking the rules of nature. But our goal is to save lives and cure diseases, and we believe we will soon be successful."[3] An unsourced *Wikipedia* article claims that Binyamin Netanyahu had endorsed Ben-Abraham and that he had won the primaries for a top seat on the Likud party list of candidates for the Knesset in 1999, but was not elected when Likud and Netanyahu were defeated by Labor party leader, former Israeli Defense Forces chief of staff Lt. General Ehud Barak.[4]

In late 2002 the Cypriot-American reproductive biologist Panayiotis Zavos, who had been collaborating with Antinori, met with Mohammed Fadlallah (b. 1935), the spiritual leader of Hezbollah, in Beirut. Fadlallah saw no Islamic objection to human reproductive cloning and gave his blessing.[5]

In 2004 The London newspaper *The Independent* wrote that "mainstream fertility scientists" had attempted to "gag" Zavos by "imploring the British media not to give him the oxygen of publicity."[6] On April 21, 2009, Zavos claimed to have cloned 14 human embryos and transferred them into the wombs of four women. He also claimed to have implanted DNA from three dead people into enucleated cows' eggs for research purposes, not for implantation. On April 29, 2009, he said: "I think we have three embryos that could be *in utero* today" in a secret laboratory in an undisclosed "Middle-Eastern country" (Aman, Jordan). [7]

A competing group is that of the biochemist Brigitte Boisselier (b. 1956) of the so-called Raëlian religion. In January 2003 Boisselier

---

[1] A-J World News, 2001.
[2] Lyman, 2002.
[3] Rogers/Follain, 2001.
[4] "Avi Ben-Abraham," *Wikipedia*, accessed July 11, 2009.
[5] http://www.zavos.org.
[6] Connor, 2009.
[7] SkyNews, 2009.

claimed in a Florida courtroom that her company Clonaid had cloned three people and that she had seen a videotape of one child living in Israel.[1]

While it seems inevitable that human clones will *soon* be a reality, the question is *how soon*. In 2001, the Bioethics Advisory Committee of the Israeli Academy of Sciences and Humanities noted that embryos outside the uterus are not regarded as human life and that human improvement of God's plans is considered laudable: "reproductive cloning may one day be a safe technology...."[2]

Israel's innate tension between its majority secular culture and minority religious culture fades away when it comes to cloning. Neither finds any intrinsic moral objection to the prospect in as much as both favor a pronatalist policy. Israel is truly a leader in this regard. The current moratorium is not a ban, but merely a requirement that the technique be developed on animals before using it to create human beings. Embryo and stem cell research are not prohibited or even regulated in Israel. The legal specialist Carmel Shalev of Tel Aviv University writes that in the parliamentary committee discussions any resistance was crushed by *government* intervention.[3]

Ethicist Asa Kasher perceives the prevailing negative attitude as excessively influenced by German Catholics and German traumatic memories of the National Socialist period:

> *We, the Jews from Israel, are the only ones in the world who can come to the Germans and tell them, you are exaggerating.[4]*

And there is the physician and specialist on halachic bioethics Avraham Steinberg:

> *Cloning, whether reproductive or therapeutic, is permitted at a basic fundamental moral level. The moral Israeli position says this is a right process.[5]*

As noted earlier, a 2006 survey conducted by members of the Department of Nursing of Tel Aviv University revealed that 16% of the respondents agreed with the statement "Cloning should be permitted for producing individuals with high IQ," and 35% believed that "cloning

---

[1] Associate Press, 2003.
[2] Prainsack, 2006, 181-182.
[3] Shalev, 2008, 327.
[4] Shalev, 2008, 329.
[5] Shalev, 2008, 329.

should be permitted for avoiding genetic diseases."[1] The specialist in biopolitics Barbara Prainsack quotes an anonymous Israeli 'expert': "*Banning* human cloning would be against human dignity."[2] Molecular biologist Michel Revel of the Weizmann Institute of Science in Revohot, Israel, complains of "irrational fantasies and fears."[3] Joshua Lipschutz of the University of California opines that "there is nothing inherently wrong with the idea of human cloning" and hypothesizes that "even if a body was cloned, the brain, which is the essence of humanity, would remain unique.... The debate should be changed from 'Is cloning wrong?' to 'When is cloning wrong?'"[4]

On July 5, 2009, a discussion appeared in *Ha'aretz* on the use of primates or cows as surrogate mothers for human clones, with Rabbi Moshe Botschko opining that such a person could be killed as non-human, and theologian J. David Bleich even advising that *kashrut* (Jewish dietary laws) would permit him to be "slaughtered" (?!); John Levica of the Columbia University Department of Medicine took an opposing point of view. What catches the eye here is the selection of primates and cows as surrogates. Other primates are the species biologically closest to man, of course, but cows have the same gestational period as women. The journalist reporting the story noted that Judaism considers in vitro fertilization to be "a superb alternative for infertile couples determined to fulfill the biblical commandment to 'be fertile and multiply.'"[5]

When the Van Leer Jerusalem Institute conducted meetings in 2002 on the prospect of Jews becoming a minority in Israel because of much higher birth rates among Israeli Arabs and Arabs in Gaza and the West Bank, *Ha'aretz* reporter Lily Galili exclaimed: "Had the Jewish underground operated with the same degree of secrecy as the team that meets at Van Leer, it is doubtful that it would ever have been exposed."[6]

Such a level of furtiveness deprives the observer of verifiable facts and requires an assessment of probability. (Think Jack Nicholson in the movie *Chinatown.*) What do we know?

According to the online *CIA World Factbook*, accessed in February, 2009, Jews officially comprise 76.4% of the Israeli population,

---

[1] 120 Israelis (68 health professionals and 52 non-health professionals; Barnoy/Ehrenfeld/Sharon/Tabak, 2006, 27.
[2] Prainsack, 2006, 195.
[3] Revel, 2000, 8.
[4] Lipschutz, 1999, 105.
[5] Ilan, 2009.
[6] Galili, 2002.

with non-Jews accounting for 23.6%,[1] and no one knows just how many non-Jews from Eastern European countries pass themselves off as Jews in Israel. In addition, there are the Jews from India and Africa, many of whom have been settled in the occupied territories to stake out territory. Even if these official figures are accepted at face value, according to Arnon Sofer, Chair of Geostrategy at the University of Haifa, the Jewish majority *within* Israel will shrink to 65% by 2020, and overall 8 million non-Jewish Palestinians will outnumber the 6.6 million Jews by the end of that very brief stretch of time.[2]

The *2009 World Population Data Sheet* published by the Population Reference Bureau in Washington, D.C. shows a total fertility rate for Israel of 2.9, a disproportionate amount of which is accounted for by Israeli Arabs (4.36 in 2004[3], and 4.6 in Gaza and the West Bank), giving a projected population in the "Palestinian Territory" of 8.8 million by mid-2050 (113% growth), as opposed to 11.2 million in Israel (only 49% growth). Israel is acutely aware of the "demographic threat," which has been referred to as a "velvet holocaust."[4] Are experiments in human reproductive cloning being observed with more than a casual eye?

## A Paradoxical Attitude

How, the reader must inevitably ask, has it come about that the fundamental pro-eugenics thrust – not just of Judaism and Zionism, but of Jewry in its essence and totality – has been so assiduously concealed, but that eugenics has been wed in the public consciousness to a tragic infatuation with the topic on the part of a political figure who put a bullet through his brain in a besieged Berlin bunker?

The Jewish attitude toward eugenics is truly paradoxical. We are dealing here with a disconnect between practice and theory. For lack of a proper understanding of the nature of eugenics as a worldview, the rage has been directed at the word, so that a number of alternative terms have been proposed – "reproductive genetics," for example (as if such a beast as 'non-reproductive genetics' actually existed). It is like writing an encyclopedia article about horses without using the word 'horse.'

Jews are a dynamic, intelligent, and resourceful people whose contribution to culture and civilization is vastly out of proportion to their

---

[1] https://www.cia.gov/library/publications/the-world-factbook/print/is.html.
[2] A pamphlet entitled *Israel Demography 2000-2020*, cited in Galili 2002; Prainsack, 2006, 187-188.
[3] Ilan, 2006.
[4] Prainsack, 2006, 188.

numbers. Jewish eugenic practices may often have been inadvertent and hugely influenced by outsiders, but their existence can no longer be denied.

Not merely our understanding of the past is at stake, but so too is the genetic future of our planet and species. We owe truth to the past, and parental love and responsibility to the future. No less.

## Some Questions

Eugenics has come to be associated with conservative or 'right-wing' politics. In point of fact its historical roots are just as much in the left as in the right. Now too, eugenics extends the full length of the ideological spectrum, and unquestionably this state of affairs will continue on into the future.

While the assault on the 'biological paradigm' of human behavior was a long time in the making, only eight years passed between the Arab-Israeli 1967 war and the 1975 appearance of Wilson's *Sociobiology* – the now accepted model of scientific inquiry. That is how long it took science to move ahead to its new synthesis without even bothering to respond to the claims of radical egalitarianism. At the same time the popular media still support this seductive view of life.

Jewish political influence has elevated the Jewish debate over eugenics from a parochial topic to a major factor in determining current and future human evolution, and this dualism has huge political implications which extend far beyond Jewish topics.

First of all, what are the implications for 'democracy' if the overwhelming majority of the population is largely ignorant of and indifferent to the sheer survival of our species, not to mention the teleology of culture?

Second, should we perhaps reconsider the question of within- and between-group variance? Is all 'diversity' desirable?

Third, we are free to do what we want to do, but we do not decide what to want. Where is there place here for 'free will' and 'morality'?

Fourth, the animal-based model of human behavior leaves precious little space for belief in a deity who will restore to us life and our loved ones in some eternal higher sphere. The religious model promised blissful eternal life. What's left now?

Fifth, what mechanism will ensure biological selection and prevent catastrophic human decline – a recognized requirement for all species?

Sixth, should we not prepare for the soon-to-arrive moment when the machine brain, created by us, will be superior to its makers? Will our place not be in some cosmic zoological garden – along with zebras, bumblebees, and speckled trout?

Seventh, should we, can we adopt rational principles of animal husbandry? Just one example: what breeder would even consider monogamy for dogs and horses? (I told you I was going to get into hot water.)

Eighth, are quality considerations the most urgent peril, or do quantitative considerations constitute the more imminent menace? Clearly, the promised 'demographic transition' is taking far too long. There are too many of us, and the twenty-first century will unquestionably witness horrible – and preventable – mass tragedies.

Lastly, and I apologize for asking a millennia-old question, what is existence all about?

## E. M. Lilien

The 'Muscle Jew' (*Muskeljude*)

# A Macro-Chronology of Jewish Eugenics

## 18th-19th Centuries

Jewish biology is pursued largely by non-Jewish scholars, with some Jewish writings by figures such as the criminologist Caesare Lombroso. Jews are self-defined by two mutually overlapping criteria: as a) descendants of ancient Jews and b) persons who practice Judaism. The major demographic event is the beginning of Jewish emigration from the Russian Empire in the 1880s. The negative reception of East European Jews by German and American Jewry stimulates interest in Zionism. Egalitarian Marxism and Lamarckian theories of acquired characteristics become popular among Jews. Frequent discussion by both Jews and non-Jews of Jewish mental illness. Concern over assimilation.

## 1894-1908

Jewish advocacy of "racial hygiene," strong overtones of Social Darwinism. The topic of declining civilizations (*Untergang*) is frequently discussed in Western society, and Jews become troubled by the growing rate of intermarriage, with Zionism still unpopular among the overwhelming majority of Jews but viewed by some as a means for combating population loss. On the qualitative side, Zionism presents itself as a biological model for race improvement. Secularization among Jews renders problematic the religious criterion for determining who is Jewish, leaving "race" as the chief determiner. Freudianism appears as an environmentally determined model of human behavior.

## 1909-1933

Enthusiastic Jewish advocacy of eugenics, but "race" begins to be replaced by "ethnicity" in determining Jewishness. Boasian cultural anthropology is advanced as an alternative to physical anthropology. World War I and the subsequent establishment of the USSR halt Jewish emigration from the Russian Empire. Coalescence of a powerful Jewish lobby in American politics and commerce, in the Weimar Republic, and within the Soviet government.

## 1933-1939

German persecution of the Jews by the National Socialist government raises some doubts in the biological paradigm of human society, but Jewish support for eugenics remains strong. The Zionist movement collaborates with the German government in pursuing the same goal –

the transfer of German Jews to Palestine. Soviet assault on genetics. Continuing low Jewish fertility and increasing rates of intermarriage. Heavy Jewish emigration from Germany.

## 1939-1945

Continuing Jewish support for eugenics, but now Jewish racialist claims must compete with anti-hereditarian views espoused by figures such as Ashley Montagu. Massive Jewish population losses disproportionately reduce the upper end of the intellectual spectrum. The first claim of six million killed is advanced by Ilya Ehrenburg in Russia in December 1944.

## 1946-1948

Continuing Jewish support for eugenics under the rubric 'genetic counseling.' A majority of American Jews still view Zionism in a negative light.

## 1948-1967

The creation of the State of Israel makes possible an "ingathering" of the Jews in Palestine. The topic of "Holocaust" is largely ignored by the Jewish community. Jewish rejection of Zionism is overcome, and Jewry is defined by its identification with Israel. Continuing support for eugenics among American and Israeli Jews. Discovery of the double helix structure of DNA makes "designer babies" theoretically possible.

## 1968-1975

The heretofore largely ignored Holocaust is cited as justification for the creation of Israel. Massive exodus of Jews from their ancestral homes in the Muslim world and Eastern Europe. The Vietnam War and the American civil rights movement radicalize American Jewish youth. The Holocaust Memorial Movement is launched. The first Jewish attacks on sociobiology and eugenics. Continued Jewish support for eugenics, but eugenics is more and more frequently presented as the ideology of genocide. Jewry comes to be defined more by its relation to the Holocaust than to Israel, leading one Jewish scholar to speak of an "anti-Zionist" period, reducing the Zionist period to the blink of an eye. Strong Jewish support for multi-culturalism and multi-racialism in America, but not in Israel.

## 1975-1982

Continuing attacks on intelligence testing, denial of the existence of human subspecies, popularization of the Khazar roots of the Ashkenazim, continued Jewish support for eugenics in the United States and especially in Israel, Jewish religious groups embrace genetic screening. The Soviet Union gives into American pressure and permits Jewish emigration.

## 1983-1993

American Reform Jews accept patrilineal descent, but Israel does not recognize the children of Jewish fathers and non-Jewish mothers as Jewish. The media continue their assault on studies of human diversity in opposition to sociobiology, creating an unbridgeable split with this now predominant paradigm within the scientific and scholarly communities, on the one hand, and popular opinion, on the other. Increasing Jewish uneasiness over low Jewish birth rates and intermarriage. Exodus of Russian Jews to the United States and Israel, unknown numbers of them simply claiming to be Jewish and destined to continue the gene infiltration that has always existed. Even Ethiopian Beta Israel members are accepted by many in Israel as Jews. Genetic screening becomes generally accepted within the Jewish community. Religious representatives and physicians in Israel are reported to label women undergoing abortion 'reproductive deviants.'

## 1994-1997

Many scientists reject media-promoted denials of human diversity, leading anti-eugenics activist Jeremy Rifkin to declare: "This is the dawn of the eugenics era." The publication of *The Bell Curve* by Charles Murray and Jewish-American psychologist Richard Herrnstein rekindles the debate about race and intelligence. Continued Jewish support for and practice of eugenic measures while avoiding usage of the word 'eugenics.'

## 1998-2001

A number of Israeli and American scholars – religious and secular – openly speak out in support of eugenics, while American Jewish pundits continue to attack it. Secular Jewish bioethics is under an in-house assault as sociobiology reduces morality to 'reciprocal altruism' and an increasing number of television documentaries focus on the near-human qualities of other species. Jewish popular attention is excited when a large proportion of contemporary Jewish Kohanim are found to share a set of Y chromosomal genetic markers. Jewish pundits ally themselves with the Christian Right and intensify their denial of human diversity. Attacks on eugenics constitute a significant Jewish cottage industry.

## 2002-2010

Donor insemination is popular in Israel. A plethora of Holocaust Memorial Museums continue to attack eugenics, ignoring its widespread *de facto* acceptance by American and Israeli Jews. Jewish-promoted immigration of non-Jews to the United States continues to grow and encounter fierce nativist opposition. France passes a law imposing criminal imprisonment and a fine of €7,500,000 for the practice of eugenics. More Israeli scholars and scientists either advocate eugenics or point out that it is openly practiced in Israel. Jewish eugenic measures eradicate Tay-Sachs Disease among Jews. Temple University's Center for Afro-Jewish Studies conducts a seminar on 'Jews and Race. As a result of high fertility, the Indian group Bene Israel numbers 50,000 in Israel. No satisfactory definition exists for the word 'Jew.'

## E. M. Lilien Drawing of Max Nordau

# A Micro-Chronology of Jewish Eugenics

*One day a labor-loving monk like me*
*will come across my diligent, nameless work,*
*and just like me, he'll light his lamp,*
*shake loose the dust of centuries,*
*and recopy onto parchment these true tales,*
*that the children of the Orthodox might know*
*the past fate of their native land,*
*and honor their great kings*
*for their feats, their glory, their good deeds.*
*As for their sins and evil doings,*
*let the people humbly beseech the Savior.*
*In old age I live again,*
*the ocean of the past surges by,*
*littered with the flotsam of events,*
*few are the faces preserved by memory,*
*and the rest has perished without retrieve.*
*But day approaches and the lamp burns down.*
*Just one more, last tale*
*and my chronicle will be finished.*

Pimen's 1598 soliloquy,
historical play *Boris Godunov*
Alexander Pushkin, 1825

## 1844

1. British statesman and novelist Benjamin Disraeli (1804-1881) in the novel *Coningsby* advances the theory of Jewish intellectual superiority. One of the characters, Sidonia, declares "Race is everything."

## 1845

### Context

1. From the *Occident and Jewish American Advocate* on "Intermarrying with Gentiles":

   - A letter to the Editor: "How ridiculous it is to see a man who has married a gentile wife, and has for her sake given up every thing which his religion demands of him, mount

the reading-desk on our most solemn days, and participate in
the religious services of the day.... [Such] men should not
be allowed to be called to the reading of the law, nor to be
reckoned to make Minyan nor in any way to be counte-
nanced or regarded as Jews. Besides this, in case of their
death no especial notice should be taken of them, they having made their selections of companions for life, let their
gentile relatives take care of their dead bodies, and inter
them in any manner they may deem proper." (Simeon Abra-
hams)

- Editor's response: "It is our doctrine לארשי אטחש יפ לע ףא
  הוא 'Though he has sinned, he is still an Israelite.' Is this li-
  berality or not? Let our readers decide.... We are glad in the
  meantime that the subject has elicited attention, and we shall
  be happy could we be assured that we have contributed a little towards correcting so great an evil as admixture with
  gentiles, through bonds of consanguinity, the greatest danger which Israel is exposed in the dispersion."[1]

## 1846

### Context

1. From a sermon by Rabbi Isaac Leeser: "they who leave the Synagogue, either through apostacy, through the neglect of circumcision, or
   through intermarriage with gentiles, become part and parcel with
   the non-Israelites among whom we dwell, and they and their descendants, except under rare circumstances, become strangers, and
   must remain so, to the worship of the God of Israel; they merge into
   the nations of the earth, and have neither right nor inheritance in the
   congregation of Jacob. Here then we have a view of the past and
   present condition of the sinners in Israel."[2]

## 1851

### Context

1. Jewish-Polish biologist Robert Remak (1815-1865) discovers cell
   division.

---

[1] Vol. II, No. 12, March.
[2] "The Lord Our Guide," *Occident and Jewish American Advocate*, vol. IV, No. 9,
Dec.

## 1851-1923

### Context

1. In Berlin Jews have lower fertility than do non-Jews.[1]

## 1851-1862

### Context

1. Italian Jews have a lower fertility rate than the total population.[2]

## 1860

### Context

1. The Society for the Amelioration of the Condition of the Jews presents its *Annual Report*, stating that it has visited 1,382 Jewish families and that "34 individuals had called for serious conversion." During the year the Society has distributed 34 Bibles, in German and Hebrew, 11 New Testaments, and 84,000 tracts, mostly in German. Doctor Hicock, of Scranton, Pennsylvania, delivers a sermon: "The universal triumph of the Church, as the events of the time proclaimed, was near at hand, and the conversion of the Jews – who were distributed over the whole earth – who circulated wherever money circulated who adhered to their faith with a tenacity which – as their religion was divine, so far as it went, would make them the better Christians – was inevitably designed to aid in that glorious work"[3]

## 1861

### Context

1. British ethnologist John Beddoe (1826-1911) in an address to the Ethnological Society of London stresses the interaction of natural selection and environment in producing Ashkenazi and Sephardic Jewish types ("On the Physical Characteristics of the Jews").[4]

---

[1] Liebman Hersch, "Jewish Population Trends in Europe," *Jewish People: Past and Present*, II, 11, Table 10, cited in Goldscheider, 1967, 200.
[2] Roberto Bachi, "The Demographic Development of Italian Jewry from the Seventeenth Century," *The Jewish Journal of Sociology*, IV, Dec., 184, Table 13; cited in Goldscheider, 1967, 200.
[3] The Society for the Amelioration of the Condition of the Jews, 1860.
[4] Beddoe, 1861.

## 1862

### Context

1. Socialist and precursor of Zionism Moses Hess (1812-1875): "The Jewish race is one of the primary races of mankind that has retained its integrity, in spite of the continual change of its climatic environment, and the Jewish type has conserved its purity through the centuries."[1]

## 1864

### Context

1. German physiologist Carl Vogt (1817-1895) "distinguishes between Eastern European and Mediterranean Jewish types."

## 1865

### Context

1. Lead article in *Allgemeine Zeitung des Judenthums*, evidently written by editor, Rabbi Ludwig Philippson: "Genetic predispositions are only a tiny beginning and an individual factor in the development of a great and long-lived race."[2]

## 1867

### Context

1. The Southern Baptist Convention, meeting in Memphis, Tennessee, resolves "that it is our duty to labor a pray more earnestly for the conversion of the Jews. The resolution gave rise to discussion, which took the widest range and consumed the greater part of the morning discussion, all prominent members participating."[3]

---

[1] *Rom und Jerusalem: Die letzte Nationalitätsfrage*; cited in Cantor/Swetlitz, 2006, 138.
[2] Philippson, Ludwig. 1865, 709.
[3] *New York Times*, 1867.

## 1868

### Context

1. Notice in *New York imes*: "BEDFORD MA. BAPTIST CHURCH. – Rev S. Kristelder, a converted Jew, now under the auspices of the Christian Brotherhood, and a recent of the Union Theological Seminary of the City, will preach at the above church (Dr. Dowling's, on SUNDAY EVENING next, a 7¾ o'clock P.M. All who are interested in the conversion of the Jews are invited.[1]

## 1870

### Context

1. The frequency of mixed Jewish and non Jewish marriages begins to climb steadily.[2] American reform rabbi David Einhorn (1809-1879) calls it "a nail in the coffin of the small Jewish race."[3]

## 1873

### Context

1. Rabbi Israel Meir Hakohen of Radin writes a book popularly known as *Chafetz Chayim* on the prohibitions in Judaism against talebearing (Leviticus 19:16), evil gossip (Psalms 34:14), in the Talmud (Yoma 4b, Sanhedrin 31 a), and in the Codes of Jewish law such as Maimonides' Mishneh Torah (Deot 7:2). These prohibitions require that professional confidences between patient and physician be maintained. In 1998 Dr. Fred Rosner interprets them to mean that "a person who is the carrier of a serious and potentially lethal genetic disorder is obligated to divulge that information to a prospective spouse."[4]

---

[1] June 20, 3.
[2] Doron, 1980, 412.
[3] Goldstein, 1997, 48.
[4] Rosner, 1998, 410.

## 1874

### Context

1. Southern physician Madison Marsh describes Jews as "the purest, finest, and most perfect type of the Caucasian race."[1]

2. Polish Jew Naftali (Naphtali) Levy proselytizes Darwinism in his Hebrew-language book *Seh Sefer Toledot Adam* (Generations of Man).[2]

## 1875

### Context

1. From Jewish World (London): "few Jewish fathers or brothers, no matter how lax in their religious observances, would think of introducing a Christian gentleman to their daughters or sisters."[3]

## 1876

### Jewish Precursor of Eugenics

1. The Jewish-Italian criminologist and physician Cesare Lombroso (1836-1909) claims to have established during autopsies certain physical stigmata characteristic of the born criminal, whom he sees as possessing a more primitive type of brain structure.

## 1877

### Context

1. *Allgemeine Zeitung des Judenthums* angrily rejects the "fraudulent racial theory" that Aryans are nobler and superior to Semites.[4]

2. From the *Pall Mall Gazette*, reprinted in the *New York Times*: "That the Jews in our day do not wish to begin the business of conversion we may rest well assured. The truth is that some of the most influential members of the Hebrew community have during the last three of four years taken unto themselves Christian wives; some noble Jewesses have made themselves happy with Gentile hus-

---

[1] Goldstein, 1997, 36.
[2] Viyen: *Bi-defus shel Shpittser `et Holtsvartah.*
[3] Goldstein, 1997, 35.
[4] *Allgemeine Zeitung des Judenthums, 1877.*

bands; and it happens that Jews rarely maintain the rites and obligations of their faith after they have married out of it."[1]

3.   A "Religious Notice" in the *New York Times*: "**Address to the Jews.** – Rev Charles E. Harris, a converted Jew, will deliver a lecture NEXT SABBATH AFTERNOON, in the M E Church, Norfolk-st, between Rivington and Stanton. Subject – *An Investigation from the Old Testament Scriptures into the Claims of Jesus Christ as the Jews' Messiah*. He will also preach morning and evening, and in the evening relate his conversion. Israelites are affectionately invited to attend. Services commence at 10½ A M, 3 and 7 P M.[2]

## 1879

### Context

1.   German writer Wilhelm Marr (1819-1904), founder of the "League of Anti-Semites," which advocates the deportation of Jews from Germany, publishes *Jewry's Victory over the Germanic Peoples (Der Sieg des Judenthums über das Germanentum)*, popularizing the word "anti-Semite" as a racial concept scientifically justifying a hostile attitude toward Jews.

2.   Chicago reform rabbi Bernard Felsenthal (1822-1908) comments on mixed marriages: "The aversion against entering into family connections is not so strong any more as it used to be, particularly if the family is in good social and financial position."[3]

## 1880

### Context

1.   Chicago reform rabbi Emil Gustav Hirsch (1851-1923): "We preserve no sympathy with a physiological Judaism. Our Judaism is rooted in our conviction and not in our blood."[4]

2.   Jewish emigration from the Russian Empire engenders both sympathy and disdain among Western Jews.

---

[1] *New York Times*, 1877
[2] Mar 25, 5.
[3] Goldstein, 1997, 34.
[4] Goldstein, 1997, 49.

## 1881

### Context

1. German geologist and ethnographer Richard Andrée (1835-1912) in *On Jewish Anthropology* (*Zur Volkskunde der Juden*) notes the stability of the Jewish physical type, which he attributes to "Semitic Blood."[1]

2. Ethnologist Eugen Dühring (1833-1921) concludes that the defining element of Jewry is common descent rather than religion[2]: "The influx of Jewish blood cannot but... lead to deterioration. This corruption assumes its worst forms when women of a superior people are constrained by fate to offer places of reproduction to the Jewish tribe and to the Jewish character.... Neither climate nor cultural environment can later modify any essential aspect of such a hybridization on either side. Half-Jews, quarter-Jews... are all a curse, for they are capable of penetrating the remainder of society more easily than can pure-blooded Jews."[3]

## 1882

### Context

1. In Russia the May Edicts enforce conscription into the Army for all first-born Jewish males, helping to trigger a massive migration to the United States that lasts until 1913.

2. German physician Bernhard Blechmann: "It is a remarkable fact, recognized by all researchers, that the Jewish tribe has remained virtually unchanged ever since its appearance roughly 4,000 years ago, and that no other racial type can be traced back in history as is the Jewish [tribe]."[4]

---

[1] Andrée, 1881.
[2] Weindling, 1989, 58.
[3] *Die Judenfrage als Racen-, Sitten- und Culturfrage*, Karlsruhe-Leipzig, 144; translated from the French, Essner, 1995, 6.
[4] *Anthropologie der Juden*, Dorpat, 1f; quoted in Lipphardt, 2008, 57.

## 1883

### Context

1. German psychiatrist Emil Kraepelin (1856-1926) claims in *Psychiatric Compendium* that race plays a role in Jewish mental illness.[1]

2. French historian Ernest Renan (1823-1892) makes famous comment that there is not a Jewish type, but rather there are Jewish types.[2]

### Jewish Advocacy of Social Darwinism

3. Zionist and eugenicist Ludwig Gumplowicz (1838-1909) writes in *Racial Struggle* that a stronger people enslaves a weaker people and takes over its land: "How does this amalgamation come about? Only in eternal racial struggle, in war and 'peace' – there is no other way. Man would have to cease being human, he would have to – if it were only possible – shuffle off everything that nature had made of him."

## 1884

### Context

1. Anthropologist R. N. Ikow: "Ultimately Russia's Jews (and probably the Karaim along with them) must be excluded from the Semites in as much as they are basically unrelated to the latter and belong to a quite different race."[3] Ikow argues that the Jews arrived in Russia, not from the West, but from the East.

### Jewish Advocacy of Eugenics

2. Editorial in the *American Hebrew*: "Mixed marriages are not only a religious evil. That they are this, is very generally conceded. No one who has studied the question will maintain that the offspring of such marriages is apt to be of much value to Judaism. There will spring up among us a large class which will be neither Jews nor Gentiles, but a hybrid – useless for all purposes of further development. But the question involves ethnological and sociological problems also. The maintenance of the Jewish race purity has caused the preserva-

---

[1] Kraepelin, 1883.
[2] Renan, 1883.
[3] R. N. Ikow, 1884, *Neue Beiträge zur Anthropologie der Juden* (*Arch. für Anthropologie* XV), Braunschweig; quoted in Judt, 1903, 9.

tion of Humanity's present system of Semitic civilization. Wherever the Aryan has stood for pillage the Semitic race has stood for peace. When the Aryan races pursued the chase, Semites cultivated letters. The law of fittest surviving, aided by the breeding of hereditary qualities in a pure race, has given the Jews a physiological and mental superiority which can be perpetuated only by the perpetuation of the race purity."[1]

## 1885

### Context

1.  Introduced by Sir Francis Galton at London's Anthropological Institute, specialist in rabbinic Hebrew Adolf Neubauer (1831-1907) and Jewish eugenicist Joseph Jacobs (1854-1916) present conflicting findings on the physical anthropology of the Jews, Neubauer arguing against the notion of purity of the Jewish race, wherein he is opposed by Jacobs, who writes: "What are the qualities, if any, that we are to regard as *racially* characteristic of Jews? Much vague declamation has been spoken and written on this subject. All the moral, social, and intellectual qualities of Jews have been spoken of as being theirs by right of birth in its physical sense. Jews differ from others in all these points, it is true, as I have partly shown elsewhere. But the differences are due, in my opinion, to the combined effect of their social isolation and of their own traditions and customs, and if they have any hereditary predisposition towards certain habits and callings, these can only be regarded as secondarily racial, acquired hereditary tendencies which cannot be brought forth as proof of racial purity."[2]

2.  At the request of Jacobs, Galton takes a series of photographs of London Jews, superimposing ten original photographs to produce four composite shots to illustrate the Jewish type. The London newspaper *Jewish Chronicle* publishes a paper by Galton on the photographs: "They are, I think, the best specimens of composites, I have ever produced... I may mention that the individual photographs were taken with hardly any exception, from among Jewish boys in the Jews' Free School, Bell Lane.... They were children of poor parents, dirty little fellows individually, but wonderfully beau-

---

[1] American Hebrew Publishing Company, 1984.
[2] Jacobs, 1886, 25.

tiful, as I think, in these composites. The feature that struck me the most as I drove through the adjacent Jewish quarter was the cool scanning gaze of man, woman and child, and this was no less conspicuous among the schoolboys. There was no sign of diffidence in any of their looks, nor of surprise at the unwonted intrusion... I felt, rightly or wrongly, that every one of them was coolly appraising me at market value, without the slightest interest of any other kind."[1]

3.  *Allgemeine Zeitung des Judenthums*[2] criticizes physical anthropologist Constantin Ikow's theory that the Jews represent three different groups (Eastern, Western-European, and Russian): "According to his theory, all the Russian Jews must have come from Inner Asia, and this is only a whim [on Ikow's part]."

## 1886

### Context

1.  The German pathologist Rudolf Virchow (1821-1902) surveys over ten thousand German children and determines that 10% of the Jewish children are blond, as opposed to 31% of the German.[3]

### Jewish Precursors of Eugenics

2.  Joseph Jacobs publishes *The Comparative Distribution of Jewish Ability*, stressing 'Jewish Genius.'

3.  Jewish-Austrian journalist Nathan Birnbaum (1864-1937); "Even when the Jews speak the purest and the most fluent German, there is no denying that Jewish spirit and Jewish mood dictates these German sounds."[4]

## 1887

### Context

1.  Boston reform rabbi Solomon Schindler (1842-1915): "it remains a fact that we spring from a different branch of humanity, that different blood flows in our veins, that our temperament, our tastes, our humor is different from yours; that, in a word, we differ in our

---

[1] Galton, 1910.
[2] *Allgemeine Zeitung des Judenthums*, Feb. 24, 1885, 139.
[3] Efron, 1994, 25.
[4] Birnbaum, 1886, 4.

views and in our mode of thinking in many cases as much as we differ in our features."[1]

2. Author and chief rabbi of Sweden Marcus Ehrenpreis (1869-1951): "The Jewish youth of our city have split into two camps and make war upon each other. The Union of Brothers calls for complete assimilation, on the one hand, and the society Mikra kodesh calls for nationalism (*leumiut*) on the other."[2]

3. Routinely referring to Europeans as 'Aryans,' Zionist physician Karpel Lippe rejects Jewish-assimilation proposals: "A little holy water, some brief nasal singing, a few church bell peals, and the Jew is transformed into a genuine Polish nobleman." The Polish nobility, he comments, has interbred to such a degree that it is "drenched in Jewish blood," and the Jewish population is equally interbred with Poles.[3]

## 1888

### Context

1. The Jewish nationalist newspaper *Serubabel* concedes that "the Jewish national idea may indeed coincide at some points with anti-Semitism."[4]

## 1889

### Context

1. A study of over 10,000 U.S. Jewish families reveals a Jewish birth rate lower than the non-Jewish birth rate.[5]

## 1890

### Jewish Advocacy of Eugenics

1. Jewish-American anarchist, sexual reformer, feminist, and eugenicist Moses Harman (1830-1910) is put on trial for expressing "ob-

---

[1] Schindler, 1887, 5.
[2] Mendelsohn, 1971, 529.
[3] Lippe, 1887, 28, 29, 30.
[4] *Serubabel*, No. 5, 35; cited in Doron, 1983, 171-171.
[5] John S. Billings, "Vital Statistics of the Jews in the United States," *Census Bulletin*, No. 19, Dec. 30, 1889, 49; cited in Goldscheider, 1967, 197.

scene" views in discussing birth control. He responds: "There is nothing referred to except a free given allusion to human conduct and different members of human anatomy. I do not deem any of these obscene. All the words that are in the article are in Webster's dictionary." Judge Cassius Foster concludes his remarks by stating: he had seen circus performers stick their heads into lions' mouths, but he had never seen them have the temerity to twist the beast's tail or kick them in the ribs while performing the risky act. After the laughter in the courtroom subsides, Foster sentences Harman to serve five years in the Kansas penitentiary and to pay a fine of $300.[1]

## 1891

### Context

1.  Polish historian J. Krzywicki: "Despite their fanatical self-isolation the Jews have much non-Semitic blood in them. In India, this side of the Ganges, they are black; in England they are blue-eyed and blond; in Russia's western *guberniyas* they have broad faces and Slavic noses."[2]

2.  *Allgemeine Zeitung des Judenthums* publishes a series of letters from London professor of history G. M. Asher entitled "On the Aryan Origin of the Jews and Their World-Historical Purpose": "the Jews do not stand alongside the European world as an alien element, but are the *undiluted continuation of the more important of the two ethnic elements that have been united in the European nations*" (emphasis in original).[3] Jewish-German anthropologist Moritz Alsberg (1840-1920) in *Racial Mixing Among the Jews* also maintains that the Jews have broadly interbred with the "Aryans."[4]

---

[1] West, 1971.
[2] Krzywicki, J. 1891. *Lyudi*. Warsaw. Quoted in Judt, 1903, 14.
[3] Asher, July 3, 1891.
[4] Alsberg, 1891.

## 1892

## Context

1. The Russian Senate having abolished in 1889 all limitations on Jewish converts to Christianity, Tsar Aleksandr III rules that Jews leaving the Russian Empire will not be permitted to return.[1]

2. President of the British group of Hovevei Zion, Elim Henry D'Avigdor-Goldsmid (1841-1895): "Anyone who believes in the Torah must resist mixed marriages and conversions.... The Jewish faith... has effectively kept the race of the Jewish people pure.... Race is inseparable from faith."[2]

3. Publisher of the satirical Zionist magazine Schlehmil, Max Jungmann: "Jewry has its purity of the blood to thank for its lasting, sturdy existence.... According to Gobineau's theory, a bastard nation from its very birth stands under the sign of degeneration."[3]

4. Austrian physician, anthropologist, and eugenicist Felix von Luschan (1854-1924) describes "Aryan" and "Semitic" as linguistic terms not applicable to race categories and points out that the Jews are as interbred as any other European group.[4] By the end of the century his view becomes predominant in scientific circles.

## Jewish Advocacy of Eugenics

5. Jewish-German philosopher Max Nordau (1849-1923) writes a treatise entitled *Degeneration* (*Entartung*), which becomes an instant best-seller. Influenced by Friedrich Nietzsche (1844-1900), Gobineau, and the French psychiatrist Bénédict Morel, Nordau predicts decline and eugenic rebirth: "Over the earth the shadows creep with deepening gloom, wrapping all objects in a mysterious dimness, in which all certainty is destroyed and any guess seems plausible. Forms lose their outlines, and are dissolved in floating mist. The day is over, the night draws on. The old anxiously watch its approach, fearing they will not live to see the end. A few amongst the

---

[1] "Evrei," *Entsiklopedicheskii slovar'*, 1893, vol. XI, 455.
[2] Reverse translation from the German, quoted by Doron, 1980 (pg. 404), who refers to him as "Elim Henry d'Avigdors."
[3] Jungmann, 1892.
[4] Luschan, 1892

young and strong are conscious of the vigour of life in all their veins and nerves, and rejoice in the coming sunrise."[1]

## 1893

### Context

1. Vittorio Hayim Castiglioni, later appointed chief rabbi in Rome, sees a harmonious relation between science and Judaism and understands the six days of the Biblical creation story metaphorically. He denounces the "false creed" of racial superiority.[2]

## 1894

### Jewish Precursor of Eugenics

1. Criminologist Cesare Lombroso in *Anti-Semitism and the Jews*: "The broad Aryan basis of Jewry received the fertile impetus of racial mixing, which, as we shall see, is an essential factor in human progress... Despite certain inferior characteristics, Jewry... has so completely adapted itself to Aryan customs, so assimilated Aryan intelligence, and in some cases even surpassed it, that the Jews have become similar to the Aryan population among which they dwell. At the same time one must concede that they have preserved their own type thanks to inbreeding."[3]

## 1895

### Context

1. In *Die Tüchtigkeit unserer Rasse* (*The Viability of Our Race*) the German eugenicist Alfred Ploetz (1860-1940) advocates the complete absorption of Jews into the Aryan race: "The Hygiene of the entire human race converges with that of the Aryan race, which apart from a few small races, like the Jewish race – itself quite probably overwhelmingly Aryan in composition – is the cultural race par excellence.... All anti-Semitism is a pointless pursuit – a pursuit whose support will slowly recede with the tide of scientific knowledge and human democracy."[4]

---

[1] Nordau, 1968, 5-6.

[2] Elucidated by Dubin, 1995.

[3] Lombroso, 1894, 29.

[4] *Die Tüchtigkeit unsrer Rasse und der Schutz der Schwachen*, 142; quoted in Adams, 1990, 17-18.

## Jewish Advocacy of Eugenics

2. Jewish-German dermatologist Alfred Blaschko (1858-1922): "Now it cannot be denied that the Darwinian theory is an eminently aristocratic theory; aristocratic on the one hand since it proclaims the inequality of everyone bearing a human face, and on the other hand, because, proceeding from this inequality, it preaches the right of the stronger, of the one better equipped for the struggle for existence."[1]

### 1871-1918

### Context

1. Despite the urgings of German eugenicists, no eugenics laws are passed in Germany during the Wilhelmine period.[2] It is only in the strongly Jewish influenced Weimar Republic that eugenics enters the popular consciousness and that such legislation begins to be proposed.

### 1896

### Context

1. Jewish-Austrian journalist Nathan Birnbaum, who coined the words 'Zionism' and 'Zionist,' maintains that "the secure foundation of nationality is always and everywhere race."[3]

2. Zionist publisher Berthold Feiwel (1875-1937), a close associate of Martin Buber, Theodor Herzl, and Chaim Weizmann and also co-founder of Jüdischer Verlag (Jewish Press), argues that "blood and tribal belonging" – not language or geography – is decisive in determining Jewish nationality."[4]

3. Black Judaism as a self-conscious religious identity arrives in America in Lawrence, Kansas, when charismatic Baptist preacher and former slave William Saunders Crowdy (1847-1908) establishes a black congregation called the Church of God and Saints of Christ, where he preaches that Africans are the true descendants of the

---

[1] Alfred Blaschko, "Natürliche Auslese und Klassentheilung," *Die neue Zeit*, 13 (1), 1894-95, 615; quoted in Weikart, 2002, 333.
[2] Adams, 1990, 29.
[3] *Die Jüdische Moderne*, Leipzig, 54; quoted in Gelber, 2000, 132.
[4] *Modernes Judentum: Tendenzrede*, 1897; cited in Gelber, 2000, 135.

Ten Lost Tribes of Israel and thus are God's chosen people. The denomination practices an eclectic "roll your own" brand of religion that combines beliefs and practices of the Old and New Testaments. Crowdy's tabernacles practice male infant circumcision, observe Saturday as the Sabbath, celebrate Passover and other Jewish holidays, but venerate Jesus Christ. More than 200 congregations are eventually established in the United States, Africa, and the Caribbean, and the group still has more than 50 affiliated congregations as of 2009.[1]

## Jewish Advocacy of Eugenics

4. Founder of modern political Zionism Theodor Herzl (1860-1904): "the strong among us were inevitably true to their race when persecution broke out among them. This attitude was most immediately apparent in the period immediately following the emancipation of the Jews. Later on, those who rose to a greater degree of intelligence and to a better worldly position lost their communal feeling to a very great extent."[2]

5. Responding to a questionnaire of the editor of *Allgemeine israelitische Wochenschrift* as to whether Jewry is in a process of decline, Jewish-German eugenicist Max Nordau prescribes a return to agriculture to rejuvenate Jewish bodies. His Lamarckian eugenics becomes popular within the Zionist movement.[3]

# 1896-1934

## Context

1. In Budapest Jews display lower fertility than do non-Jews.[4]

---

[1] Chavets, 2009.

[2] Cited according to English translation, *A Jewish State: An Attempt at a Modern Solution of the Jewish Question*, The Maccabaean Publishing Company, New York, 5.

[3] Falk, 1998, 594.

[4] Liebman Hersch, "Jewish Population Trends in Europe," *Jewish People: Past and Present*, II, 11, Table 10, cited in Goldscheider, 1967, 200.

This drawing by E. M. Lilien indicates that it was chosen as the so-called 'Congressional Postcard' at the Fifth Zionist Congress in Basel, December 26-31, 1901. The lower inscription reads: "Let our eyes witness your loving return to Zion," which inverts the words from the Amidah prayer, waiting not for God's return but for "yours." Note that the goal indicated in the drawing itself – without the border – is that of a man plowing a field, not of any specific geographical area.

## 1897

## Context

1. According to the 16-volume Russian-language *Jewish Encyclopedia* (*Evreiskaia èntsiklopediia*), Jews constitute 11.35% of the population of the Russian Empire, but account for 21.1% of government officials and white-collar workers. While they constitute only 0.6% of those engaged in agriculture, they account for 72.8% of those engaged in trade.[1]

2. Jewish-German financier and political figure Walther Rathenau (1867-1922) advocates assimilation of Jews into Western culture: "'A Jew is a Jew.' That is the simple principle of the State. Strict and exceptionless expulsion from the army, the administration, and the universities. The goal – to act as a counterweight to Jewification of society – is justified."[2]

### Jewish Advocacy of Eugenics

3. At the First Zionist Congress in Basel, Max Nordau inserts the concept of degeneration into the Zionist cause: "It is a great sin to let a people degenerate…"[3]

4. Arthur Ruppin writes in his diary: "A renewal of Jewry is possible only on the basis of racial belonging."[4]

## 1898

## Jewish Advocacy of Eugenics

1. At the Second Zionist Congress in Basel Max Nordau proposes creating a 'muscle Jew' (*Muskeljude*) – physically strong, sexually potent, and morally fit. Modeled after images of Hellenic athletes, the new ideal is intended to overcome frequent images, internalized even by many Jews, of the Jew as scrawny, weak, and inferior.[5]

---

[1] "Rossiia," vol. 13, 650.
[2] Rathenau, 1897.
[3] Presner, 2003, 281.
[4] *Tagebuch*, Jan. 6, 1897; quoted by Doron, 1980, 414.
[5] Presner, 2003.

## 1899

## Context

1.  Despite lower fertility than in Europe as a whole, Jews have increased their population from 2.7 million in 1800 to 8.5 million thanks to lower mortality.[1]

2.  Historian Houston Stewart Chamberlain (1855-1927) publishes his widely read two-volume book *The Foundations of the Nineteenth Century* (*Die Grundlagen des neunzehnten Jahrhunderts*), popularizing the idea of an Aryan race stemming from Indo-European culture and led by the Nordic or Teutonic peoples. The book is dedicated to his Jewish mentor Julius Wiesner (1838-1916), professor of botany at the University of Vienna.

3.  American economist William Zebina Ripley (1867-1941): "The modern Jews are physically more Aryan than Semitic, after all. They have unconsciously taken on to a large extent the physical traits of the people among whom their lot has been thrown."[2] Joseph Jacobs, President of the Jewish Historical Society, responds: "Professor Ripley, as a student of anthropology, declares, as the result of his inquiries, that there has been so large an admixture of round skulls with the (hypothetically assumed) original long skulls of the Hebrews that all signs of racial unity have disappeared. I, on the other hand, who have approached the subject as a student of history, see no evidence of any such large admixture of alien elements in the race since its dispersion from Palestine, and have come, therefore, to the opposite conclusion – that the Jews now living are, to all intents and purposes, exclusively the direct descendants of the Diaspora."[3]

4.  Jewish-German essayist and novelist Robert Jaffé (1870-1911) describes the poet Heinrich Heine as a 'Rasse-Jude' (racial Jew).[4]

5.  Jewish-Austrian lawyer and president of the Jewish community Emil Byk (1845-1906): "We prospered in Poland so long as Poland

---

[1] Efron, 2001, 118-120.
[2] W. Z. Ripley, *The Races of Europe*. Quoted by Judt, 1903, 16, after 1890 edition, London.
[3] Jacobs, 1899, 502.
[4] "Ghettodichter," *Die Welt*, 28, 269; quoted in Gelber, 2000, 143.

prospered, and well may we understand the writers of bygone years who wrote: Poland is paradise for the Jews [*Polonia Judaeorum paradisus*]."[1]

## 1900

### Context

1. Jewish-Austrian philosopher Martin Buber (1878-1965) develops *völkisch* concepts, referring to the Jews as a *Blutstamm* (Blood Tribe).[2]

2. Both Jewish and non-Jewish psychiatrists at the turn of the century take it as a given that Jews are more inclined to suffer from mental illness than are non-Jews.[3]

3. Prominent Russian Zionist and ophthalmologist Max Mandelstamm (1838-1912) insists that one can locate "Jewish degeneration," but attributes their "decrepit, miserable, weak bodily condition" to ghetto conditions, rather than to inherent racial characteristics.[4]

### Jewish Advocacy of Eugenics

4. *The Jewish Gymnast's Newspaper* (*Jüdische Turnzeitung*) is founded, featuring images of the new 'muscle Jew'[5] (see pg. 107).

5. American historian Mitchell B. Hart looking back in 2007 at the first half of the twentieth century: "If the Jews 'became white' over the course of the early twentieth century, then surely one way in which this occurred was that they became imbued with the spirit of racialism, eugenics, and colonialism. Not all of them were so imbued; but many, and many more than Jews and others after the end of World War II have acknowledged. And the fact that Jews were one of the main targets of this racialized worldview does not obviate the fact that, at certain points, they participated in it."[6]

---

[1] Mendelsohn, 1971, 526.
[2] Martin Buber, "Jüdische Renaissance," *Ost und West*, 8; cited in Gelber, 2000, 134.
[3] Efron, 2001, 152.
[4] Hart, 1999, 276; citing *Stenographisches Protokoll der Verhandlungen des IV Zionisten-Congresses*, Aug. 15, 1900, Vol. 4, 117-131.
[5] Presner, 2003.
[6] Hart, 2007, 141-142.

## 1900-1930

### Context

1. The Jewish birth rate is lower than the general birth rate in Romania, Hungary, Prussia, Vienna, Amsterdam, and Leningrad.[1]

2. Radical intellectuals frequent the 'red salon' of the mathematician Leo Arons [1860-1919]. Socialist doctors such as Ignaz Zadek [1858-1931] (the brother-in-law of the revisionist politician Eduard Bernstein [1850-1932]) form influential groups campaigning for extension of social medicine. Bernstein befriends a group of young radical Jewish doctors, who include Hermann Lisso [1856-1926], Alfred Blaschko [1858-1922], Paul Christeller [1815-1915] and Mieczslaw Epstein [1868-1931], who move from eastern towns like Danzig and Posen to Berlin as an intellectual center. In the words of contemporary historian Paul Weindling, such figures "transposed biology to the broader realm of the social organism."[2]

3. In Warsaw Jews have lower fertility than do non-Jews.[3]

4. Contemporary American historian Donald K. Pickens in *Eugenics and the Progressives*: "Progressivism, heir to the nineteenth century with its concern about Darwinism, naturalism, revolution, class struggle, industrialization, and the multitude of urban problems, was not a pure substance; rather, it was an alloy through which ran sizable streaks of conservatism and, on occasion, a vein of reaction. The progressive theme ran from optimism founded on utopian assumptions to deep naturalistic despair. Little wonder then that eugenics from 1900 to 1929 was a synthesis of those moods."[4]

## 1901

### Context

1. The Fifth Zionist Congress is held in Basel.

---

[1] Uriah Z. Engelman, "Sources of Jewish Statistics," in Louis Finkelstein (ed.), *The Jews: Their History, Culture, and Religion*; cited in Goldscheider, 1967, 200.
[2] Weindling, 1989, 35.
[3] Liebman Hersch, "Jewish Population Trends in Europe," *Jewish People: Past and Present*, II, 11, Table 10, cited in Goldscheider, 1967, 200.
[4] Pickens, 1968, 102.

## 1902

## Context

1. Viennese physician Martin Engländer explains the physical inferiority of Eastern European Jews relative to Western European Jews as stemming from environmental rather than from hereditary factors.[1]

2. Jewish-German painter Max Liebermann (1847-1935): "Race remains race" (*Rasse bleibt eben Rasse*).[2]

3. Zionist Alfred Waldenburg (1873-1942) argues that intermarriage leads to "racial suicide" and calls for "superior racial inbreeding" in Palestine.[3]

4. Alb. Lucas, Secretary of the Union of Orthodox (Jewish) Congregations of the United States and Canada, protests an article published in the *New York Times*: "As to whether the conditions into which the Royal Commission is inquiring as to the effect of alien immigration upon English labor will be found to be as 'grim' as you appear to think remains to be seen when the Royal Commission publishes its report. But the testimony of the Registrar of Births and Deaths, (of Stepney,) even if it does contain the statement 'that that parish had been utterly ruined and the standard of living of the whole neighborhood lowered,' is an unsupported expression of opinion entirely at variance with the rest of the evidence which I have seen. Even Mr. Arnold White [1848-1925], the arch-enemy of Jewish immigration into England, said nothing like it. Our own experience stamps as utterly untrue the next statement that 'half of the aliens apply for medical charity and their defective (?) children become public charges.' The final paragraph of your editorial shows that you are as biased against Jewish aliens, and with as little reason, as is Mr. Arnold White and the British Brothers' League. You did not mention the word 'Jew' openly in your creed, and your sneering references to 'these not yet Americanized New Yorkers,' is 'stranger than any nightmare,' when it is published on the page of a newspaper that gives more than a column to a report of a public meeting of these very 'not yet Americanized New Yorkers.... Fair

---

[1] Efron, 2001, 170-171.

[2] Reported by Adolph Donat, "Max Liebermann über den Zionismus," *Die Welt*, 43, 3; quoted in Gelber, 2000, 156.

[3] Hart, 1999, 280.

play is a jewel, and we expect, nay, we demand, fairer treatment at the hands of our fellow-citizens than is expressed by the innuendo and sarcasm of your leader. An explanation and apology is in order."[1]

## 1903

### Context

1. Jewish anthropologist J. M. Judt (Ignacy Maurycy Judt): "The classification of the Jews as race among the Semitic tribes is not rationally grounded.... The false identification of philological classifications with racial breakdowns has led to confusion.... When I speak of Aryans, I by no means have in mind their blood, their hair, or their skulls. I mean, quite simply, those who make use of Aryan languages."[2]

2. Jewish-Austrian philosopher and convert to Christianity Otto Weininger (1880-1903): "The real Jew, like a real woman, lives only within his species, and not as an individual."[3] Weininger reflects the current image of the Jew as effeminate.

### Jewish Advocacy of Eugenics

3. Jewish eugenicist Arthur Ruppin: "In our view, social opinion is fully capable of fulfilling its intended function of restraining the psychically or physically hindered from having children."[4]

## 1904

### Context

1. From a letter to the Editor of the *New York Times*: "[Israel] Zangwill's mission here, to prepare the way for the Jews to go to Uganda, lacks full knowledge of conditions, and therefore does not deserve our sympathy. Not Uganda, but America, is the land for the final preparation of the Jew which will befit him to return to the Land of Israel."[5]

---

[1] Lucas, 1902; referring to *New York Times*, 1902, g166.
[2] Judt, 1903, 224-225.
[3] Weininger, 1920, 412.
[4] Ruppin, 1903, 94.
[5] S.P.F., 1904.

## Jewish Advocacy of Eugenics

2. Zionist leader Ze'ev Jabotinsky (1880-1940): "The source of national feeling...lies in a man's blood...in his racio-physical type, and in that alone... Autonomy in the Golah [exile] is likely to lead...to the complete disappearance of the Jewish nation as such from the face of the earth.... Just imagine...when our offspring will be living at peace among a strange people.... This will mean the inception of complete assimilation.... A preservation of national integrity is impossible except by a preservation of racial purity, and for that purpose we are in need of a territory of our own.... If you should ask me in a sense of revolt and outrage: but surely in that case you want segregation at all costs! I would answer that one must not be afraid of words and not of the word 'segregation.'"[1]

3. *The Journal of Jewish Demography and Statistics* (*Zeitschrift für Demographie und Statistik der Juden*) is founded, one of its areas of interest being Jewish biology.

4. The *Jewish Quarterly Review* publishes an article by Jewish-British journalist Lucien Wolf (1857-1930) entitled "The Zionist Peril," in which Wolf characterizes Zionism as "an attempt to turn back the course of modern history, which hereto, on its political side, has had for its main object to secure for the Jewish people an equal place with their fellow-citizens of other creeds in the countries in which they dwell, and a common lot with them in the main stream of human progress. It is essentially an ignorant and narrow-minded view of a great problem – ignorant because it takes no account of the decisive element of progress in history; and narrow-minded because it confounds a political memory with a religious ideal."[2]

5. Physician, prominent Zionist, member of the governing board of the Berlin Jewish Community, and Council National League of Jewish Communities Aron Sandler (1879-1954) in *Anthropology and Zionism* argues that Jewry over the ages has remained racially pure (*rassenrein*) by means of inbreeding and even the "separating out" (*Entmischung*) of infusions of alien blood "into their original components! A second factor acting as a 'guardian of the races' is constituted by the further elimination of racially alien children whom

[1] Meyer, undated, citing letter.
[2] Vol. xvii, 22-23; quoted in Fishberg, 1911, 499.

sexual selection determines shall henceforth be expelled by virtue of their birth. I cannot dwell here on this difficult point.... In observing the struggle for existence, particularly in the animal world, we encounter acts that far outshadow in viciousness and unfairness the most extreme spawn of the phantasy as judged by our ethical concepts. But seen from the lofty viewpoint of the teachings of evolution, this 'viciousness' reveals itself as the expression of the deepest wisdom.... When a surgeon calmly amputates a limb, intending to preserve the body of a patient from a general infection who is ill because of his heredity so as to preserve his life and spiritual development, a fool observing the operation would experience horror over the 'viciousness' of the physician. Such expulsion mechanisms...are probably the main factors preserving that individuality which we have focused on as the chief characteristic of a race."[1]

6.  In response to the condemnation of feminism as dysgenic by the Dutch sociologist Sebald Rudolph Steinmetz (1862-1946), the Alliance for the Protection of Mothers (Bund für Mutterschutz) is created to add a radical feminist voice in support of the eugenics movement, including such Jewish feminists as Henriette Fürth (née Katzenstein, 1861-1938), and Adele Schreiber (1872-1957).[2] Fürth counters the mainstream-eugenics view that promiscuity is itself a negative genetic trait, maintaining that such behavior demonstrates health and even... good heredity.[3] Schreiber argues that female economic and social equality would encourage eugenic improvement by enabling women to choose healthy and fit fathers for their

---

[1] Because of the historic importance of this claim, I provide here the German original: *Die Mischung zerlegt sich wieder in ihre Componenten! Einen zweiten Factor, der als "Wächter der Rassen" auftritt, bildet die weitere Eliminierung dieser von Geburt zur Ausscheidung bestimmten Kinder der fremden Stamrasse, die sich nunmehr durch die geschlechtliche Auslese vollzieht. Ich kann indes auf diesen schwiegeren Punkt an dieser Stelle nicht eingehen. Wenn der Chirurg in der Absicht, den Körper des Erbkranken vor der Allgemein-Infektion zu bewahren, ihn dem Leben, der weiteren geistigen Entwickelung zu erhalten, ein Glied kühlen Blutes amputiert, so wird ein Narr beim Anblick der Operation vor der 'Graumskeit' des Chirurgen Ensetzen empfinden. Derartige Ausscheidungsmechanismen... sind es wohl hauptsächlich, die uns den Fortbesitz einer bestimmten individualität sichern, die wir als das Haupt-Characteristicum einer Rasse aufgefasst haben.* Sandler, 1904, 27-28.
[2] S.R. Steinmetz, "Feminismus und Rasse," *Jahrbuch für Sozialwissenschaften*, 1904. vol. 7, 752; cited in Taylor-Allen, 1988, 31.
[3] "Bund für Mutterschutz," *Archiv für Rassen- und Gesellischaftsbiologie*, I, 164; cited in Taylor-Allen, 1988, 39.

children rather than be forced to enter into economically advanta-
geous but biologically disadvantageous marriages.[1]

## 1905

### Context

1. American inventor Max Levy (1857-1926): "I see no reason why
   the Jews, any more than any other man or animal, should be above
   the conditions which govern the development of species. The influence of the environment upon the origin and development of species
   has been clearly set forth by Spencer, Huxley, Darwin, and other
   exponents of the theory of evolution, and the same class of considerations that effects the physical development of species is clearly
   shown to exert an equally powerful influence on the development of
   civilization."[2]

2. Anthropologist and physician Samuel Abramowitsch Weissenberg
   (Samuil Abramovich Vaisenberg, 1867-1928) argues that the Sephardim and the Ashkenazim are "racially different."[3]

3. Zionist, artist, and publisher Ephraim Moses Lilien (1874-1925)
   writes to his wife about the nature of Jewishness: "Not because we
   are brothers in faith, but rather because we are members of the same
   tribe…. We are one people, one race" (*ein Volk, eine Rasse*).[4]

4. Jewish eugenicist Arthur Ruppin in his *Jewish Demography and
   Statistics Newspaper* (*Zeitschrift für Demographie und Statistik der
   Juden*) publishes an article by physician Curt Michaelis: "World
   history… must now be constructed according to the general laws of
   biology in their specific application and anthropological interpretation…. Thus, Jewish history is no longer the product of some extra-
   terrestrial God influencing the masses, who are merely passive
   plasma for a moral and self-defining divine artist, but rather are the
   product of the abilities and demands that individual people brought
   with them and kept alive by virtue of inheritance…. Immeasurably
   huge Jewish racial pride took on a specific form – the idea of cho-

---

[1] Schreiber, speech, reported in *Dresdener Nachrichten*, Nov. 29; cited in Taylor-
Allen, 38.
[2] "Jewish People and the Laws of Evolution," 187; quoted in Hart, 2007, 118.
[3] Weissenberg, 1905.
[4] E. M. Lilien, *Briefe an seine Frau 1905-1925*, ed. Otto M. Lilien and Eve Strauss,
Königstein, Ts: Jüdischer Verlag, 1985; quoted in Gelber, 2000, 136.

senness, which was then stubbornly inherited because it titillated the vanity of the community.... It was passed on all the more easily thanks to the racial hatred of the Jew for the non-Jew, and its reaction – the racial hatred of the non-Jew for the Jew."[1]

5. Jewish eugenicist Alfred Nossig responds to Michaelis with an article entitled "The Chosenness of the Jews in the Light of Biology" in the same issue of *Zeitschrift für Demographie und Statistik der Juden*: "If it was the goal-oriented efforts of innumerable generations of Jewish thinkers and statesmen that bred a people of pure blood not poisoned by venereal disease or alcohol but stamped with a sense of family, a deep-rooted tradition of intellectual liveliness, and ideal spiritual orientation, it was also their natural measure of completion that refused to hand over these most lofty ethical treasures to be destroyed along the path of mixing with less carefully bred races. The prohibition of mixed marriages was the chief race-structuring factor that activated inheritance in its supreme efficacy even as these advantages were passed on from generation to generation and steadily accrued thanks to inbreeding. It was in this fashion that a people could arise that Ibsen termed 'humanity's nobility.'"[2]

6. Editors of the American Zionist journal *The Maccabbean* assert that "the mixed marriage is more tolerable from a religious than from a racial standpoint."[3]

### Jewish Advocacy of Eugenics

7. The London *Jewish Chronicle* reports comments by Dr. A. Eichholz at a London meeting of the Maccabeans on the Jewish population of the United Kingdom: "There was among the Jewish population a great deal of material to investigate from the point of view of deterioration, and also from the opposite and more important standpoint of amelioration, and he hoped some investigation would be made either by the Maccabeans or the Statistical Society, or by a joint committee of both. There was a field for useful enquiry of an eugenic nature among the Maccabeans themselves. They were likely to be asked shortly to make such an enquiry. The University of London had recently instituted a lectureship in National Eugenics, and they would probably be asked to assist the lecture or Mr. Francis Galton

---

[1] Michaelis, 1905.
[2] Nossig, 1905, 4.
[3] Hart, 1999, 279.

in their researches. He would like to learn how far such an enquiry was possible."[1]

8.  One of the members of the German Society for Racial Hygiene (Deutsche Gesellschaft für Rassenhygiene) is the Jewish physician, anthropologist and eugenicist Heinrich Wilhelm Poll (1877-1939). The Society's finances are held with the Goldschmidt-Rothschild Bank.[2]

## 1906

### Context

1.  Vienna physician Leo Sofer notes in the Jewish journal *Zeitschrift für Demographie und Statistik der Juden* theories on heightened Jewish resistance to contagious disease developed in the Middle Ages by inbreeding, but also the greater Jewish vulnerability to nervous diseases.[3]

2.  The *Jewish Encyclopedia* publishes an article entitled "Purity of Race": "The general arguments hitherto advanced against the purity of the Jewish race are: (1) The evidence that in Bible times the Jews intermarried with surrounding nations; (2) the frequent reference to proselytes in early Christian literature; (3) the prohibition of inter-marriage repeated in many of the councils of the Church implies frequent infringements; (4) the conversion to Judaism of the Cha-zars, a Turanian tribe in South Russia, from whom, it is suggested, most of the Russian Jews, who form about half of contemporary Israelites, are descended; (5) the marked difference in type to be ob-served among contemporary Jews. To these arguments the uphold-ers of the purity of the race reply: (1) The intermarriages mentioned in the Bible are few in number and with cognate tribes; (2) prose-lytes were the chief sources from which the early Christian Church drew its members, thus removing them from contact with Judaism; (3) the severity of the punishment attached by the Church to inter-marriage proves how infrequent intermarriages must have been; (4) the conversion of the Chazars was merely nominal, and it has left traces on only the few Karaites of South Russia: the other Russian Jews came from Germany, as is shown by the German dialect they

---

[1] Anonymous, 1905.
[2] Braund/Sutton, 2008, 13.
[3] Cited in Hart, 2007, 121.

use; (5) the differences of type may have been produced by social differences and are not so great when a series is taken into consideration. The upholders of the purity point out: (6) That cohanim, or members of the priestly caste, were and are not allowed to marry a proselyte, and must, therefore, have preserved their purity of descent; (7) that the marked resemblance of Jewesses throughout the world, showing as they do less variation among the females of the race, conforms to the biological test of purity of breed; (8) that mixed marriages in the present day are markedly infertile, which would reduce the influence of such intermarriages in an increasing geometrical ratio; (9) the rarity of instances in historical sources of proselytism in mixed marriages since the Middle Ages; (10) the prepotency of Jewish blood, as shown by the marked Jewish type of even the remoter offspring of Jews that have intermarried; (11) the stringent social separation, which can be historically proved throughout the Christian centuries; (12) the existence of marked Jewish type in the features and bodily measurements of contemporary Jews wherever found."[1]

3. Jewish-Viennese writer Stefan Zweig (1881-1942) in an early poem: "I hear my blood through the midnight."[2]

### Jewish Advocacy of Eugenics

4. Eugenics-Society member Edgar Schuster co-authors with Sir Francis Galton Noteworthy Families: An Index to Kinships in Near Degrees between Persons whose Achievements are Honourable, and Have Been Publicly Recorded.

5. Zionist art critic Lothar Brieger (1879-1949) in his novel *René Richter*, described by historian Mark H. Gelber as "the seminal fictional text of racialist Zionism," maintains that racial admixture is the common enemy of both Aryans and Jews and objects against "a foreign, Semitic race mixing itself in, changing its [the Germanic] racial character and usurping its rights, without having any justification."[3] [4]

---

[1] http://www.jewishencyclopedia.com/view.jsp?artid=1573&letter=A
[2] "Das singende Blut."
[3] London: J. Murray.
[4] Lothar Brieger-Wasservogel, *René Richter: Die Entwicklung enes modern-en Juden*, Richard Schröder Verlagsbuchhandlung, Berlin, 5; quoted in Gel-ber, 2000, 148.

## 1907

## Context

1. From an article entitled "Negro Eugenics" (*New York Times*): "most negroes who excel in the arts and professions do possess an infusion of white blood. But so, also, do nearly all negroes on this hemisphere, and in most instances a previous admixture is discoverable through African ancestry with the ancient Hamitic and Semitic stocks."[1]

2. Jewish-French philosopher Henri Bergson (1859-1941) in *Creative Evolution* accepts biological determinism while at the same time defending a limited scope for free will.[2]

3. Halachic literature on sex organ transplants: "If we follow the sex organs, then in any case of prohibited sex, it is possible that they have put [in the woman] generative organs from a woman who is not prohibited, and the witnesses [to the sex act] could not know this. *The Torah surely prohibited* [only] *the woman herself, and it makes no difference where the genitive organs are from.... Aside from this, sex prohibitions do not apply to an organ, which has no life of its own, and is more like a piece of meat; there is not even a rabbinical prohibition....* The story of the Arab who bought a haunch in the market, made a hole in it and performed a sex act with it (Avodah Zarah 22a)... is cited only to show that they were so bound in sexual lust that he performed a sex act with a mere haunch... But it is obvious that sexual prohibitions do not apply to a piece of meat. If so, in our case there is no possibility of a sexual prohibition" (Emphasis in original).[3]

4. From the *Jewish Chronicle* (London): "The Reform Congregation has always had at its head fine types of men, and the three leading families associated with its fortunes – those of (Frederic David] Mocatta [1828-1905], [Elim Henry D'Avigdor-]Goldsmid [1841-1895] and [Alfred Gutteres] Henriques – would provide interesting

---

[1] Jan. 28, 6.

[2] Bergson, 1907.

[3] Halachic periodical *Vayelaket Yosef*, edited by Joseph Schwartz, vol. 10, nos. 3, 4, 6, and 9; quoted in Rosner/Bleich, 2000, 439.

material for Galton and his coadjutors in investigating the new science of eugenics."[1]

## Jewish Advocacy of Eugenics

5.  Jewish-Austrian biologist Paul Kammerer (1880-1926), a socialist and outspoken proponent of Lamarckian eugenics: "Are we slaves of the past or master-workmen forging the future?"[2]

6.  Jewish-German anthropologist and eugenicist Elias Auerbach (1882-1971): "The Jews are a classic object of racial research, because we can better work together in their case in measuring and studying their history than with any other race.... Their varying fates and environments have not managed to wipe out their common, sheerly indestructible type. It is precisely the Jews who, more clearly than any other race, demonstrate the overwhelming power of heredity over adaptation in the fate of a race."[3]

## 1908

### Context

1.  Zionist Israel Zangwill (1864-1926) writes "The Melting Pot" – the most popular play on Broadway: "America is God's crucible, the great melting pot, where all the races of Europe are melting and re-forming. Here you stand, good folks, and your 50 groups with your 50 languages and histories, and your 50 blood hatreds and rivalries, a thing for your feuds and vendettas. Germans and Frenchmen, Irishmen and Englishmen, Jews and Russians, in the crucible with you all. God is making the American."[4] Four years later Zangwill advocates the exact opposite for the Jews: "But if dissolution would bring degeneracy and emancipation dissolution, the only issue from this dilemma is the creation of a Jewish state or at least a Jewish land of refuge..."[5]

2.  Arthur James Balfour (1848-1930), who will later write the Balfour Declaration supporting a Jewish homeland in Palestine, delivers a speech at Newnham College, promoting a Lamarckian view of 'de-

---

[1] Anonymous, 1907.

[2] Lenz, review of paper delivered by Kammerer, 1913; cited in Lipphardt, 2008, 101.

[3] *Jüdische Rassenforschung*, 1907, 333; quoted in Lipphardt, 2008, 72.

[4] http://en.wikipedia.org/wiki/Israel_Zangwill.

[5] Zangwill, 1912?, 20-21.

cadence': "if civilization wear out, and races become effete, why should we expect to progress indefinitely, why for us alone is the doom of man to be reversed?" He is criticized by eugenicist Caleb Williams Saleeby (1878-1940) for paying insufficient attention to eugenic solutions.[1]

3.   Formerly an ardent Zionist, eugenicist Alfred Nossig founds the Jewish Settlement Association (Allgemeine Jüdische Kolonisations-Organisation), lobbying on behalf of Jewish emigration to parts of the world other than Palestine.[2]

## Jewish Advocacy of Eugenics

4.   Jewish-American anarchist and feminist Moses Harman moves his *American Journal of Eugenics*, which emphasizes the "right genera-tion of human beings," from Chicago to Los Angeles. Harman de-votes his publication to natural selection through freedom of mo-therhood, self ownership of women in the realm of sex and repro-duction, and intelligent and responsible parenthood with the woman being dominant in the home. Harman calls the eugenics movement an almost forgotten science once openly taught by the Greeks and Egyptians, and refers to it as the "Science of Right Borning."[3]

5.   In his book *Religion and Socialism* the Jewish-Russian future Soviet Minister of Education Anatoly Lunacharsky promotes the *bogo-stroitel'stvo* movement (literally "Constructing God") which seeks to link socialism with religion – a school of thought later dismissed by Vladimir Lenin as "the narcissistic navel gazing of stupefied fra-gile shopkeepers and the dreamy self-contempt of shysters and *pe-tite bourgeoisie* drowning in despair and exhaustion."[4]

## 1909

## Context

1.   Zionist and eugenicist Ignaz Zollschan writes to Nordic racial theo-retician Stewart Houston Chamberlain: "It would be of the greatest interest to me to hear from one of my chief opponents, whom I can in advance assure of my loyalty [!], his opinion of my work."

---

[1] Saleeby, 1910, 327.
[2] "Alfred Nossig," http://www.holocaustresearchproject.org/ghettos/nossig.html.
[3] June; cited in West, 1971.
[4] Letter to Russian writer Maksim Gorky, Nov. 14, 1913.

Chamberlain politely responds that he is too busy to occupy himself with racial questions at the present time, but asks Zollschan "not to view [his refusal] as an unfriendly gesture."[1]

## Jewish Advocacy of Eugenics

2. Zollschan again: "There is no denying a 'shortage of genius' among the Jews, even among the most outstanding of them…. The under-developed consciousness of an entire people must break through to the surface with the force of the elements…. Inheritance must be favorable to develop a brilliant mind, but no less crucial to development of this so sensitive seed is the soil and other external relationships, the environment, education, etc. Only when all these complicated relationships are harmonized can a true genius develop."[2]

3. From an interview granted by philanthropist Alice Isabella Model (née Sichel, 1856-1943) to the London *Jewish Chronicle*: "Women are better educated than they used to be, and their views are broader. They are studying economics and eugenics."[3]

4. Jewish-German feminist and eugenicist Henriette Fürth points out that her advocacy of legalized abortion does not imply approval of abortions sought for merely "selfish" reasons.[4]

5. Jewish-German physician and, later, Minister of the Interior Affairs Eduard David (1863-1930) argues that eugenics is the proper social response to Darwinism and promotes abortion as one eugenics measure among others.[5]

---

[1] Doron, 1980, 420.

[2] Zollschan, 1920 (identical to 1909 edition, with the exception of new prefaces), 273-276.

[3] Model, 1909.

[4] See Taylor-Allen, 1988, 45, 55.

[5] In Max Apel (ed.), *Darwin: Seine Bedeutung im Ringen um Weltanschau-ung und Lebenswert, Berlin'* cited in Weikart, 2002, 340.

# 1910

## Context

1.  German Jews have increased their numbers from 512,158 in 1871 to 615,021, but this represents a percentage decline from 1.25% to 0.95% of the total German population.[1]

2.  The American Jewish Committee successfully lobbies against efforts to label Jews as a separate race in the U.S. census. Later it will pay anthropologist Franz Boas to promote this view.[2]

3.  Jewish-German-American banker Jacob Schiff (1867-1940): "In the early eighties, when they [Jewish immigrants] began coming, we could take them as they came. We have continued to take them ever since. Now, we must put our foot down. We must learn to regulate for our own good. We must insist that only those Jews who are strong, and who are able to earn their own living, shall be allowed to enter the country. Only a certain number, too, should be allowed to come to North Atlantic seaports. The rest should be deflected to the country sections, to the Gulf ports, or to west of the Mississippi River. West of the Mississippi, especially, opportunity is great, and labor is in demand. Two or three million more Jews can be absorbed. But we feel that we have already reached the limit of absorption in this city [New York]. We must set you recent immigrants thinking. And you must advise your friends at home. You must do this for your own good, for your health and prosperity, and for the future of your children. Do say: 'We want all our friends here.'"[3]

## Jewish Advocacy of Eugenics

4.  The London newspaper *Jewish Chronicle* publishes an interview with Francis Galton, praising him for having "devoted a long life to the pursuance of a high ideal – that of improving the fitness of the human race"[4]

5.  Jewish-British Marxist political scientist Harold Laski (1893-1950): "The different rates of fertility in the sound and pathological stocks

---

[1] Efron, 2001, 119-120.
[2] Popper, 2005.
[3] *New York Times*, 1910.
[4] Galton, 1910.

point to a future swamping of the better by the worse. As a nation, we are faced by racial suicide. It is to this problem that eugenics supplies the solution. It believes that the time has now come when man can consciously undertake the duties that have heretofore been performed by nature. Natural selection must be supplemented by reproductive selection. The parentage of the fit must be encouraged, the propagation of the unfit must be prevented. Such people, the opposition of whom eugenists have to face, assert that marriage is purely a private affair, and that the State has no right of interference. Eugenists maintain that such a view is anti-social, and productive of infinite harm. Whatever action is fraught with national consequences rightly comes within the cognisance of the State.... As Galton has so finely said, we must hold the eugenic ideal of parenthood with the fervour of a new religion. The advance of modern science, and the insight it has given us into life, make us realize more vividly, and with greater truth, the possibilities and limitations of our civilisation."[1]

6.  From an exchange of letters to the editor of the London *Jewish Chronicle*:

'Jewvenis': "The question of the object of discussion [mixed marriages] lies quite outside the discussion. It may be the propagation of the species, in which case eugenics should supersede every other consideration."[2]

'M.' objects: "What ultimate good would arise from a temporary improvement in the physique of a small proportion of the human race, supposing for a moment that improvement should result? What high ideals would be served? What great purpose accomplished? The world would still have to continue its long evolution, but lacking the aid of Judaism and its adherents. The progress of mankind towards a 'perfect generation' would be seriously, perhaps fatally retarded by the loss of Israel and its mission.... To take 'Jewvenis' on his own ground, does he dare to predict that a union of his own pure blood... with a lady of unknown origin and haphazard, untraceable descent, will produce offspring of the highest possible quality? No matter that he has no sense of difference, he cannot vouch for hers."[3]

---

[1] Laski, 1910.
[2] Jewvenis, *Jewish* Chronicle, 1910, Sept. 30, 19.
[3] Nov. 18, 27.

'Jewvenis': "Words, words, words. I feel inclined to retort, 'Israel and its grandmother's cat.' Define this mission. (Again I repeat.) Show how we, to-day with our snobbery and mechanical orthodoxy, are doing one bit more for the world than the Gentiles. Our fine men, for whom three of your correspondents can find nothing better to say than that they are 'far more mercenary than the Christian men!' O, 'M,' pause and reflect! As for her last paragraph, words fail me. We are to assume then, that, rather than unite myself to a perfectly honourable, intelligent, healthy and beautiful Christian lady, I may marry the most miserable specimen of fifth-rate Jewry, however objectionable physically, morally and intellectually. And why? Because the Jewess is necessarily of 'pure blood'; the Christian necessarily of unknown origin and haphazard, untraceable descent.' Oh religion. Oh history. Oh eugenics. Oh commonest of common sense!"[1]

B. Felz: "'Jewvenis' probably sneers at faith and all that is associated with it. He is a disciple, apparently, of the 'advanced thinkers.' These thinkers have thought so rapidly that they have thought themselves to a standstill. They have debased faith and exalted reason.... Hypothetically we all have a right to commit murder, but we do not act on that hypothesis.... He wants to marry for love but he wants others to be married according to the laws of eugenics, whatever they may be.... From internal evidence it can also be gathered that Jewvenis is a Fabian and that is the crux of the matter. What he needs is a little simple thinking. Hitherto he has had too much of the advanced variety."[2]

'Jewvenis': "Mr. Felz... says that I desire a marriage of love for myself, a marriage of eugenics, for others. This is true, so far as I desire (under the present regulations) a marriage at all. It is similarly true that I desire a marriage of eugenics for myself, and a marriage of love for others. In short I believe love to be a practical quantity in eugenics. If I didn't, I should consider I had ample justification for blowing up the world."[3]

7.  Jewish-Austrian sociologist Ludwig Gumplowicz (1838-1909): "To comply with the obvious will of nature is the highest morality: With a perceptible voice nature calls back into its bosom those who are sick and weary of life. To follow this call and to make space for

---

[1] Nov. 25, 30.

[2] Dec. 2, 23.

[3] Dec. 16, 32-33.

healthy people filled with zeal for life is certainly no evil deed, but rather a good deed, for there are not too few people on the earth – rather too many." Before these words are published Gumplowicz, who is ill with cancer, and his blind wife commit suicide by taking cyanide.[1]

8.   Anthropologist, eugenicist, and author of many articles for the *Jewish Encyclopedia* Joseph Jacobs: "Nietzsche appears to have anticipated the Jewish idea of eugenics, by which, through mating together, the superior specimens of humanity, it is hoped to develop a higher form of man. But with Nietzsche, as with Galton, such higher forms cannot constitute the whole of humanity. They are the exception and upper limits of the human race, and are predestined to lead it and form its ideals....Nietzsche is aristocratic to the core, America is democratic to the depths. Nietzsche is Hellenic, aesthetic. America is Hebraic, ascetic. He directed his fiercest onslaughts against Christianity, which he appears to have regarded as a cunning and necessary invention of the Jews to preserve themselves amid a world of enemies.... There is a crude, rough vigor in all this which is immensely attractive amid all the overturning of ideals. His virile thought ranges over the whole field of modern culture.... His works are as stimulating as a storm by the seaside; the salt spray lashes but invigorates you."[2]

9.   Eugenicist Caleb Williams Saleeby: "the most conspicuously persistent of all races in the historic epoch, the Jews, have survived one Empire of their oppressors, but have never had an Empire of their own. Thus, so far as the historian is concerned, it is not races at all that die, but civilizations and Empires."[3] The London *Jewish Chronicle* comments: 'Dr. Saleeby's new book on "Parenthood and Race culture" (an outline of eugenics) should be widely read."[4]

10.  London Jewish Working Men's Club: "An interesting lecture was delivered, on Sunday evening by Dr. S. Herbert, before a large audience, the subject being 'Race Progress,' the three main principles of which the lecturer enumerated ethics, economics and eugenics. A

---

[1] Ludwig Gumplowicz, *Sozialphilosophie im Umriss* (1910), in Emil Brix (ed.), *Ludwig Gumplowicz oder die Gesellschaft als Natur* (Vienna, 1986), 272-73; quoted in Weikart, 2002, 337-338.
[2] Jacobs, 1910.
[3] Saleeby, 1910, 297.
[4] M.H., July 2, 1909, 25.

spirited discussion followed. Cordial thanks were accorded to the lecturer on the motion of Mr. Levy Davis. Mr. Louis Katz presided."[1]

## 1910-1920

### Context

1.  In St. Petersburg Jews have lower fertility than do non-Jews.[2]

## 1911

### Context

1.  The *New* York *Times* quotes an anonymous woman who is very successful as a writer of faction and verse, but who complains that writers looked with disfavor on "stories about sick people, about sex relations, about eugenics, about Jews, Catholics, Episcopalians, as such, about women who have broken social laws, about artists, as such."[3]

2.  German eugenicist Alfred Ploetz: "All these races (alpine, Jewish, etc.) are seldom found pure here."[4]

3.  Jewish-Russian-American anthropologist and eugenicist Maurice Fishberg (1872-1934) in his fundamental study *The Jews*: "The causes of this failure of the [Zionist] movement are manifold. In general it is due to the fact that the bulk of the modern Jews are entirely opposed to repatriation. The strictly orthodox class in Eastern Europe is against it because they do not believe in 'forcing the hand of Providence'; they believe that Messiah will come sooner or later to redeem the scattered children of Israel without any assistance of mortal man. The reformed Jews in Western countries insist that the Jews are not, like the Turk, only encamped in Europe, ready to retreat to Asia at the first favourable or unfavourable opportunity. Their Rabbis teach that Jews are only a religious community, and condemn all attempts at repatriation. In England the Chief Rabbi warned several Rabbis against preaching Zionism in their synagogues, and the Haham of

---

[1] *Jewish* Chronicle, Dec. 30, 14.
[2] Liebman Hersch, "Jewish Population Trends in Europe," *Jewish People: Past and Present*, II, 11, Table 10, cited in Goldscheider, 1967, 200.
[3] *New York Times*, 1911.
[4] *Ziele und Aufgaben der* Rassenhygiene, 190; quoted in Adams, 1990, 24.

the Spanish and Portuguese synagogue in London was admonished from touching upon this subject from his pulpit. In the United States the Conference of American Rabbis declared itself officially against the Zionist movement because 'America is the Jews' Jerusalem and Washington their Zion'.... The assimilated Jews in Western Europe, and many also in the East, oppose this movement vigorously.... 'The German Jew who has a voice in German literature must, as he has been accustomed to for the last century and a half, look upon Germany alone as his fatherland, upon the German language as his mother-tongue, and the future of the nation must remain the only one upon which he bases his hopes,' says Ludwig Geiger [Jewish-German historian, 1848-1919].... These are the views held by the majority of the cultured Western Jews. They recognize that the oppression of their co-religionists in Russia and Roumania is an important problem, but it cannot and must not be met with repatriation."[1]

4.  The *New York Times*: "He [Fishberg] is at pains to prove that the Jews, so far from being a pure race, have throughout their history intermingled with the races.... Whereas, until quite recently, they increased much more rapidly than other people, so much so that many statesmen were alert at the prospect of a Semitic world within a few centuries, recent inquiry, especially in Germany, shows that they are fallen behind the general births by fewer births, by considerable conversions.... [He] strongly advocates such assimilation, though, by his own showing, this would practically result in the disappearance of the Jewish race within a comparatively few generations."[2]

5.  Theologian, rabbi, and opponent of Nietzsche's and Spengler's "cultural pessimism" Ludwig Stein (1859-1930): "History is not a simple extension of nature, as Herder and Spencer assume, but a kingdom in and of itself. It is the creation of the conscious human spirit rising above the dumbness of the unconscious.... We take our fate in our own fist. We ourselves are the blacksmiths of our own happiness. We shall not permit any fate to be forced upon us by the mystique of blood."[3]

---

[1] Fishberg, 1911, 498-500.

[2] Anonymous, 1911a.

[3] *Allgemeine Zeitung des Judenthums*, 1911, 103, 104; quoted in Doron, 1980, 395.

6.  Supposedly, mixed marriages produce far fewer children than do marriages in which both partners are Jewish: 12.4:1, and only 10% of the children of such mixed marriages retain their Jewishness.[1]

7.  Physician Sir James Barr (1849-1938), Vice President of the local Eugenics Education Society, delivers an address before the Liverpool Jewish Literary Society:

    • "I think the time has now come when even the Jews might consider the question of quality as well as quantity. It is quite true that the Jews have not kept themselves pure and undefiled and unspotted from the world; but on the whole their strong religious persuasion has kept them A FAIRLY PURE RACE. Personally, I have a warm admiration for the Jews. Apart from any admixture of Gentile blood, there are large numbers of degenerates among the Jews. It is perhaps a question for the Jews themselves as to where the elimination should begin, but from a Gentile point of view I would be inclined to start with those who charged fifty percent, and upwards for the loan of their money."[2]

    • Livingstone: "...there is one quality which Sir James Barr cannot deny the Jews, and that is that they listen to their critics with much tolerance – or rather suffer their admirers with much fortitude. I am not hyper-sensitive, but I fancy that this tolerance and sufferance can go a little too far."[3]

8.  Jewish-German physician and eugenicist Felix Theilhaber (1884-1956) writes *The Decline and Fall of the Jews*, in which he diagnoses a "degenerative process" caused by mixed Christian-Jewish marriages, which are increasing in frequency and, to make matters worse, often childless.

---

[1] Doron, 1980, 412, quoting several sources.
[2] Anonymous, 1911b.
[3] Anonymous, 1911c.

9. **Total number of Jews placed in mental institutions in Prussia**

|  | 1880-1882 | 1883-1885 | 1886-1888 | 1889-1891 | 1892-1894 | 1895-1897 | 1899-1900 |
|---|---|---|---|---|---|---|---|
| 1. Direct genetic handicap | 81 | 62 | 86 | 116 | 160 | 139 | 120 |
| 2. Family predisposition | 158 | 144 | 203 | 226 | 268 | 261 | 281 |
| 3. Direct genetic handicap or family predisposition | 51 | 80 | 93 | 113 | 156 | 161 | 149 |
| 4. No heritability | 351 | 352 | 378 | 396 | 444 | 636 | 724 |
| 5. Unknown | 321 | 797 | 757 | 509 | 542 | 417 | 522 |
| TOTAL | 962 | 1,085 | 1,277 | 1,360 | 1,566 | 1,614 | 1,796 |

(Theilhaber, continued): "If we compare institutionalizations of the Jewish mentally ill with persons of other faiths, we see the following figures":

|  | 1892-94 | | 1895-97 | | 1898-1900 | |
|---|---|---|---|---|---|---|
|  | Christians | Jews | Christians | Jews | Christians | Jews |
| A. Simple mental disturbance | 23,693 | 988 | 25,746 | 1,058 | 31,121 | 1,243 |
| B. Paralysis | 5,603 | 301 | 6,071 | 237 | 6,149 | 202 |
| C. Epilepsy | 3,907 | 80 | 4,355 | 107 | 5,034 | 117 |
| D. Idiots | 4,837 | 135 | 4,777 | 159 | 5,320 | 169 |
| E. Alcoholism | 3,080 | 14 | 3,688 | 21 | 3,545 | 21 |
| F. Not sick | 940 | 38 | 1,024 | 32 | 1,381 | 32 |
| TOTAL | 42,060 | 1,566 | 45,651 | 1,614 | 52,650 | 1,796 |

10. (Theilhaber continued): "According to these figures during the years 1892-1900 Jews constituted 3.5% of all institutionalized mentally ill – 3 ½ times as the numbers in the general population. The

number of Jews suffering from paralysis is particularly high: 12-25%, moreover the percentage is ten times higher among Jewish men than among Jewish women." Theilhaber is a Lamarckian eugenicist and recommends a healthy life style to promote 'regeneration.'[1]

11.  Manchester psychoanalyst Solomon Herbert (1874-1940) in the *Jewish Review*: "We have above drawn attention to the extraordinary mental activity of the Jews. Good as this is in many ways, it appears to be by no means an unmixed blessing. For the close connection between genius and insanity, which has often been remarked upon, seems to be verified in this case. Not only have the Jews a very high percentage of lunacy – about two to three times larger than that of their neighbours – but mental and nervous disorders of all kinds are very common among them.... But, while we would not advocate Spartan measures against these victims of nature, it becomes incumbent upon us for the future welfare of our people to restrict the unlimited propagation of the unfit.... No scientific researches of any extent have been made on the subject of race blending.... While, on the one hand, it would seem undesirable to lower the standard of inbred qualities of the Jews by a large infusion of extraneous blood, on the other hand, a judicious admixture from other stocks cannot be rejected as biologically wholly bad, as long as care is taken that the Jewish unity is not thereby jeopardised. Indeed, it would appear an open question, for instance, whether such selective mating would not tend rather to temper down the somewhat highly strung, nervous disposition of the Jews."[2]

12.  Max Besser (1877-1941) in *Jews in Modern Racial Theory*: "We wish to peer into the future with joyous eyes, confident that the genius of the Jewish people, once the iron chains of oppression are removed, will in its free unfolding make a gift to mankind of new, eternal values."[3]

13.  Prominent Zionist Redcliffe Salaman (1874-1955) in *Heredity and the Jews*: "The deductions that might be drawn... strengthen the view that complex as the origin of the Jew may be, close inbreeding for at least two thousand years has resulted in certain stable or homozygous combinations of factors which react in accordance with the laws of Mendel

---

[1] Theilhaber, 1911, 139-140.
[2] Herbert, 1910-1911, 453-454.
[3] Besser, 1911, 29.

and which may explain the occurrence of the peculiar facial expression recognized as Jewish."[1]

14. Jewish-German dermatologist, sexologist, and eugenicist Max Marcuse (1877-1963): "We are not at all interested in the preservation of a Jewish cultural or racial community but believe, on the contrary, that... the decline of German (in general West-) Jews should, rather, be accelerated. But all the greater is our interest in preserving the lofty health- and cultural values that have been stored up within German Jewry that they might not be lost simultaneously with the death of their biological bearers." Marcuse views the "Jewish problem" as an "instructive, serious, unique case of eugenics" whose resolution is feasible "only along the pathways of an All-German population policy oriented toward racial hygiene that might preserve the physically and psychically valuable, culturally gifted germ plasma, awakening and nurturing a cultural will favorable to the establishment of families."[2]

15. Biologist Jacques Loeb (1859-1924) wishes to create a "biological engineering art" which would transform life.[3]

16. Theologian Ludwig Stein reflects the majority Zionist belief in Lamarckian eugenics: "If we wish to dialogue with [August] Weismann [advocate of German plasm theory, 1834-1914], that which is nowadays purported to be unalterable Jewish racial characteristics do not lie preformed in our germ plasm, in our blood as preordained fate. Rather, these are acquired characteristics that have been shaped over the course of history and been inherited."[4]

17. Lewis Morris, B.A., of Jews College, delivers a paper on "eugenics" before the Great Garden Street Torah Old Boys (London): "The lecturer referred to the bearing of Judaism on the general problems of race culture. A discussion followed in which messrs. M. Braun, B.A., H. A. Kassilovitch, A. Zeitlin, I. Portugal, and the Chairman [A. Plaskow] took part."[5]

---

[1] Salaman, 1911, 290.

[2] Marcuse, Max, Rezension zu Theilhaber Untergang der deutschen Juden, 1911, in *Zeitschrift für Sexualwissenschaft* 14
(1927), H. 7, S. 280; cited in Lipphardt, 2006.

[3] Loeb, 1911.

[4] *Allgemeine Zeitung des Judenthums*, 1911; quoted in Doron, 1980, 399.

[5] *Jewish Chronicle*, Mar. 17, 1911, 32.

## 1912

### Context

1. Arthur James Balfour, author of the Balfour Declaration, delivers the opening speech at the First International Congress of Eugenics in London: "the study of eugenics is one of the greatest and most pressing necessities of our day."[1]

2. The final lines of the sonnet "The New Colossus" by the Jewish-American poet Emma Lazarus (1849-1887) are engraved on a bronze plaque in the base of the Statue of Liberty: "Give me your tired, your poor/Your huddled masses yearning to breathe free,/The wretched refuse of your teeming shore./Send these, the homeless, tempest-tost to me,/I lift my lamp beside the golden door!" In sharp contrast to her recommendations for the United States, Lazarus is considered to be a "proto-Zionist": "Brethren, my cup is full!/Oh let us die as warriors of the Lord./The Lord is great in Zion. Let our death/Bring no reproach to Jacob, no rebuke/To Israel. Hark ye! let us crave one boon/At our assassins' hands; beseech them build/Within God's acre where our fathers sleep,/A dancing-floor to hide the fagots stacked."[2]

3. At the First International Congress of Eugenics, London, a statistician from the Prudential Insurance Company of America reports that married Jewish women are the group most likely to have had children – 88.0%, the other categories being native-born (71.6%), foreign-born (82.5%), Protestants (72.7%), and even Catholics (82.3%).[3]

### Jewish Advocacy of Eugenics

4. Reform rabbi Emil G. Hirsch (1852-1923) writes that, without invoking the "jargon of eugenics," rabbis had done and continued to do their part to prevent unwise marriages.[4]

---

[1] http://www.eugenics-watch.com/briteugen/eug_babh.html, accessed May 12, 2008.
[2] Project Gutenberg Etext of The Poems of Emma Lazarus, Vol. II,
http://www.gutenberg.org/dirs/etext02/2mlaz10.txt.
[3] http://www.pubmedcentral.nih.gov/picrender.fcgi?artid=2334093&blobtype=pdf, accessed May 30, 2008.
[4] Rosen, 2004, 58.

5.  Eugenicist Ignaz Zollschan in *The Racial Problem*: "Without Zion-
    ism there are merely two possibilities: the dissolution of the race or
    physical degeneration."[1] Alexander Schüler writes that he regards
    Zollschan's work as the latest in scientific knowledge and that he
    draws from it an understanding of "the demand for preserving the
    race and the racial purity of the Jewish nation... of the need for a
    movement emanating from the entire people (*vom ganzen Volk*) and
    to which the entire people would turn." He feels that Zollschan's
    book provides a new, scientific, understanding of Zionism.[2]

6.  Anarchist Emma Goldman (1869-1940) travels to Denver in the
    hope of teaching eugenics there, but the classes are canceled for
    lack of interest.[3]

7.  Feminist and eugenicist Henriette Fürth argues that population qual-
    ity is best served by maximum freedom of reproductive choice, in-
    cluding abortion.[4]

8.  Physicist Arthur Schuster and Francis Galton's co-author Dr. Edgar
    Schuster attend the First International Congress of Eugenics, Lon-
    don.[5]

9.  From a *Jewish Chronicle* editorial column entitled "In the Com-
    munal Armchair," signed by "Mentor":

    > "For the Jewish race is, among the races of the world, as Dr.
    > Lindsay in the course of his paper at the Eugenics Congress
    > pointed out, a remarkable testimony to the value of Eugenics.
    > Our survival to this day is living proof to the truths which euge-
    > nicists are enforcing....That the Pentateuch raised Eugenics into
    > a matter of religion goes to show only either that thousands of
    > years ago the Jewish people regarded Eugenics as a supreme
    > value to man, or that it was feared that only as a series of Divine
    > commands would Eugenics be practiced. We note it in every di-
    > rection; in the laws of segregation as in the connubial prohibi-

---

[1] *Das Rassenproblem unter besonderer Berücksichtigung der theoretischen
Grundlagen der jüdischen Rassenfrage*, Vienna/Leipzig, 1912, third edition, 494;
quoted in Gelber 2000, 139.
[2] Schüler, 1912, 56-58.
[3] Falk/Cole/Thomas.
[4] *Staat und Sittlichkeit*, 14; cited in Taylor-Allen, 1988, 43.
[5] *Eugenics Review, Bulletin of the Eugenics Society, Biology and Society,* Eugenics
Watch.

tions the Jew was taught Eugenics as his religion. The much despised Shadchan or marriage broker as an institution had many obvious faults. Yet, in a quiet, unscientific manner he has been the means of curing mere sentiment and passion in the mating of sons and daughters of Israel. The Shadchan, when he was conscientious – and who will say he never was? – made it his business to bring about marital unions that should be happy in the sense of being fit, proper and healthy. His reputation was at stake if his 'introductions' did not show a clean bill of family health. His art consisted in 'matching' those who were to be joined in matrimony, so that he became an agent in multiplying marriages of the fit. We have laughed consumedly at the Shadchan's vagaries, and have been shocked at his turning what we instinctively feel out to be a matter of love and affection into one of barter and bargain. But the Shadchan is distinctly on the side of Eugenics, in 'regulating' the union of men and women, and he must have contributed a trifle to the preservation of the race. If Eugenics has its way, the Shadchan in every land which cares for the preservation of its race looks like being nationalized into a state department."[1]

10. A response from 'Infelice': We women are helping to change many things…. We require to change the status of marriage altogether. If the injudicious love of marriage is to be deplored, the coldly-arranged union for gold is to be more so. We want to learn more about eugenics. We want to learn more about what really makes a happy wedded life. Better than all the riches is the man who comes from a good stock…."[2]

11. A correction in London's *Jewish Chronicle*: "Eugenics Congress, – In reference to "Mentor"'s comment in last week's issue that Jews were not specifically represented at the Eugenics Congress, it is pointed out that Mr. Ernest L. Willard represented the Jews' Free School at the Congress."[3]

## Jewish Resistance to Eugenics

12. Mr. N. S. Burstein of Cardiff. "To say that in bringing children into the world the question of quality and not quantity ought to be most

---

[1] "Mentor," *Jewish Chronicle*, Aug. 2, 1912, 6-9.
[2] Jewish Chronicle, Feb. 21, 44.
[3] Aug. 9, 18.

carefully considered, is, to my mind, simply preposterous. Who has not seen mentally and physically weak children grow into clever men and women?... There is, to my mind, only one tangible and concrete way for the eugenists to do something toward bettering the coming race and that is – to teach assiduously to future parents the grammar of virtue, the grammar of chastity, and the grammar of morality."[1]

## 1913

### Jewish Advocacy of Eugenics

1. Rabbi Ephraim Frisch of Far Rockaway attacking the Eastern Council of Reform Rabbis in a letter published by the *American Hebrew*. Frisch maintains that the Council is superfluous and is seeking to disrupt the Central Council of American Rabbis, from which, he asserts, it sprang in part. "It [the Eastern Council] ought never to have been born, and if we had a good system of theological eugenics it never would have seen the light."[2]

2. Eugenicist Felix Theilhaber in *Das sterile Berlin* proposes taxing families with few or no children so as to subsidize large families.[3]

3. In Berlin Jewish venerologists Magnus Hirschfeld (1868-1935) and Iwan Bloch [1872-1922] cofound the Medical Society for Sexology and Eugenics (Die ärztliche Gesellschaft für Sexualwissenschaft und Eugenik), but have a falling out with one of their Jewish co-founders, neurologist Albert Moll (1862-1939), who promptly founds a competing institute. Moll, who also is a bitter enemy of Freud's, later dies the same day as Freud.[4]

4. The British Society for the Study of Sex Psychology is founded in London.[5] It advocates eugenics and attracts a number of Jewish intellectuals, some of whom are homosexuals, establishing the historical nexus homosexuality/Jews/eugenics, but also including marriage counseling and birth control.

---

[1] Burstein, 1912, 24.
[2] Cited in *New York Times*, 1913.b
[3] Theilhaber, 1913a.
[4] "Moll, Albert" German *Wikipedia*.
[5] Crozier, 2001, 304.

5.  From the *New York Times*: "Rabbi Maurice H. Harris of the Eastern Council of Reform Rabbis… urged a study of eugenics, saying: 'All recognize that the time has come when such vital issues must be discussed. In this relation the Eastern Council, like the Central Conference, should be prepared to indorse the requirement already adopted by some states and churches of a physician's certificate of health prior to the solemnization of a marriage."[1]

6.  North London Jewish Literary Union: "Yesterday week Dr. J. Snowman read a paper on 'Jewish Eugenics.' A discussion followed, in which Dr. Mary Seharlieb (who presided), Miss Denhoff, Dr. Goitein, the Messrs. J. Brodsky, A. Rabinstein, H. Sperling, F. S. Spiers, and S. Wallach took part."[2]

7.  The Liverpool Jewish Literary Society: "The final meeting of the session was held on Sunday at the Royal Institution. Mr. Bertram B. Benas, B.A., LL.B., Vice-President, presided over a numerous attendance. The lecturer was Dr. Glynn Whittle, M.A., M.R.C.P., who chose as his subject 'Another Aspect of Eugenics.' The author, who at the onset mentioned that from a perusal of the interesting pages of the JEWISH CHRONICLE he concluded that Anglo-Jewry was concerned with the eugenic aspects of legislation., gave a critical consideration of the marriage laws of England, tracing their ecclesiastical origin, and reviewing at length the sources of their principles, and the interpretation placed upon these sources by the canon law and its exponents. Thanks were accorded to Dr. Glynn Whittle on the motion of the Rev. I. Raffalevich, seconded by Mr. L. Collins, and supported by Mr. M. Jacobson and the Rev. J. Bach."[3]

## 1914

### Context

1.  The leading promoter of orthodox Marxism after the death of Friedrich Engels, Karl Kautsky (1854-1938) publishes *Jewry and Race*, debunking racial theories about Jews and conceding that while only Zionism can save Jewry from assimilation, Zionism amounts to "abandoning the colors" [of socialism].[4]

---

[1] *New York Times,* 1913a.
[2] *Jewish Chronicle, Feb. 21, 18.*
[3] *Jewish Chronicle*, April 4, 30.
[4] Kautsky, 1914.

2.  The world Jewish population increases from about 4¾ million in 1850 to 13½ million in 1914, i.e., by 180% or by 16 per 1,000 annually. In comparison, during 1850-1900 the total world population is estimated to have grown by 67 per 1,000 annually, and the population of Europe, North America, and Oceania by 11 per 1,000 annually.[1]

## Jewish Advocacy of Eugenics

3.  From the *New York Times*: "Rabbi M. Hyamson told of a... nursery established by London Jews in East End. He referred to the Jewish interest in eugenics, and said that the work of day nurseries provided for better children of the future as well as for better children in the present."[2]

4.  Physician and author of *A Short History of Talmudic Medicine* Jacob Snowman (1871-1959) argues in an article entitled "Jewish Eugenics" (published in *Jewish Review*) that "Judaism and eugenics are in complete accord in encouraging the marriage of the fit, which, indeed, forms the more important aim of the new science – Eugenics is an ultra-modern form of הלכ תסנכה [the appearance of the bride] conducted on principles of careful selection. The value of the eugenic movement will not be gauged so much by the prevention of marriage among the degenerate and diseased, but by its success in promoting marriage, in the prime of life, among the best types of men and women.... Of all races, our interest in the future is probably the keenest. Judaism proclaims the bond of mutual responsibility which unites contemporary generations, but it also recognizes the duty it owes to succeeding generations, and imposes the religious obligation of carrying on a vigorous, intellectual, and self-reliant race."[3]

5.  The magazine *Naturopath and Herald of Health* on the contribution to the eugenics movement made by Moses Harman: "He gave the start and spur to this effort. Through his journals, *Lucifer, the Lightbearer*, later rechristened *The Eugenic Magazine*, encouraged by a small circle of earnest men and women, he dug down below

---

[1] Schmelz/DellaPergola, 2007.
[2] *New York Times*, 1914.
[3] Snowman, 1913-1914, 171-172, 174.

the surface endeavoring to bring forth a stronger and better type of man."[1]

6.  From the *Jewish Chronicle*: "We hear much nowadays of the new science of eugenics. But there is nothing new under the sun, and the principles of this science were formulated thousands of years ago in the inspired pages of the Pentateuch, anticipated in the words of Deuteronomy: 'Hear, O Israel, and observe to do it, that it may be *well* with thee, and that ye may increase mightily.'[2]

## 1915

### Context

1.  Psychologist and President of Clark University G. Stanley Hall in *Menorah Journal*: "There is much in common between the Yankees, whom I represent, and the Jews, and this alone ought to give us a friendly feeling toward one another.... I realized very keenly how closely related were the Jews to the Yankees, – with this tremendous difference, that you are increasing in numbers while we are decreasing."[3]

### Jewish Advocacy of Eugenics

2.  Emma Goldman is arrested on morals charges for distributing a 4-page pamphlet in English and Yiddish entitled *Why and How the Poor Should Not Have So Many Children*.

3.  The *New York Times* reports that Nathan Rabinowitz has proposed a "drastic" eugenics law to the Health Officer's Association of New Jersey: "According to the proposed law no marriage would be permitted without the presentation by both bride and bridegroom of a health certificate made in the form of an oath made by their physicians. If, after the marriage, it shall be found by either party that the other's health certificate was false, the physician who gave it is made liable to a damage suit by the injured party and is also liable to prosecution by the State for perjury. The law also provides that in the case of persons going out of the State to marry in order to escape its consequences their marriage shall be void.... 'The serious-

---

[1] March; cited in Wikipedia, http://en.wikipedia.org/wiki/Moses_Harman, accessed Dec. 26, 2008.

[2] Jan. 9, 20.

[3] Hall, 1915, 87-88.

ness of the situation is such that New Jersey can no longer afford to neglect it,' said Mr. Rabinowitz. 'It was first brought to my attention by a case in my practice in which the bride was told by the bridegroom on their honeymoon that he was suffering from a dread disease. She got a divorce, but now she is a social outcast...' This indorsement was given without a dissenting vote."[1]

4.  From a commentary by London's *Jewish Chronicle*: "The war has raised many problems for us, and not the least among them are: What is to become of the future of the race, how are we to replace those who are lost and maimed, and whereby shall we improve the quality of the stock left behind? These and kindred questions all belong to the province of the eugenicist. He more than anyone else can surely claim to speak with authoritative knowledge on these matters and offer some solution which should serve as a basis for free and impartial discussion and be of particular help to us Jews in this terrible time of 'blood and fire.'... The idea [of eugenics] is not new fangled, but was clearly laid down by Plato and Theognis about twenty-two centuries before him. The modern expression of it is now nearly a quarter century old and it has already passed the stage of ridicule except by the ridiculous."[2]

5.  The Young Men's Debating Society (London): "Last Friday the Rev. L. Morris, B.A., delivered his presidential address. He chose as his subject 'The Jews and Eugenics.' Mrs. S. L. Lipschitz, an Hon. Vice-President, presided over a record attendance."[3]

## 1916

### Context

1.  *Im deutschen Reich*, a publication of The Central League of German Citizens of Jewish Faith, comments that if the 1914 claim by the German Society for Racial Hygiene really avoided involvement in political and confessional aspirations, it would not ask: "My confessional belonging."[4]

---

[1] "Eugenic Marriages Urged..." 1915.
[2] Introduction to an interview on eugenics with C. W. Saleeby, April 16, 6.
[3] *Jewish Chronicle*, Nov. 12, 26.
[4] *Im deutschen Reich,* 1916.

2.   Two Yiddish plays – *Birth Control or Race Suicide* (*Geburth Kon-*
     *trol, oder, Rassen zelbstmord* by Harry Kalmanowitz) and *A Wom-*
     *an's Duty in Birth Control* (*Di Flikhten fun a froy in geburt control*
     by Samuel B. Grossman) – are submitted for copyright deposit at
     the Library of Congress, both written in the same year that Margaret
     Sanger opens America's first birth-control clinic in Brooklyn. Five
     thousand flyers are printed in English, Italian, and Yiddish to in-
     form women of the clinic's opening, but it is closed by police within
     ten days. *Birth Control or Race Suicide* is performed at New York's
     Roof Garden Theater.[1]

3.   The Glasgow Literary and Social Society, as reported in the *Jewish*
     *Chronicle*: "'The Jew – A Eugenic Factor' was the title of a highly
     interesting lecture delivered last Sunday by the Rev. Louis Morris,
     B.A. Mr. Morris introduced the subject by reviewing briefly the
     aims and purpose of the science of Eugenics, and concluded by
     showing that the Jews approximated closer to the application of that
     science than any other people. Among the Jews the rate of infant
     mortality was lower, in some cases less than one third, than that of
     non-Jews, but not so organic derangement to which Jews were less
     liable than other peoples. The Jew was a good Eugenic subject an-
     ywhere, but better in the ghetto. Mr. Morris was thanked for his lec-
     ture by the Chairman, ex-Bailie M. Simmons, D.L., and J.P."[2]

### Jewish Advocacy of Eugenics

4.   Rabbi Joel Blau in a paper read before the New York Board of Jew-
     ish Ministers: "If the Rabbis of the Aggadah have a philosophy of
     Subnormality looking towards its cause and origin, the Prophets
     have an Eschatology of Subnormality, looking towards its end and
     final extinction. On the day when the crooked will be made straight
     and the desert bloom as a rose, both cause and effect of Subnormail-
     ity will be done away with, both soul and body will be made whole.
     In the meantime, the Rabbis of the Halachah, being practical men,
     were right in dealing with a knotty human problem in a practical
     way…. Thus our final word about the *Defective in Jewish Law and*
     *Literature* is, that if the Aggadists point the way to deep speculation

---

[1] http://www.jewishvirtuallibrary.org/jsource/loc/birth.html.
[2] Feb. 25, 25.

and the Prophets to sublime inspiration, the Halachists point the way to effective service."[1]

5.  N. S. Burstein delivers a lecture at the Cardiff Jewish Institute on "Eugenics, and the Jewish Daughter."[2]

6.  The Jewish Association of Arts and Science: "At Toynbee Hall, on Sunday evening last, the Rev. Dr. Abelson read a paper on 'Some Disorders of Personality and Their Significance to the Race,'... The Association has resolved to devote a special sub-section to Disease, specially affecting the Jews. Dr. Abelson has consented to act as secretary for this purpose, and all those interested are invited to co-operate. Invitations to co-operate will be sent to all the leading Jewish physicians in London and the provinces, to the Eugenics Education Society, medical students, social workers, ministers of religion, etc., and it is intended later on to publish a book on the results achieved." [3]

7.  *New York Times* obituary for D. A Gorton, founder of the Eugenic Society of America: "Married His Secretary. Although he had long been prominent for his advocacy of eugenics and was a fellow of the New York Academy of Medicine, in addition to being widely known as a writer on medical subjects, Dr. GORTON came into public notice when he married his secretary, Miss Bertha REHBEIN, in 1911. She was less than half his age, and before marrying her he carefully studied her family history. At the time it was said that the unusual wedding was a romance, but Dr. GORTON discussed the union scientifically with his wife before marriage, and they decided that it should be a practical test of his eugenic theories. On April 25, 1912, about a year after they were married, the twins were born.... It was said to-day that the 'eugenic twins' would be brought up in the strictest observance of their father's theories, which would be followed until they were fully grown."[4]

---

[1] Blau, 1916, 26-28.
[2] *Jewish Chronicle*, Feb. 4, 27.
[3] *Jewish Chronicle*, Dec. 22, 24-25.
[4] *New York Times*, 1916.

## 1917

## Context

1.  Civil war in Russia prevents a resumption of the massive pre-war Jewish emigration.

2.  Jewish-Romanian-American civil-rights activist Henry Moskowitz (1875-1936), a co-founder of the National Association of Colored People: "This movement [Zionism and Nationalism] became the refuge of many Jews to whom Judaism as a religious message had ceased to appeal. Zionism as a political weapon against anti-Semitism has always represented a minority in Israel.... The American Jew is nationally an American only. He is thrilled by one flag, which symbolizes the ideals of a democracy in which men and women of all races have joined together to make government of the people, for the people, and by the people a success. The first tenet of American democracy is the separation of church and State. It has demonstrated that citizens can have one political allegiance and as many religious allegiances as their conscience may dictate. The American Jew will not tolerate any other political loyalty. Any movement which emphasizes racial groups in America inconsistent with the spirit of this democracy.... Responsible, organized English Jews are fearing the establishment of a free republic in Palestine upon the basis of special Jewish rights. What they want is equal rights for their people wherever they live. This is a much sounder position for the Jews than the dangerous political experimentation of Zionism. But to the Jews in America Zionism has no positive message."[1]

## Jewish Advocacy of Eugenics

3.  Jewish-Russian-American anthropologist Maurice Fishberg (1872-1934) in an article entitled "The Racial Breeding of Jews": "The study of the history of the Jews, of their marital laws, customs, and traditions over the course of the last two thousand years clearly shows that ghetto life was efficient in promoting positive eugenic tendencies with favorable outcome, so that we now find among European Jews a greater percentage of persons with special abilities than among the remaining population. On the other hand... it is a commonly observed fact that Jews are physically weak, shorter than

---

[1] Moskowitz, 1917.

normal, their musculature is feebly developed, the chest narrow, flat, and inferior; in brief, they create the general impression of a weakling people. Moreover, not only physical defects, but also mental shortcomings are more frequent among them than among other civilized peoples.... Blindness, color blindness, deaf-mutism, idiocy, feeblemindedness, weakness, etc. are found among them two to five times more often than among Europe's Christians.... Practiced for centuries, the most influential anti-eugenic factor was the institution of the *Hachnassath-Kallah* – societies that occupy themselves with providing for poor Jewish brides and their dowry. In the ghetto there were practically no unmarried persons of either sex. Every Jew was supposed to marry and have children. When young people were too poor to venture marriage or if they suffered from such mental and physical defects that they could find no partner willing to link their fates together, then these societies provided the means and opportunity to find a partner for life. Thus every physical and even mentally fragile person was induced to marry and bring legitimate offspring into the world. Frequently such persons had to be forced into marriage and parenthood. A blind youth was matched with a lame girl. The community fool was given a paralytic wife, etc. The blind and the dumb paired off. Socially minded Jews, Jewesses in particular, collected charity to provide these unhappy persons with furniture and money. These unhealthy persons undoubtedly left behind a disproportionate number of degenerate offspring, who in turn remained a burden for the community until they came of age, at which time pious Jewesses and Jews once again collected for them the means to marry. ¶ A superstition among Eastern European Jews ran that an epidemic, particularly cholera, would be immediately stopped if two cripples married. The ceremony had to take place in a Jewish cemetery. In times of pestilence Jews never neglected to employ this method.... In many instances matchmakers played a eugenic role, as when they brought together a rich man's daughter with a promising scholar, but a large number of invalid Jews also owe their existence to their efforts. Since a matchmaker was legally entitled to demand compensation for his labors, more often than not he had no scruples in pursuing his line of business. Jewish folktales and literature tell of many unscrupulous such persons who brought together cripples and criminals. ¶ They're active even now in Eastern Europe, and their trade flourishes particularly among London's and New York's Jewish immigrants. The wealthy and publicly assimilated Jews in Berlin,

Paris, Vienna, London, etc. especially favor their services.... ¶ On the whole charitable societies among the Jews of Vienna, Berlin, London, and particularly New York continue to support invalids, pay their rent, give them welfare payments, so that these disadvantaged members of society reproduce themselves to a degree that is frightening when one realizes that most of the children produced in this fashion display mental or physical defects."[1]

4.  Fishberg on Jewish physical attractiveness: "Wealthy persons and scholars were little concerned with the physical appearance of their future sons-in-law. Intellectual abilities were the main thing. If the bridegroom was a significant, promising scholar, even a physical defect was ignored. By contrast, great importance was attached to the physical appearance of the bride. The Talmud praises a woman's beauty, and every Jew is supposed to marry a beautiful woman. And it is indicated that a large man should give preference to a small woman, and vice versa, a person of dark complexion should seek out someone of light complexion. This may to a certain degree explain why one sees in the ghetto many beautiful women, while handsome men are encountered only seldomly. Selection may have exercised its effect."[2]

5.  Fishberg on race: "I have elsewhere demonstrated that there exists no homogeneous Jewish race and that we can speak just as little of racial purity among the Jews as we can speak of racial purity among other communities. In the sense of ethnicity we find among contemporary adherents of Jewry all sorts of racial elements. Thus we must conclude that it is not race that is the cause of these social and pathological distinctions between Jews and other peoples."[3]

6.  The *Journal of Heredity* publishes "Jewish Eugenics: Perpetuation of the Race Explained by Application of Sound Biological Principles – Marriage Held in High Esteem and Its Success Measured by Eugenic Standard," an anonymous review of Rabbi Max Reichler's *Jewish Eugenics and Other Essays*.[4]

---

[1] Fishberg, 1917, 71-83.
[2] Fishberg, 1917, 80.
[3] Fishberg, 1917, 74-75.
[4] Anonymous, 1917.

7.  The German Psychiatric Research Institute (Deutsche Forschung-sanstalt für Psychiatrie – DFA) in Munich, one of the most important research institutes in the field of theoretical and clinical psychiatry and a bulwark of the eugenics movement, is founded by Emil Kraepelin. Its financial existence between the World Wars is guaranteed by large donations from the Jewish-American philanthropist James Loeb (1867-1933).[1]

8.  Founder of the science of biometrics William Moses Feldman (1880-1939): "Recognizing the relative importance between heredity and environment, the Rabbis formulated certain rules and principles of selective breeding, or, as Galton has named it, 'eugenics,' for the deliberate purpose of permanently raising the standard of the Jewish race. 'Eugenics,' says Sir Francis Galton, 'deals with what is more valuable than money or lands – namely the heritage of a high character, capable brains, fine physique, and vigour... and deserves to be strictly enforced as a religious duty.' And such was also the opinion of the Jewish sages in the time of the Talmud."[2]

9.  Sioux City, Iowa, 'progressive' rabbi and prominent social leader Emanuel Sternheim (1882-1942) advocates a planned eugenics program to cope with the general degeneration of the 'lower classes': "No more noble task can be undertaken than to go among the very poorest girls in our big towns with the view of instilling into their minds both from a moral and a material point of view the necessity of greater self-restraint and of the advantage of looking before they leap into marriage. Human nature is such that it has placed in the hands of women far greater practical opportunities of doing good in this portion of the eugenics field than is the case with men; and this is a force which ought to be used to the utmost."[3]

## 1918

### Context

1.  A Wiesbaden physician with the surname Ratner writing in *Hygie-nische Rundschau* sees the prophet Abraham as having practiced eugenics. "There is a constant leitmotif in the ancient history of the

---

[1] Weber, 1991.
[2] Feldman, W., 1917, 43.
[3] Emanuel Sternheim, *A Sociological Reverie*, Sioux City, Iowa, 6-11; cited in Burger, 1974-1975.

Jewish people: racial choice and eugenic selection intended to strengthen and preserve the tribe. Later we see that any mixing and purposeless outbreeding that could lead to a decline is severely discouraged, even threatened with punishment."[1]

## Jewish Advocacy of Eugenics

2.  Psychiatrist Shneor Zalman (Zygmunt) Bychowski (1865-1934), an ardent Zionist and founder of the 'Maccabi' organization in Poland, is also an enthusiast of the eugenic idea and preaches explicitly harsh measures to prevent degeneration of the Jews and also to uphold their viability, although he categorically denies that specific Jewish neuropathies are of a hereditary nature: "The Russian revolution has abolished all our restrictions – this is a wonderful physician to the nerves of the sons of Israel. The 'hidden complexes' of Freud, those mental wounds with painful thorns in them have all been abolished. In the free Russia there will be no room for 'our nervousness.' The 'Wandering Jew' will remain only in the world of stories and fantasies."[2]

3.  A group of physicians and social workers in Warsaw found the Section for Social Hygiene and Eugenics (Sekcya Higieny Społecznej i Eugeniki) within the Society for the Health Protection of the Jewish community in Poland (Towarzystvo Ochrony Zdrowia Ludności Żidowskiej w Polsce – TOZ). Gerson Lewin (1867-1939) is appointed its first President, and it is officially recognized by the Polish Ministry of the Interior in 1923.[3]

4.  Bychowski again: "The resurrection of the nation in its homeland will be possible only if the 'human material' that goes there is healthy. In this respect it will be necessary to employ from the beginning strict means, like the 'law' against immigration that has been instituted in the United States. It is of special significance that the Zionists should learn to view marriage not as a personal act that a person my handle as his heart may desire, but rather as an important public act, on which depends the future of the race.... If there are reasons to believe that the marriage may produce sick children,

---

[1] Ratner, 1918, 250.
[2] Falk, 2003-2004, 44-45.
[3] Turda/Weindling, 2007, 289.

these must be strictly forbidden. The Zionists must be especially careful, when they come to rebuild anew the life of the nation."[1]

5.  Zionist psychiatrist Raphael Becker (1891-ca 1943) rejects the view that Eastern European Jews are psychopathic and instead maintains that it is deracinated Western European Jews who are inferior. In the words of American historian John Efron, this is a "paradigm shift."[2]

## 1919

### Context

1.  *The Nation* publishes in its December issue a letter by Franz Boas, a declared pacifist who had opposed America's entry into World War I, accusing four other anthropologists of having conducted espionage in Central America during the conflict, and the American Anthropological Association (AAA) votes to censure him for using his professional position for political ends. [3]

2.  Jewish eugenicist Elias Auerbach: "[Fritz] Lenz was very perceptive in noting that most representatives of this theory [Neo-Lamarckism] are Jews (I name [Richard] Semon, [Paul] Kammerer, [Ignaz] Zollschan, [Friedrich] Hertz). One will now understand why this is the case: a misdirected apologetics of Jews opposing racial theories."[4]

3.  Jews' College Union Society: "Last Sunday, Lieut.-Col. C.S. Meyers, F.R.S., R.A.M.C. read a paper on 'Eugenics.' Dr. A. Buchler presided.

    "The lecturer outlined the problem of a practical program of eugenics, since much mental talent and genius is found in the physically unfit. After an exposition of the Mendelian laws of heredity, the lecturer dealt with the problem from a Jewish standpoint, with reference to inter-marriage. Dr. Redcliffe Salaman's statistics had shown that 93% of the children born from such marriages lost Jewish facial characteristics. On the other hand, Jewish

---

[1] "Nervous Diseases and Eugenics of the Jews" (in Hebrew), *Ha-Tekufah* 2, 289-307, on 289; cited by Falk in Cantor/Swetlitz, 2006, 149.
[2] Efron, 2001, 175.
[3] "Scientists as Spies."
[4] *Rasse und Kultur*, 1919, 15; cited in Lipphardt, 2008, 102.

blood was far more widely distributed among the non-Jewish population than people suspected. Again, though the Semites were a long-headed race, the Jews are broad-headed. Perhaps they had derived this character from inter-marriage with the Hittites. In conclusion, Lieut.-Col. Meyers referred to the enormous increase in intermarriage, especially in Scandinavia, and its disastrous effects on the predominance of a Jewish type.

"A discussion followed, in which Misses Regina Miriam Bloch and Lizzie Hands, and Messrs. Robinson, Max Footerman, and G. Weber participated.

"Mr. James Solomon, in proposing a vote of thanks to the lecturer, maintained that more stress ought to be laid on environment.

"The Rev. E. M. Levy, B.A., who seconded the vote, said that politics, in which compromise between two competing forces is the rule, presented an interesting contrast with the Mendelian rule. 'All or none.'

"The CHAIRMAN, referring to a remark of the lecturer that inter-marriages must have been frequent in the early centuries of the current era, since edicts were made by the Church against them, remarked that inter-marriage did not occur in the Ghetto, except in Italy. The problem of the inheritance of qualities was observed by the Rabbis as early as the second century A.D. Dr. Buchler heartily thanked the lecturer, and expressed the hope that he would come to the college as well as the Union Society."[1]

4. Mordechai Brachyahu, physician and head of Hadassah, the largest Jewish health organization in British-mandate Palestine: "The material from which we must take the cornerstones for the new edifice is in large part flawed and defective in several aspects, flawed and defective in itself, and in relation to the 'race.'"[2]

5. Jewish physician and eugenicist Magnus Hirschfeld (1868-1935), together with dermatologist Friedrich Wertheim and psychiatrist-eugenicist Arthur Kronfeld (1886-1941), establishes in Berlin the

---

[1] *Jewish Chronicle*, Jan. 28, 20.
[2] "Organ for the Health of the Nation," part II, *Ha'aretz*, Oct. 30, 1919; cited in Hirsch, 2009, 599.

Institute for Sexual Science (Das Institut für Sexualwissenschaft). Hirschfeld, who coined the words "transsexualism" and "transvestism," wishes to prove that homosexuality is an inborn trait and advocates its decriminalization. Later the Institute will open a Eugenics Department.[1]

6.  Jewish-German physician Max Marcuse publishes "Eugenics and the Psyche" in the journal *Zeitschrift für Sexualwissenschaft und Sexualpolitik.*[2]

## 1920s

1.  The Zionist Jewish Agency enacts a medical selection policy for immigration to Palestine.[3]

## 1920

### Context

1.  Hitler sets forth a list of 25 points, none of which deal with eugenics.

2.  *New York Times*: "Felix M. Warburg of New York, Chairman of the Joint Distribution Committee for American Jewish relief funds, who is here, is endeavoring to impress Jewish leaders in Europe with the necessity of discouraging European Jews from flocking to the United States in order to keep Jewish emigration [sic] within reasonable limits."[4]

3.  German psychiatrist Alfred Hoche (1865-1943) in *Permitting the Annihilation of Life Unworthy of Living* (*Die Freigabe der Vernichtung Lebensunwerten Lebens*): "There was a time, now considered barbaric, in which eliminating those who were born unfit for life, or who later became so, was taken for granted. Then came the phase, continuing into the present, in which . . . preserving every existence, no matter how worthless, stood as the highest moral value. A new age will arrive – operating with a higher morality and with great sacrifice – which will actually give up the requirements of an exagge-

---

[1] Pretzel, 1997.
[2] "Die Eugenik und das Psychische."
[3] Shvarts *et al.*, 2005, 6.
[4] *New York Times*, 1920.

rated humanism and overvaluation of mere existence."[1] In 1933 Hoche resigns his professorship at the University of Freiburg so as to avoid being dismissed for having a Jewish wife.[2]

4.  Albert Einstein: "It may be thanks to anti-Semitism that we are able to preserve our existence as a race."[3]

## 1920-1960

### Jewish Advocacy of Eugenics

1.  According to the National Library in Jerusalem, some 200 Hebrew-language parents' manuals are published. These publications contain a coherent worldview, of which eugenics forms an integral part, subjecting Jewish mothers to an unremitting program of education, indoctrination, and regulation. During the British mandate, Jewish physicians in Palestine actively promote eugenics.[4]

## 1921

### Context

1.  Secretary of the Lemberg Jewish Community and eugenicist Alfred Nossig (1864-1943) proposes that Jewish groups end their bickering and resolve the "Jewish question" by concentrating their efforts on raising money for the Jews to relocate to Palestine.[5]

2.  Jewish physician Benno Chajes (1880-1938): "In as much as sterilization and castration must necessarily be limited to a small number of persons, their eugenic value is minor. How much more humane and more appropriate to the task at hand is the prevention of births."[6]

3.  Psychiatrist and eugenicist Shneor Zalman (Zygmunt) Bychowski: "For years I have been occupying myself with that problem [of Jewish nervous diseases and Jewish degeneration]. It interests me as a

---

[1] *Die Freigabe der Vernichtung lebensunwerten Lebens. Ihr Maß und ihre Form*, Binding, K. Hoche, A. 1920, 1922 Felix Meiner Verlag, Leipzig.
[2] German *Wikipedia*, http://de.wikipedia.org/wiki/Alfred_Hoche, accessed June 29, 2008.
[3] Quoted by Entine, 2007, 240.
[4] Stoler-Liss, 2003.
[5] Nossig, 1921.
[6] Chajes, 1921, 163.

doctor, it irritates me as a Jew, and it torments me as a Zionist.... It is deplorable that we must mention here the habit common among Polish and Lithuanian Jews not to let a man remain a bachelor, even when he is sick, and may transmit the disease to his progeny.... We must fight with all our means such a prejudice, which may cause a great loss to the nation. This must be especially noted by those who construct the future of the nation – the Zionists. The resurrection of the nation in its homeland will be possible only if the 'human material' that will go there will be healthy. In this respect it will be necessary to apply from the beginning strict means, like the 'law' against immigration that has been introduced in the United States. It is of special significance that Zionists should learn to view marriage not as a personal act that one may handle as one's heart may wish, but rather as an important public act, on which depends the future of the race, the flourishing of the nation, and its hopes – the next generation. If there are reasons to believe that a marriage may produce sick children, this must be strictly forbidden. Zionists must be especially careful when they come to rebuild anew the life of the nation. It is necessary to make much propaganda in Palestine also against the notion of [having only] 'one or two children,' which leads to the annihilation of the race."[1]

4. Arthur Ruppin writes from Jerusalem to the Berlin police that 10,000 persons have their fingerprints checked to determine typical Jewish characteristics. Fortunately, the request is denied for lack of funds.[2]

5. Two papers are delivered at the Second International Congress of Eugenics, is held in New York on the inheritance of mental disorders:

- Long Island psychiatrist Aaron J. Rosanoff (1873-1943): "It would seem part of a healthy conservation to refrain from the employment of any eugenic measure which is irrevocable – such as sterilization – at least for the present.... Sterilization is a measure which requires no forcing. Existing institutions are greatly crowded and many of them have long waiting lists."[3]

---

[1] "Yiddische Nerven Un Yiddishe Degeneratie"; quoted in Falk, 2007, 133, 135.
[2] *Ruppin Archive (Nachlass)*, Central Zionist Archive, A 107/V/16; cited in Doron, 1980, 416.
[3] Rosanoff, 1923, 229.

- Boston physician A. Meyerson: "The most fertile work that can be done at present in the study of the inheritance of mental disease is to look for the agents that injure germ-plasms rather than try to link up the transmission of mental disease with the phenomena of Mendelism of other great biological laws."

- Other persons attending the Congress: Dr. William Bierman; Dr. Hugo A. Freund; Benjamin Gruenberg; Frederic W. Simonds; Dr. Abraham Leo Wolbarst (b. 1872).[1]

6.  London physician and eugenicist William Moses Feldman on "rabbinical and contemporary eugenics": "A study of the Bible, Talmud and later Jewish writings reveals, amidst a great deal of eugenic lore, a very considerable amount of sound knowledge not only regarding the influence of heredity on offspring, but also concerning the application of such knowledge for the purpose of raising the standard of the Jewish race. When Abraham instructed his servant not to choose a wife for Isaac among the daughters of the Canaanites he carried out a eugenic act. The same was the case when, in view of Rebeka's anxiety regarding a possible *mesalliance* between Jacob and one of the daughters of Heth, Isaac commanded Jacob to contract a consanguineous marriage in Padan-Aram rather than enter into an undesirable union at home."[2]

## 1922

### Context

1.  Arthur Ruppin estimates that nearly 90% of American Jews have no interest in Zionism.[3]

2.  Jewish-German gynecologist, artist, and popular science writer Fritz Kahn (1888-1968), polemicizing with Maurice Fishberg: "All former attempts to present Jewish racial types anthropologically have failed because they ignored the simple fact that not all adherents to Judaism are Jews, but rather are 'proselytes at the gates.... One might just as well juxtapose a Hottentot, a Boer, and an Indian and

---

[1] Eugenics Watch.
[2] Feldman, 1921.
[3] Elchanan Friedlander, *Ruppin and the Zionist Policy of Building the Land*, 1989, dissertation in Hebrew; cited in Bloom, 2007, 193.

proclaim to the entire world that 'the racial homogeneity of the Jews is nothing but a myth.'"[1]

3.  Mordechai Brachyahu, pediatrician and head of the Department of Hygiene at Hadassah Hospital: "the idea that the greatest sin that humans can commit against the God of life is to give birth to sick children marked with the seal of degeneration; the public that preserves the assets of the collective and the individual demands, in this matter, adherence to rational means, so that helpless offspring will not be born; and in the war of nations, in the secret, 'cultural' war between one people and another, the victor is he who sees to the improvement of the race, to the elevation of his descendants' biological worth ..."[2]

4.  Biologist Raymond Pearl (1879-1940), a proponent of eugenics about to become its opponent, writes to Lawrence J. Henderson, professor of biological chemistry at Harvard, that discrimination against Jewish applicants "is a necessary move in the struggle for existence on the part of the rest of us." Jews have a "higher survival value" because they do not let morals or decency get in the way of their personal advancement, he writes. "The Jewish mind has developed in the direction of versatility and superficiality. In the immediate struggle for existence, these traits will win out, I think, always over thoroughness and depth.... The real question seems to me to come to this. Whose world is this to be, ours, or the Jews?"[3]

### Jewish Advocacy of Eugenics

5.  Jewish-Polish eugenicist Zewy Parnass:

    "Our religious regulations indicate that hygiene, and particularly racial hygiene, is what we were aiming for in social life. Let us revive old rules in accordance with the spirit of the past; revive them and we will get the solution to all the problems, solutions which are an ideal for the European eugenicists. They dream of the time when the necessity of race hygiene will be so deeply rooted in social consciousness that it becomes a kind of social religion. We have had this religion for a long time; it arose in the Jewish tradition in Palestine.

---

[1] Kahn, 1922, 164.
[2] Karpel, 2006.
[3] Hendricks, 2006.

"The whole legislation of Babylonian and Jerusalem Talmud, in the chapters relating to national and racial life, forms the greatest book of eugenic laws.

"In the course of time each incurably ill patient will voluntarily undergo sterilization. And those that oppose it will be stigmatized by public opinion as social outcasts who dared to contradict nationally sacred values."[1]

6.  The Eugenics Department for Mother and Child is established in Magnus Hirschfeld's Institute for Sexual Science (Berlin).

7.  Union of Jewish Literary Societies: "The [morning] annual meeting of the Society will take place on Sunday, May 7th.... The afternoon meeting will be held at the Francis Galton Eugenics Laboratory, University College, Gower Street, at half-past four. Professor Karl Pearson (Galton Professor of Eugenics at London University) will lecture on 'Alien Jewish Children.'"[2]

8.  In the early 1920s British Jewish institutions begin to take an interest in eugenics, creating the Jewish Health Organization of Great Britain (JHOGB), which exists until 1946.[3]

## 1923

### Context

1.  Fritz Lenz (1887-1976), chair of Racial Hygiene in Munich, holds to a hierarchal view of human races, with Nordics and Jews at the top.[4]

2.  Jewish-German Lamarckian eugenicist Julius Tandler (1869-1936): "Today at 3:00 pm a group of students stormed into the Institute corridor. The students screamed at my assistants that they wanted to search the Institute for Jews. The commotion drew me out into the corridor as well, and a student, who should have known me, immediately came up to me and asked – calmly, I must admit – if there were Jews here. I responded to him: I am a Jew and the Director of

---

[1] *Kwestia żydowska w świetle nauki*, Beth-Israel, Lwov, 78; cited in Turda/Weindling, 2007, 288, 294.
[2] *Jewish Chronicle*, April 28, 36.
[3] Endelman, 2004, 75-77.
[4] Adams, 1990, 31.

this Institute, which seemed to perplex him highly, and he went back to the noisy crowd in the background."[1]

3. When eugenicists Karl Pearson and Margaret Moul test the children of Jewish immigrants in Great Britain and recommend that such immigration be limited because of poor showing – mental and physical – on the part of these children, the Rev. S. Levy comments in *Jewish Chronicle*: "One is more than once reminded of Thackeray's story of the English traveler who was so deeply impressed by the remarkable fluency with which the children in a French village spoke French and was so mortified when he contrasted the ignorance of French displayed by children in an English village."[2]

4. Contemporary Israeli-American sociologist Elazar Barkan: "The predominantly German and Jewish ethnic background and the liberal-left ideology that characterized the Boasian school had only inflamed the xenophobia of the old-guard antagonists, who were primarily Anglo-Saxon conservatives. Beyond that lingered the question of cultural anthropology. Was the business of anthropology a racial classification of humankind, or was it cultural relativism? The sides were drawn around the heredity-environment debate. The controversy was dynamic.... The source of the egalitarian conviction lay outside of scientific discourse...."[3]

### Jewish Advocacy of Eugenics

5. Physician Solomon Samoilovich Vermel' (Vermelia?) (1860-1940) makes a presentation on Jewish criminality for the Jewish Commission of the Russian Eugenics Society. Vermel' maintains that, contrary to popular opinion, Jews commit fewer crimes than do Christians, and that the crimes which they do commit tend to petty theft and fraud. Vermel' employs the phrase "Jewish nation," but defines it as a "cultural-psychological complex" with some "purely biological factors," but concedes that Jews have to a significant degree interbred with non-Jews. Vermel' also makes the observation that for

---

[1] From a letter to the Dean; quoted in Lipphardt, 2008, 224.
[2] "The Problem of Alien Immigration into Great Britain," Pearson/Moul, *Annals of Eugenics*, vol. I, parts I and ii; Levy article: "Children of Jewish Immigrants," *Jewish Chronicle Supplement,* Jan. 29, viii.
[3] Barkan, 1992, 92-92, 346..

centuries the Jews have not participated in wars, which is of interest from the perspective of selection.[1]

6.  Georg (George) Chaym in *Socialist Monthly* (Sozialistische *Monatshefte*) writes: "Socialism certainly does not take a negative position toward race hygiene insofar as race hygiene concerns theoretical and practical measures for the improvement of race or avoiding its debasement."[2]

7.  Dr. Joseph Meir (1890-1953), for whom the hospital in Kfar Sava, Israel, is named: "Who should be allowed to raise children? Seeking the right answer to this question, eugenics is the science that tries to refine the human race and keep it from decaying. This science is still young, but it has enormous advantages.... Is it not our duty to insure that our children will be healthy, both physically and mentally? For us, eugenics in general, and mainly the careful prevention of hereditary illnesses, has a much higher value than in other nations. Doctors, athletes, and politicians should spread the idea widely: Do not have children unless you are sure that they will be healthy, both mentally and physically."[3]

## 1924

### Context

1.  The Immigration Restriction Act, which pursues the goal of preserving the ethnic composition of the United States, is passed with support from the eugenics lobby. It has the effect of hindering immigration of East European Jews, Southern Italians, and Central Europeans. Gedalia Bublick, Editor of the *New York Daily News*, protests at a hearing before the Committee on Immigration and Naturalization: "Now some gentlemen want to... create a new America, with no equality, and they say instead that the man of the Mediterranean race is not born equal to the man of the Nordic Race.... This new literature... will remain a shame to America in her history."[4]

---

[1] Vermel', 1923.

[2] Graham, 1977, 1141: George Chaym, *Sozialistische Monatshefte*, No. 10, 638.

[3] Y. Meir and A. Rivkai, *The Mother and the Child*, 1934, Tel Aviv: Kupat Holim, 63-64, Stoler-Lis, 2003, 110. Date approximate, Shvarts *et al.*, indicate early 1920s (pg 14).

[4] 68th Congress, 394; quoted in Merkel, 1997.

2.   German racial theoretician Hans F. K. Günther (1891-1968) in *Racial Features of the German People* classifies Jews as a race "of a second order," that is, one that consists of an interbred mixture of several primary races.[1]

3.   The Vienna Society for Racial Nurturing (Racial Hygiene) is founded. Membership is limited to ethnic Germans, evidently thus excluding Jews.[2]

### Jewish Advocacy of Eugenics

4.   The Society for Jewish Family Research is founded in Warsaw. According to Zionist psychiatrist Rafael Becker, it pursues the explicit goal of "acquainting the broad Jewish masses with the basic premises of eugenics."[3] Becker stresses the need for negative eugenics for contemporary Jewry so as to prevent unhealthy marriages between the mentally and physically ill, or between cousins.

5.   Jewish Director of the Society for Jewish Family Research and eugenicist Arthur Czellitzer (1871-1943 or 1945) writes that Jewish charitable works can be "a good deed in the present but a crime in the future." Thus he advocates assistance for deaf mutes, but accompanied by a marriage ban.[4]

## 1925

### Context

1.   The United States introduces the National Origins Act, setting the annual nationality immigration quota at two percent of the number of foreign-born persons of such nationality resident in the continental United States in 1890. The Act's consequences for Jewish immigration is limited by the fact that, despite massive Jewish representation in the Soviet government, emigration from the Soviet Union is impossible.

---

[1] Günther, 1924.
[2] Die Wiener Gesellschaft für Rassenpflege (Rassenhygiene). Hofmann *et al.*, 2005, 43.
[3] Lipphardt, 2006.
[4] Lipphardt, 2006.

2.   Zionist eugenicist Ignaz Zollschan views assimilation as "racial suicide."[1]

3.   Writing in *Annals of Eugenics,* demographers Karl Pearson and Margaret Moul discuss immigration in its eugenic perspective, pointing out the positive contribution of Jewish immigration, as opposed to the negative role of certain other ethnic groups, but nevertheless argue forcefully against free immigration: "the English Jew has been theoretically a free man for a century, and practically one for a much longer time. In the case of the Russian and Polish Jews there has been more or less continuing oppression, nay a veritable selection going on for a much longer period. Such a treatment does not necessarily leave the best elements of a race surviving. It is indeed likely to weed out the mentally and physically fitter individuals, who alone may have had the courage to resist their oppressors. We can sympathize with a man who has suffered hard treatment, but that in itself is not an adequate eugenic reason for granting him citizenship in a crowded country. For that citizenship we demand physical and mental fitness; we need the possibility of an ultimate blending and we need full sympathy with our national habits and ideals. Those of us who had occasion to travel during air-raids on London will not lightly forget the sights and sounds we encountered among the Yiddish-speaking population who sought refuge in the tube stations. But that is only an isolated aspect of the problem; we know also of acts of great courage among Jews of Russo-Polish origin. We know further of brilliant achievements and university distinctions gained by recent immigrants or their children. No satisfactory conclusions can be reached by citing individual instances which may tell one way or the other. There is only one solution to a problem of this kind, and it lies in the cold light of statistical inquiry.... ¶ It would seem to follow... that the Jewish alien children are not superior to the native Gentile. Indeed, taken all round we should not be exaggerating if we asserted that they were inferior in the great bulk of the categories dealt with.... But while the characteristics we have dealt with are very essential, there remains a distinct possibility that our unfavourable judgment might be largely reversed if we should find that these alien Jews are markedly superior in intelligence to the native Gentiles. We might pardon a poor phy-

---

[1] Hart, 1999, 279; citing *Das Rassenproblem unter besonderer Berücksicht-igung der theorethischen Grundlagen der jüdischen Rassenfrage,* 5th edition, Vienna.

sique and even uncleanliness if these characteristics were accompanied by a dominating intelligence."[1]

## Jewish Advocacy of Eugenics

4.  Jewish neurologist Oskar Vogt (1870-1959) dissects Lenin's brain into 30,000 sections as part of an effort to study the brains of super geniuses.

5.  In Russia Jewish geneticist Tikhon Ivanovich Iudin (1879-1949) defines eugenics as a union of genetics and sociology and predicts it "will become a biology of social types."[2]

6.  Jewish-British geneticist and eugenicist Redcliffe Salaman gives a lecture in London in which he counsels against permitting "mental defectives" to marry and advocates the segregation of their children.[3] [4]

7.  Some members and officers of the American Eugenics Society (renamed Society for the Study of Social Biology in 1973): Isaac M. Abt (1867-1955), MD, Dr. Herman M. Adler (1925, 1930), Professor H. F. Bergman (1925, 1930), Professor H.L. Bruner, Benjamin Gruenberg, biologist A. H. Hersh (1925, 1930), Robert G. Leavitt (1925, 1930), Mrs. R. Mayer, Rabbi Louis Leopold Mann, Max Schrabisch, political scientist Edwin R. A. Seligman (1925, 1930), University of Chicago Botanist H. S. Wolfe (1925, 1930), Professor L. B. Wolfenson.[5]

## Boasians

8.  Jewish-American cultural anthropologist Melville Herskovits (1895-1963), a student of Franz Boas: "The invidious comparison, between races as between individuals, is always odious, and it is to be hoped that the present craze for the Nordic myth will go the way

---

[1]Pearson/Moul, 1925, 8, 50-51.

[2] Adams, 1990, 170.

[3] Endelman, 2004, 76-77.

[4] Salaman was, at various times, President of the Union of Jewish Literary Societies, President of the Jewish Historical Society, Chairman of the Jewish Committee for Relief Abroad, Governor of the Hebrew University in Jerusalem, Trustee of Jews' College, and Founder and Chairman of the [eugenicist] Israel Zangwill Memorial Fund (Smith, 1955, 242-243).

[5] *Eugenical News, Eugenics Quarterly, Social Biology*, Web site Eugenics Watch.

of all crazes. The problem of who is to settle in this country is a real one; our need is to put it on the economic and social basis where it belongs, and to leave out of it vague hypotheses concerning racial intelligence."[1]

## 1926

### Context

1.  Chair of Anthropology Department in the Jewish Medical Society for Sexology and Eugenics Hans Friedenthal (1870-1943) argues that while individual Jewish populations may have common features, such characteristics are not shared by Jewry in its entirety and thus Jews cannot be considered a race.[2]

2.  Canadian census data show a Jewish birth rate only 70% that of the overall population.[3]

3.  The Jewish Health Organisation of Great Britain awards a grant to psychometrician A. G. Hughes to do a study under the direction of Cyril Burt: "Jews and Gentiles: Their Intellectual and Temperamental Differences, A Psychological Study Which Reveals the Innate Superiority of Jewish Children over Their Gentile School-Mates." The results are published in the *Eugenics Review*, with the reservation that they are not necessarily equally true for Jews living in a different environment."[4]

### Jewish Advocacy of Eugenics

4.  On the occasion of Mother's Day, the American Eugenics Society announces a competition among churches and synagogues for the best sermons on eugenics. A number of churches and synagogues participate in the competition, including:

    *   Rabbi George Benedict of Temple, Emanu-El in Roanoke, Virginia: "How well he [Moses] understood the great moral and religious principles which, when a nation remains true to them

---

[1] Herskovits, 1925, 141.

[2] *Zeitschrift für Demographie und Statistik der Juden* (Neue Folge), 1926, Lieff. 1-3, 4-6.

[3] Mortimer Spiegelman, "The Reproduction of Jews in Canada, 1940-42," *Population Studies*, IV, Dec. 1950. 299-313; cited in Goldscheider, 1967, 199.

[4] Hughes, 1928

means its life and well-being, and which, when a people betrays them means death and disaster! He beheld very clearly that principle in nature so well known to modern eugenists by which vice and disease purify a race by destroying the vicious and the immoral. Repeatedly he warns Israel of the punishment which will overtake those who break the laws of purity and morality. Such a one, he warns, 'shall be cut off from among his people.' Unworthy to have his life continue in the life of the family of Israel, disease shall cut his career short; like a stream that is disconnected from its source, and dries up. 'Venichresoh hanefesh hahe meameo;' – 'And that soul shall be cut off from among his people,' he warns.... Defectives and degenerates are linked by the same invisible bonds; and so are those of noble inheritance. 'Heredity explains nearly 90% of the rough outline of the character and intelligence,' is the startling statement of a eugenist of authority."[1]

- Rabbi Harry H. Mayer, Kansas City, Missouri: "there is urgent need to remind ourselves and have those best qualified to speak on the subject remind us in discussions in the pulpit and the press how deeply it concerns the future of the whole human race that every child be born into the world descended from a family stock that is healthy and vigorous both in body and in mind.... [N]othing that a father or mother can do for us in the way of education, in the way of endowment of wealth is comparable to the gift that a parent bestows on a child in handing over to it a good ancestry.... How vital it is therefore that when a couple mate they should remember the eugenic factors underlying matrimony. However superior one of them may be, if the other is not an equal, physically, mentally and morally, the offspring will be likely to be inferior."[2]

- Pastor Frederick Franklin Adams of The United Church, Hinesburg, Vermont, speaking "before three denominations in one church, a union meeting": "The people of Israel were told to beware of marriages with the heathen; they were told to be fruitful and multiply; to have children and to bring them the rules and laws of health and of sanitation, so that the health of the child and of the nation might be protected. In other words

---

[1] Benedict, 1926.
[2] Mayer, 1926.

they were admonished to have their children 'well born'. Abraham and Jacob knew the laws of eugenics and of proper breeding as did many of the ancients. Jacob, we know, applied the laws of selection in breeding, to his sheep and goats to win a victory over his father-in-law. Moreover we notice that the mother of Samson was warned by the angel to drink no strong drink, in order that she might give birth to John the Baptist was hinted at when it is said of him that he would use neither wine nor strong drink. Israel did not neglect the physical and spiritual preparation of her children."[1]

5.  Jewish political theorist Karl Kautsky (1854-1938) reviews fellow socialist Oda Olberg's book *Degeneration and Its Cultural Conditioning*, agreeing with her on society's need for eugenics.[2]

6.  In Palestine the Jewish Agency champions immigration of only healthy Zionists and prevents the immigration of others who do not meet Zionist criteria. The Immigration Department of the Palestine Zionist Executive issues "Instructions for the Medical Examination of Immigrants." The instructions are in part dictated by eugenic considerations, and they engender a fierce debate over selection.[3]

7.  Anthropologist Hans Friedenthal (1870-1943): "This author would welcome the application of racial teachings and eugenics within contemporary religious communities.... Future racial teachings shall be assigned the task as science, not of separating and disrupting mankind, but of uniting it to the highest degree."[4] (Friedenthal uses the word "racial" with reference to the human race in its entirety.)

8.  The famous Jewish-Austrian biologist Paul Kammerer, a socialist and outspoken proponent of Lamarckian eugenics, is accused of committing scientific fraud. The midwife toad, a terrestrial species which lacks the pigmented "nuptial" thumb pads used by aquatic males to grasp females during mating, was supposedly injected with India ink to fraudulently demonstrate that its male offspring would inherit the pads if the toad was returned to its original aquatic envi-

---

[1] Adams, 1926.

[2] Graham, 1977,1141-2: review of Oda Olberg's "Die Entartung in ihrer Kulturbedingtheit: Bemerkungen und Anregungen," *Die Gesellschaft*, 3, 1926, 567-573.

[3] Shvarts *et al.*, 2005, 9-10.

[4] Friedenthal, 1926, 91; Lipphardt, 2006.

ronment. Kammerer is disgraced but accepts a position in a still-receptive Moscow. Falling into a deep depression, suffering not only from the assaults on his character, but from poor finances and his wife's refusal to accompany him to Russia, he maintains his innocence but commits suicide en route to Russia. For almost three decades, however, his work remains current in the Soviet Union, where his theories harmonized with the principles of Trofim Lysenko, head of the Institute of Genetics of the Soviet Academy of Sciences.[1]

9. Genetic psychiatrist Nathaniel David Mttron Hirsch (1896-1984): "Externally, our [American] immigration policy in the future is of prime importance; internally, a negative and positive eugenics program that will involve birth-control, and avoidance of miscegenation, the crossing and blending of certain Natio-Races, and the increased fertility of the intelligent healthy and beautiful of the older stock, is imperative.... The free intermarriage of the Jews with other Natio-Racial groups would probably produce a stock containing a high variability of mental qualities, resulting in the birth of many men of exceptional capacities in varying directions."[2]

## 1927

### Context

1. German racial theoretician Hans F. K. Günther posits a decisive struggle between Nordics and world Jewry for control of the world.[3]

2. Jewish-American geneticist and eugenicist Herman Muller (1890-1967) discovers the artificial induction of mutations through X-ray radiation.

3. When the anthropologist George Pitt-Rivers asks Jewish anthropologist Charles Seligman what he thinks of C. G. Jung's characterization of Jews as a subtle, scheming race, Seligman replies that he detects two strains in contemporary Jews: "the one, a desert strain, by which I mean a strain of thought and action more or less akin to that of the Arab as one knows him in fact and history, and another strain, which I tend to associate with Armenoid blood, which is characte-

---

[1] Paul Kammerer papers, American Philosophical Society.
[2] Hirsch, 1926, 378-379, 394, 403.
[3] Günther, *Der nordische Gedanke unter den Deutschen*, 129; cited in Morris-Reich, 2006b, 134.

rized by the showy efflorescent style one certainly finds among some Jews."[1]

## Jewish Advocacy of Eugenics

4.  Hugo Iltis (1882-1952), a Jewish biologist and proponent of eugenics, criticizes German race hygienists for mixing politics and science in drawing conclusions about race that are premature, dangerous, and "barbaric." Iltis advocates a universalist race hygiene once decades and perhaps centuries of work will have passed.[2]

5.  Architect of Jewish settlement in Palestine Arthur Ruppin writes in his diary: "I am becoming *increasingly* aware of the extent to which the Jews' return to Eretz Israel and agriculture should be seen as a primary eugenic phenomenon."[3]

6.  Jewish Supreme Court Justice Louis Brandeis (1856-1941) joins Justice Oliver Wendell Holmes, Jr. (1841-1935) in upholding Virginia's sterilization program for the feeble minded ("Three generations of imbeciles is enough.")[4]

7.  Rabbi Louis Mann of Chicago's Sinai Temple joins the Advisory Council of the American Eugenics Society.[5]

8.  Physician Max Marcuse publishes a marital manual on "hygenics and eugenics."[6]

## 1928

### Context

1.  Prominent German geneticist and eugenics proponent Fritz Lenz (1887-1976) attempts to persuade the Jüdischer Verlag to publish Samuel Weissenberg's *Jewish Lineage Studies* (*Jüdische Stammeskunde*), even though he disagrees with Weissenberg's thesis that Jewry is more a social than a biological phenomenon. Lenz has

---

[1] Endelman, 2004, 83.

[2] Graham, 1977, 1142: "Rassenwissenschaft und Rassenwahn," *Die Gesell-schaft*, 4, 1927, 97-114.

[3] Central Zionist Archive, A 107/592; cited in Morris-Reich, 2006a, 28.

[4] *Buck v. Bell* (274 U.S. 200).

[5] Bozeman, 2004, 424.

[6] *Die Ehe, ihre Physiologie, Psychologie, Hygiene und Eugenik: Ein biolo-gisches Ehebuch*, Verlag A. Marcus & E Weber, Munich.

been accused of "anti-Semitism," but the collegial spirit of Lenz's intervention is typical of relations between Jewish and gentile thinkers prior to Hitler coming to power.[1]

## Jewish Advocacy of Eugenics

2. Former head of the 1919 Munich Soviet Max Levien (1885-1937) rejects the racial theories of Hans F. K. Günther and Adolf Basler (b. 1878) and advocates a universalist socialist eugenics.[2]

3. Jewish-German gynecologist and eugenicist Ludwig Fraenkel (1870-1951): "In times of overpopulation and crowding the common good cannot countenance even as exceptions that the quantity of population be augmented, but it must constantly seek that the best human elements be multiplied with the greatest vigor and that cohabitation of a healthy person with an infertile individual not render fallow the improvement of the race or allow it to be steered along unhealthy paths."[3]

4. Physician and sexual reformer Max Hodann (1894-1946): "Who [nowadays] concerns himself with eliminating harmful genes?... Ultimately it will fall to a socialist society to take eugenic measures for protecting society from the burden of low-grade posterity."[4]

5. Magnus Hirschfeld (1868-1935) organizes the first congress of the World League for Sexual Reform.

6. Physician Alfred Marx: "We agree with such demands of racial hygiene as *pre-marital health certificates; the prevention of marriage between relatives suffering from heritable diseases; non-marriage for persons whose families are strongly burdened by genetic illnesses, for habitual criminals, drunkards, the mentally ill, etc.* [stress in German original].... Nowadays one frequently hears that Nordic man is the most valuable in the European racial mix, and if it could be proven, I would be quite prepared to believe it, for I am not one to automatically consider his own race to be the most valuable. I would even possess the courage of an Otto Weininger to ne-

---

[1] Lenz, 1928.

[2] Graham, 1977, 1144: Levien, "Stimmen aus dem deutschen Urwald (Zwei neue Apostel des Rassenhasses," *Unter dem Banner des Marxismus*, 2, 1928, 150-195.

[3] *Soziale Geburtshilfe und Gynäkologie*, 1928, 73; quoted in Hommel/Alexander, 1998, 477.

[4] "Max Hodann."

gate my own spiritual existence, but for now it seems to me that such views, which are shared by some of the racial hygienists, spring less from objective facts than from political tendencies. I have no way of knowing if men of the Mediterranean, Dinaric, and Alpine race are so much less valuable than is the Nordic race, but I do know that there are men of Jewish ancestry capable of making a contribution to research and culture and also that the intellectual gifts of these men are often doomed to be wasted in Germany. In conclusion, we can see that the demands of racial hygiene are justified to the degree that they are derived directly from the findings of hereditary research, but that these demands are largely based on improper preconceptions, unproven claims, and unfounded prejudices. Today's racial hygiene is a political struggle decked out in scientific conclusions."[1]

## 1929

### Context

1.   Bacteriologist, Indiana University lecturer on eugenics, and Chairman of Indiana Eugenics Committee Thurman B. Rice (1888-1952): "The principle of keeping the race pure is nowhere better illustrated than in the history of the Jewish people; sex hygiene began with the Jews: race hygiene was almost a fetish with them…. Although scattered to the four winds the Jews remain Jews; races rise and fall but Israel is immortal. In every line of progress the Jew stands at or near the head of the list and has done so for forty or more centuries. In science and medicine, in philosophy and literature, in music and art, in statesmanship, business and finance, investigation will show that a large percentage of the men at the very top are Jews. There is no better argument for the universal practice of the principles of eugenics than the marvelous success of the Jewish race – the only race of importance to have a history of progress extending over a period as long as a thousand years…. The race has been subjected to a rigorous selection because of the hardships which they have endured…. The above is not meant as a eulogy of the Jew. The writer comes of a stock that has long held the Jew in contempt – with less reason than prejudice, it must be admitted, however. It is quite true that the Jews as a race have many undesirable traits, some real and many more traditional than real, but no one can deny that the Jews as a

---

[1] Marx, 1928.

class are highly proficient in the art and science of gaining their ends. Whatever he may or may not be from others' standpoint, from the Jew's standpoint he is usually a great success. It is not claimed that the Jewish race is a pure race, or that it is the best race, but just the same, through the ages, ancient, medieval and modern, the seed of Abraham is continuous and is still going strong. The record of no other people can approach it in this respect."[1]

2.  Jewish-German mathematician Felix Bernstein (1878-1956) publishes *Variation and Heritability Statistics* (*Variations- und Erblichkeitsstatistik*), in which he searches for racial genetic markers, the goal being to detect the racial mixing of population groups.[2]

3.  German geneticist Fritz Lenz explains the Jewish inclination to Lamarckism as a denial of unalterable racial differences: "Jewish intellectual elites who feel that they are part of the German people and German culture have told me that it is tragic for them to be perceived as alien. If acquired characteristics could be inherited, the Jews by virtue of their life in a Germanic environment and their attachment to Germanic culture could become authentic Germans."[3]

### Jewish Advocacy of Eugenics

4.  William Grossman of Passaic, New Jersey, at a joint meeting of the Eugenics Research Association and the American Eugenics Society held at the American Museum of Natural History: "Based upon... statistical figures, my own investigations, personal experiences and observations, I come to the conclusion that the sex and mental hygiene advocated in the Talmud and post Talmudic literature are responsible for Jewish superiority.... The Talmud has scattered among its many pages numerous suggestions on sex, mental and social hygiene. All these sayings and warnings would make up a fair volume on these subjects. Some quotations in these subjects may awaken the interest of research workers and statisticians to conduct a careful investigation on a basis where the Sabbath observing Jew should form a separate group."[4]

---

[1] Rice, 1929.
[2] Publisher: Borntraeger.
[3] "Der Fall Kammerer und seine Umfilmung," 1919, 316; quoted in Lipp-hardt, 2008, 142.
[4] Grossman, 1929, 105.

5.  Rabbi D. de Sola Pool: "At the meeting of the American Eugenics Society, I tried to indicate that the most noteworthy emphasis of Talmudic ethics was its setting up breeding for character as the paramount eugenic consideration. I recall that other peoples have more or less consciously bred for the physical qualities of the warrior, marriage within the caste, the Samurai, the modern popular conception of 'a good marriage' as a marriage for money, the eugenic optimum as indicated by *Who's Who*, etc. But so far as I know, the eugenic emphasis placed by the Old Testament (including Amos), and by the Talmud, on moral character as the eugenic aim is unique.... Perhaps in a complete and scientific history of eugenics... Moses, Amos and the rabbis of the Talmud will be recognized as intuitive and extraordinarily influential eugenists even before the days of Galton and Karl Pearson."[1]

6.  In Brazil geneticist André Dreyfus (1897-1952) concedes that the heretofore popular Lamarckian eugenics has "sadly to be abandoned."[2]

7.  Jewish-Russian geneticist Aleksandr Sergeevich Serebrovsky publishes an article proposing a socialist eugenic state[3], provoking a response in the newspaper *Komsomolskaia Pravda*: "The Class Enemy in the Scientific Institutes."[4]

8.  Jewish-Australian-British physician Norman Haire (né Zions, 1892-1952) organizes the second congress of the World League for Sexual Reform. Eugenics is envisaged "in the Nietzschean sense of not merely a perpetuation of the race, but its improvement: *Der Übermensch*."[5] The congress's platform also advocates women's rights, birth control, and the decriminalization of sexual acts between consenting adults.

9.  Jewish-German feminist Henriette Fürth (1861-1938): "Let us display love and humanity toward those who are born handicapped, but no tolerance for their unrestricted multiplication! Let us protect fu-

---

[1] Grossman, 1929, 105-106.

[2] "O estado actual do problema da hereditariedade," *Actos e Trabalhos*, Premeiro Congresso Brasileiro de Eugenia, 91; quoted in Adams, 1990, 132.

[3] "Antropogenetika i evgenika v sotsialisticheskom obshchestve," *Mediko-biologicheskii zhurnal*, vyp. 4-5, 1930, 447-448.

[4] "Klasssovyi vrag v nauchnykh institutakh," July 31.

[5] Crozier, 2001, 312.

ture generations!... There are a number of factors that indicate that sterilization is either essential or at least desirable from the point of view of a responsible eugenics.... We need not depend on geniuses. Genius is always an exception. The broad supportive foundation of the people's development is the masses.... It is from their midst that outstanding individuals appear."[1]

## 1930s

### Context

1. American writer, actress, and believer in eugenics Mina Loy (1882-1966), herself the daughter of a Jewish-Hungarian father and a British Protestant mother, writes in her unpublished autobiographical novel *Goy Israels* that Jews are the "necessary intellectual bridge to the mystic dimension" and that her own "mongrel" ancestry is superior to a pure genetic strain.[2]

2. A great slump in Jewish fertility occurs in North America and Europe.[3]

3. Jewish eugenicist Arthur Ruppin: "The difference between a blond, tall, long-skulled northern German and a southern German of the type *homo alpinus* is greater than between a southern German and a Jew of the Western Asian type."[4]

### Jewish Advocacy of Eugenics

4. Contemporary Israeli-American historian Rakefet Zalashik maintains that the concept of social engineering was part of the psychiatric mainstream in Israel from the 1930s through the 1950s: "Eugenics was part of the national philosophy of most of [the local] psychiatrists. The theory was that a healthy nation was needed in order to fulfill the Zionist vision in Israel."[5]

---

[1] Fürth, 1929, 45-50.

[2] *Goy Israels*, Mina Loy Papers. Yale Collection of American Literature. Beinecke Rare Book and Manuscript Library, New Haven, Connecticut; cited in Vetter, 2007.

[3] Schmelz/DellaPergola, 2007.

[4] Ruppin, 1930, 37.

[5] Feldman, 2009; referring to Zalashik's 2008 book *Ad Nefesh: Refugees, Immigrants, Newcomers and the Israeli Psychiatric Establishment*, (*Hakibbutz Hameuchad* in Hebrew).

## 1930
## Context

1.  Rabbi and eugenicist Louis L. Newman (1893 or 1897-1975) in *Eugenics: A Journal of Race Betterment*: "Anti-Semitism, it is said, begins on Wall Street after five o'clock. Business friendships between men are rarely allowed to enter the domain of their womenfolk. The walls which Jews and non-Jews erect against each other are built on the foundations of their respective family integrity. The question of racial superiority does not enter in the slightest. It is purely a question of group instinct and the insistence upon the self-preservation of the social unit into which the members of each community are born.... [I]n the main there is little peril that Americans will take the slogan of the Melting Pot too literally. Historic, long-established racial integers will be preserved with little impairment, and America will be unified through means other than racial fusion."[1]

2.  In a sermon given before the congregation of the Free Synagogue at Carnegie Hall, Dr. Sidney E. Goldstein maintains that it is not biology and eugenics that invalidate mixed marriages, but psychology: "The relationship of marriage is profoundly influenced by the intangible social elements of instinct, emotion, temperament, interest, ideals.... In mixed marriage the difference in background, in psychological constitution, in attitude toward life and its fundamental problems is so great that love is not able to consume the barriers in its flame."[2]

### Jewish Advocacy of Eugenics

3.  Psychiatrist Abraham Rabinowitz: "We should also note our own primitive races (Bokharan, Georgian, Persian, etc.). Their consciousness, with its meager contents, poses no special claims to life, it surrenders in a slavish manner to external conditions, and therefore does not experience collisions and produces a small percentage of functional nervous and mental diseases. The progress of civilization and its penetration into these groups will undoubtedly affect

---

[1] Newman *et al.*, 1930, 61-62.
[2] *New York Times*, 1930b.

them, and they too will produce a considerable percentage of neurotics, psychotics, etc."[1]

4.  The Dresden anthology *Hygiene and Jews* contains an article by rabbi and folklorist Max Grunwald (1871-1953) entitled "Biblical and Talmudic Sources of Jewish Eugenics," recommending on the basis of the halachic work *Sefer Chassidim 1097* that a man would do better to marry the daughter of a well-behaved proselyte than a Jewish woman from a bad family, for she will give him good children. Particularly to be recommended is marriage with the daughter of a sister (Jeb. 62b). Grunwald points out that Jewry followed the principles of eugenics far before Galton discovered them and quotes Nietzsche: "Only he who follows the voice of blood and conscience and is dedicated to the fate of the community will perceive his life's goal as providing valuable future society members and constructive elements."[2]

5.  Writing in *Eugenics: A Journal of Race Betterment,* Austrian born Jewish-American biologist Nathan Fasten of Oregon State Agricultural College (1887-1953) provides a brief summing up of the eugenics movement: "The goal of the eugenicist is not to create a new race of people, but rather through educational, civic, and other means, to develop a social consciousness which will result in the humane treatment and eventual elimination of the hopelessly crippled, diseased, and mentally incompetent and at the same time, increase the number of children perpetuated by the normal individuals constituting our present civilization. The eugenicist, contrary to the opinion so often proclaimed by many popular writers and orators, does not aim to establish a race of supermen, but rather, a race of sturdy, intelligent and healthy individuals similar to the large proportion of the human family now in existence. Most eugenicists believe that the factors of heredity and environment are both essential to the development of the physical and mental traits of human beings."[3]

6.  Arthur Ruppin (1876-1943), head of the World Zionist Organization office in Palestine, writes in his 1930-31 book *The Sociology of the*

---

[1] "The Eretz Israel Reality and Help to Mental and Nerve Patients," HaRefu'a, No. 3, 10-11 Iin Hebrew); cited in Hirsch, 2009, 602.
[2] Grunwald, 1930.
[3] Fasten, 1930.

*Jews* that "in order to preserve the purity of our race," Jews display-ing signs of genetic defects should not have children. He maintains that Ashkenazic Jews are not Semites and considers Oriental and Sephardic Jews, whose Semitism he recognizes, as inferior. Ruppin finds a book by racial theoretician Hans Günther (1891-1968) in a Tel-Aviv bookstore and comments in his diary: "It brings on many thoughts that I had in mind for my own book."[1]

7.   The American Eugenics Society announces the appointment of Rabbi Stephen S. Wise of New York as one of the judges in its ser-mon contest. The contest is open to any minister of any faith and is to deal with the relation of the churches to eugenics.[2]

8.   Hans Goslar (1889-1945), Jewish Director of the Press Section in the Prussian Ministry of State, praises the "hygienic and eugenic ef-ficacy of Jewish religious law preserving the body of the people" (*die volkserhaltenden, hygienisch und eugenisch wirksamen Bes-timmungen des jüdischen Religionsgesetzes*).[3]

9.   Jewish-German anthropologist and eugenicist Wilhelm Poll, who this year steps down from his long-held position of Secretary of the *Journal for National Eugenics and Heredity* (*Zeitschrift für Volk-saufartung und Erbkunde*), reports having identified 121 sets of twins in Hamburg schools.[4]

10.   Biologist and animal breeder Leon F. Whitney and author of *Eugen-ics in the Talmud* William Grossman: "Each of the authors, one an Anglo-Saxon and the other a Jew, likes his own race best. They are not going to argue race superiority but frankly admit that just as there were Anglo-Saxons who came here with the true pioneering spirit and who have contributed so greatly to American ideals and institutions, so there have come Jews who have also made valuable contributions.... The Old Testament is full of good eugenic lessons, the Talmud likewise. The chief lesson to be derived from the Old Testament is the effect of morality upon the preservation of the race.... Race purity was stressed. Among the admonitions which have helped in this regard one finds striking and emphatic com-

---

[1] *Tagebuch*, Jan. 31, 1930; cited in Doron, 1980, 421.
[2] *New York Times*, 1930a.
[3] Hans Goslar (ed.), *Hygiene und Judentum*, 7; quoted in Hart, 2007, 29.
[4] Braund/Sutton, 2008, 14-15.

mandments. For example, Ezra, a Jewish sage living in 400 B.C., forced the Jews who married Gentiles in Persia to divorce their Persian wives."[1]

11. Jewish-Ukrainian physician and eugenicist Max Danzis (1878-1953) at the Men's Club of Temple B'nai Abraham, Newark, New Jersey: "The Talmud, as well as the Old Testament, has many interesting and valuable lessons. Talmudic sages were early to recognize the effect of a clean, moral life upon preservation of the race. Sex, hygiene, and eugenics, almost modern in its sense, received careful consideration in the Talmud. Marriage into epileptic and leprous families was forbidden. Imbeciles were not permitted to marry. The value of heredity, particularly in its relation to marriage, was stressed…. 'A girl with a good pedigree, a daughter of a scholar who leads a life in accordance with the regulations of the Torah, even if she be poor and an orphan, is worthy to become the wife of a king. If one sees a girl that has all the necessary qualities, he should not delay the engagement, because she might be snatched up by another."[2]

12. Jewish-American financier and advisor to Woodrow Wilson (and, later, Franklin D. Roosevelt) Bernard Baruch (1870-1965) is shown as a member of the American Eugenics Society.[3]

---

[1] Whitney/Grossman, 1930, 54, 56.

[2] Danzis, 1930, 766-767.

[3] Meehan, 1997/2001. Just *some* other members and officers of the American Eugenics Society (renamed Society for the Study of Social Biology in 1973): Mrs. Frank Abell, S. J. Appelbaum, August Belmont (né August Schonberg), Shirley C. Bierman, Philip G. Bronstein, M. Cohen, Dr. Hugo A. Freund, Mrs. Herbert Goldschmidt, William Grossman, M. Robert Guggenheim, Murray Guggenheim, Professor Charles Winthrop Gould, Professor Samuel B. Heckman, August Hekscher, Dr. N. D. Hirsch, Mrs. Otto Kahn, Samuel Kasakoff, Samuel Henry Kress, Albert J.. Levine, Ruben Liskey, Julius Manger, Albert Z. Mann, Paul B. Mann, future U.S. Secretary of the Treasury Henry Morgenthau, Jr., Nobel Prize winner (genetics) Herman J. Muller, Emma Goldman's lover and manager Ben L. Reitman, Dr. Aaron Joshua Rosanoff, H. H. Rubin, Jacob Saposnekow, Frederic W. Simonds, Alfred K. Stern, geneticist Abraham Stone (1930, 1938, 1956), Ellen A. Stone, geneticist Solomon Thieberg, Max Thorek, Dr. and attorney Felix Ferdinand Tietze, Dr. Abraham Wolbarst, Samuel Zuckerman, Wolf Zuelzer, Adolph Zukor. (*Eugenical News, Eugenics Quarterly*, web site *"Eugenics Watch")*

## 1931

## Context

1.  The Eugenics Publication Company (New York City) brings out Sigmund Freud's (1856-1939) *Modern Sexuality, Morality and Modern Nervousness*.

2.  From the posthumous memoirs of Michael Hainisch (1858-1940), second Federal President of Austria: "I wanted to see the number of Jews in leading positions limited, because the mentality of the Jews is different from that of the Aryans. In so stating, I do not wish in the least to imply that I consider the Jews to be an inferior people. On the contrary, in intelligence and intellectual activity they are a finely bred race, and in this respect are unquestionably superior to the Aryan population.... But the mistake of the Jews is that they are completely lacking in that sense of the irrational in the life of a people (fatherland, homeland, mother tongue, Christianity), without which the survivability of a people cannot be insured.... I was not only of the opinion that the university should not become a Jewish school, but I considered it intolerable that 85% of the lawyers and 70% of the doctors were Jews. For that reason I always favored the introduction of a *numerus clausus*, although I wanted to formulate it in my own way...."[1]

3.  Prominent German eugenicist Fritz Lenz: "That National Socialism truly aspires to a recovery of the race cannot be doubted. Of course, the lopsided 'Anti-Semitism' of National Socialism must be regretted."[2]

### Jewish Advocacy of Eugenics

4.  Eugenicist Wilhelm Nussbaum (1896-1985) graduates from the Kaiser Wilhelm Institut für Anthropologie, menschliche Erblehre und Eugenik, his certificate signed by the prominent eugenicist Eugen Fischer (1874-1967).[3]

---

[1] Michael Hainisch: 75 Jahre aus bewegter Zeit. Lebenserinnerungen eines östereichischen Staatsmannes. Bearbeitet von Friedrich Weissensteiner. Wien-Köln-Graz 1978, 306; quoted in Hofmann, 2005, 87.
[2] *Menschliche Auslese und Rassenhygiene (Eugenik)*, (=Erwin Baur, Eugen Fischer, Fritz Lenz), 3rd edition, Munich, 417; quoted in Schmuhl, 2003, 152.
[3] Efron, 1994, 19.

5. Jewish-German eugenicist Arthur Ruppin: "A faulty genetic predisposition of the father or the mother is the most frequent cause of degenerative phenomena in the children. The only means to reduce this degeneration is for persons with degenerative traits to refrain from having children.... Already in Babylonian times the Talmud instructed: 'Convert all that you have into money and get your son the daughter of a scholar, or your daughter a scholar as a spouse.'... But this one-sided emphasis on intelligence or scholarship had... as a consequence that even today we find among the Jews more highly gifted persons, but also more psychologically or physically disadvantaged individuals than among Christians."[1]

### Jewish Rejection of Eugenics

6. Rabbi Sydney E. Goldstein of the Jewish Institute of Religion of New York testifies in Congress as a "moral theologian" before a subcommittee of the Senate Judiciary Committee, evidently in opposition to the Gillett Bill (S. 4582), which would exempt licensed medical practitioners from federal laws forbidding circulation of contraceptive information. The American Eugenics Society supports the legislation.[2]

### 1932

### Context

1. The Jewish-Russian poet Osip Mandelstam (1891-1938), who eventually perishes in the purges, muses about consequences for people living in the Soviet Union if Jean-Baptiste Lamarck was right about acquired genetic characteristics:
   *From my lips will grow tentacles,*
   *My trunk will be hooped in rings,*
   *My suckered fingers will thrash an ocean floor,*
   *And I will disappear like Proteus behind a horny mantel...*[3]

2. In Germany Jewish eugenics proponent Hans Goslar is dismissed from his government post of Director of the Press Section.[4]

---

[1] Ruppin, 1931, 92-94.
[2] Anderson *et al.*, 1931.
[3] Glad/Weissbort, 1992, 86.
[4] Kramer, 2003.

3.  Jewish-German mathematician Felix Bernstein, who has been working on detecting genetic traces of racial mixture, emigrates for fifteen years to the United States, where he had earlier worked at the Eugenic Records Office at Cold Spring Harbor.[1]

4.  Not only in Warsaw, but in other Polish towns Jews have lower fertility than do non-Jews.[2]

5.  The Great Soviet Encyclopedia publishes an article by Jewish-Russian physician Grigory Aleksandrovich Batkis (1895-1960) calling eugenics "bourgeois" and "fascist," even castigating it as "Menshevizing idealism."[3]

## Jewish Advocacy of Eugenics

6.  In Russia Jewish geneticist Vladimir Pavlovich Èfroimson (1908-1989) is sentenced to five years imprisonment for advocating eugenics and for refusing to give false testimony against fellow eugenicist Nikolai Konstantinovich Koltsov.[4]

7.  Some participants in the Third International Congress of Eugenics, New York: Dr. Rudolph M. Binder, Professor Charles Winthrop Gould, biologist A. H. Hersh (1925, 1920), Dr. N. D. Hirsch, Mrs. Earnest Schuster, Dr. Abraham L. Wolbarst.

8.  Jewish-German physician and eugenicist Julius Tandler: "Given the attitude of mankind today, or even in a hundred years, I am not of the opinion that the physician will have the right to kill the inferior; but I am of the opinion that we have the right to prevent their birth. This is the place to state openly that the inferior should be sterilized. To speak out against the breeding of the inferior is an act of emergency self-defense on the part of human society, which must know that it itself is in danger, and the first to come to this realization are the doctors.[5]

---

[1] Schappacher, 2005, 6.
[2] Liebman Hersch, "Jewish Population Trends in Europe," *Jewish People: Past and Present*, II, 11, Table 10, cited in Goldscheider, 1967, 200.
[3] Adams, 1990, 185.
[4] Vergasov. Undated.
[5] Hofmann, 2005, 61.

## 1933

## Context

1. In Berlin, Jewish eugenicist Magnus Hirschfeld's Institute for Sexual Science and Eugenics is plundered and the building confiscated. Most of the film footage of the books burned in Opera Square is from the Institute's library, particularly his own writings. A student carries a bust of Hirschfeld impaled on a pole. In a Paris movie house Hirschfeld happens to see scenes of the event.

2. Forced as a Jew to resign his position of curator in the Ethnographical Museum at Hamburg, Otto Samson emigrates and devotes his efforts to rescuing the notion of stable racial classifications.[1]

3. There are 503,000 Jews in Germany, constituting 0.76 percent of the total population,[2] of whom approximately 350,000 emigrate.[3]

4. Just 2½ months after German President Hindenburg names Hitler Chancellor, an April 19 article in London's *Jewish Chronicle* entitled "Germany: Ghastly Hell of the Concentration Camps" makes an early claim of "Human Slaughter Houses." The article also expresses anxiety over sterilization legislation: "It is to be noted that although the Bill for Compulsory Sterilization makes no direct references to Jews, there is a provision of very grave implication which authorizes governors of penal establishments (e.g. concentration camps) to propose the sterilization of a prisoner. The proposal comes before the 'Eugenics Court," which is composed of a magistrate, a medical officer, and a physician whose special province is the study of heredity hygiene. Its decision, taken after a secret hearing, is subject to appeal by a 'High Court of Eugenics.' The surgical operation, if the decision is confirmed, can be carried out against the will of the prisoner."[4]

5. *Bayerische Ärztezeitung*: "As an example for the influence of lifestyle on the characteristics of children emerging from interbreeding, Professor [Eugen] Fischer used the example of crosses between

---

[1] Barkan, 1992, 161.

[2] Esta Bennatha, "Die demographische und wirtschaftliche Struktur der Juden," *Entscheidungsjahre*, 1932; *Zur Judenfrage in der Weimarer Republik*, 966, 89, 95.

[3] Richard Albrecht, *Exil-Forschung: Studien zur deutschsprachigen Emigration nach 1933*, Frankfurt/Main, 1988, 11; cited in Mildenberger, 2002, 184.

[4] Pg. 17.

Nordic races and Jews. He is of the opinion that it makes an enormous difference whether Nordic people cross with the offspring of
old, cultivated Jewish families or those from recently immigrated
Eastern Jewish families." The *Völkischer Beobachter* quotes Fischer
with the words: "What position should one take on commingling
with the Jewish race? It is a matter of course that the Jewish race is
not more inferior than many other races. But one thing is certain:
that it is different, and this difference is also the reason why it is
completely unsuitable for a crossing of cultures with the German
nation."[1]

6.  Johns Hopkins University zoologist Alexander Weinstein: "...the
    advance of science, which formerly inspired mankind with confi-
    dence, has in recent years resulted in diffidence and despair."[2]

7.  Sigmund Freud: "When one thinks that ten or twelve percent of the
    Nobel Prize winners are Jews and when one thinks of their other
    great achievements in sciences and in the arts, one has every reason
    to think them superior."[3]

8.  Jewish eugenicist Arthur Ruppin describes in his diary a startlingly
    collegial meeting with Hans F. K. Günther (1891-1968), who is de-
    scribed by historian Amos Morris-Reich as "Hitler's mentor"[4]: "The
    conversation lasted two hours. Günther was most congenial but re-
    fused to accept credit for coining the Aryan concept, and agreed
    with me that the Jews are not inferior but different, and that the
    Jewish question has to be solved justly."[5] Günther informs Minister
    of the Interior Wilhelm Frick (1877-1946, executed by American
    occupation authorities), who responds positively. Morris-Reich
    makes two important observations: "My impression is that in the
    1920s Ruppin came closer to the deterministic branches of anthro-
    pology – those branches that Boas so bitterly contested. It is not en-
    tirely unlikely that Ruppin came upon Boas's work through
    Günther's references to him.... Ruppin, like many others, sees a
    problem where Jews live in the midst of peoples whose racial ma-

---

[1] *Bayerische Ärztezeitung*, No. 12, 1933; quote taken from Niels C. Lösch, *Rasse als Konstrukt: Leben und Werk Eugen Fischers*, Frankfurt/Main, 1977, 244f; both passages taken from Schmul, 2003, 121 and 136 respectively.
[2] "Palamades," *American Naturalist*; quoted in Glass, 1986, 146.
[3] Quoted in Wortis, 1984, 145. See also Gilman, 1996, 123.
[4] Morris-Reich, 2006a, 1.
[5] Central Zionist Archive, A 107/954; cited in Morris-Reich, 2006a, 1.

keup is very different from theirs. This seems to be Ruppin's and Günther's common ground: that a solution to the Jewish problem must include the Jews' removal from Northern Europe."[1]

9. Jewish-German anthropologist and eugenicist Wilhelm Poll is the first "non-Aryan" member of the Hamburg Medical School to be pensioned off under the newly passed Law for the Restoration of the Professional Civil Service (Gesetz zu Wiederherstellung des Berufbeamtentums).[2]

10. Jewish-German geneticist and eugenicist Curt Stern (1902-1981): "It is terribly difficult for my wife and me to separate ourselves externally from Germany. You know that I have always considered myself fully German."[3]

11. Physician Max Hodann, who has worked largely on birth control counseling in Hirschfeld's institute, is arrested and, upon release, emigrates and at one point fights in the Spanish civil war.

### Jewish Advocacy of Eugenics

12. Although Jews account for only 0.8% of the German population, at least 16% of physicians are Jewish.[4] Eugenics is a popular cause among them.

13. Rabbi Rudolph Coffee of Oakland, California, is among the 25 Charter Members of the Human Betterment Foundation, Pasadena, California.[5]

14. The physician Abraham Matmon publishes a booklet in Hebrew entitled *The Improvement of the Human Species and Its Significance to Our Nation*: "...we must always remember the assertion that in order for a nation not to degenerate, it must take care not only of its quantitative values but also of its quality. The best material in a nation is that which always marches forward, or more precisely, drags behind itself the weaklings."[6]

---

[1] Morris-Reich, 2006a, 26, 11.
[2] Braund/Sutton, 2008, 23-24.
[3] Deichman, 1996, 21.
[4] Adam, 2007, 194.
[5] Human Betterment Foundation announcement. The archives of the Human Betterment Foundation are in Special Collections at Caltech in Pasadena.
[6] Tel-Aviv, Biological-Hygienic Library; quoted in Falk, 1998, 597-598.

15. Future President of the American Eugenics Society and Curator of the American Museum of Natural History and Chairman of its Anthropology Department Harry Lionel Shapiro (1902-1990): "...it is conceivable, even inevitable, in the future society of which man will be a part that the population will be mated as carefully as the animal breeder now controls his stock."[1]

16. Jewish-Austrian eugenicist and physician Ignaz Zollschan proposes the creation of a Jewish racial studies center at Jerusalem University.[2]

17. In Berlin, Jewish physician Wilhelm Nussbaum (1896-1985) founds the Jewish Genetic Research and Eugenics Cooperative (Die Arbeitsgemeinschaft für jüdische Erbforschung und Eugenik/Erbpflege). The project will consist of anthropological and genealogical surveys and employ twins studies. More than 2,000 Jewish families sign up as subjects for a eugenics survey.[3] The indicated range of Jewish political orientation is broad: "neutral, assimilationist, Zionist, Orthodox."[4]

18. A second list is to be found in the Nussbaum Archive – of persons and organizations slated to regularly be informed about the cooperative's activities.[5]

---

[1] From *Natural History*, Nov-Dec. 1933 quoted in *Current Biography* 1952 "Harry Shapiro"; cited in *Eugenics Watch*, http://www.eugenics-watch.com/aeugensoc/aeoff.html, accessed May 12, 2008.

[2] "Jüdische Rassenforschung: Ein Vorschlag an die Universität Jerusalem, " *Jüdische Rundschau,* Aug. 11, 1933; cited in Lipphardt, 2008, 255.

[3] Letter to *Jüdischer Arbeitsnachweiss*. Nussbaum Archive, Leo Baeck Institute, New York.

[4] Founding members: Dr. and Senator Karl Abel, gynecologist; Professor Hans Aron, physician; Professor Carl Birnbaum, psychiatrist; Dr. Arthur Czellitzer, ophthalmologist; Professor A. Gutmann, physician; Werner Haberland, profession not indicated; Dr. E. Holländer, District Court official; Dr. Ludwig Holländer, lawyer and notary; Dr. Kasten, psychiatrist; Professor Arthur Kronfeld, physician; Dr. Curt Singer, Kulturbund; Dr. Leo Löwenstein, R.J.F.; Professor E. Mathias, pediatrician; Professor L. Meyer, pediatrician; Dr. W. Nussbaum, gynecologist; Dr. Felix Reich, Institute for the Deaf and Dumb; Professor W. V. Simon, orthopedist; Dr. Else Wolfson, ophthalmologist (Nussbaum Archive, Leo Baeck Institute, New York, box 3, folder 19).

[5] Rabbi Dr. Baeck; Dr. A. Czellitzer; Dr. Driesen, The Free Jewish University; Dr. Goldschmidt; Director Gutman; Dr. Hirsch; Mrs. Falkenberg; Ms. Kaminski; Dr. Klein; Wilhelm Levisohn; Professor Dr. Stefan Martin-Openheim; Dr. P. Mayer; Dr. Elise Morgenstern; Rabbi Dr. Prinz; Dr. Nether; Dr. A. Ruppin; Director Stahl;

19. The Nussbaum Archive contains still a third list, indicated as "private" and written in longhand. Next to each name is the "part" of the organization they are advising.[1]

20. In addition to Nussbaum's institute, two Jewish-German anthropologists plan – unsuccessfully – to create their own institutes: eugenicist Ignaz Zollschan attempts to found an "Institute for Racial Studies" (Institut für Rassenforschung), and Franz Weidenreich (1873-1948) tries to launch his own institute for the study of Jewish biology. In 1933 Weidenreich writes (in English) to Charles Seligman: "I may assume you know the conditions in Germany very well, and that you also know that I am a Jew, and so am in great danger, not only on the scientific and intellectual side, but also on the material side.... But as a representative of the Science of Race, I am nevertheless in great danger. As you probably know, the Nazis have their own Science of Races which is impossible to be accepted by me or by any cultured person whatever."[2] Seligman pursues his doomed plan even after he judges it wisest to leave Germany and emigrates to the United States.

21. Jewish-American eugenicist Hermann J. Muller on the belief in Nordic superiority: "There is not one iota of evidence from genetics for any such conclusions, and it is too bad to have them issued with the apparent stamp of genetic authority. They form just the sort of ground which reactionaries desire, on which to raise a pseudoscientific edifice for the defense of their system of sex, class and race exploitation."[3]

---

Chancellor Dr. H. Stern; Dr. H. Strauss; Professor Dr. Weidenreich; Rabbi Dr. Weinberg Rosenthal; Dr. Zollschan; Medical Intern Ruth Bamberger; Professor Berliner; Dr. Fritz Bloch; L. Feintuch; Professor Dr. Bruno Heymann; Mrs. Kamberg; District Judge Dr. Robert Care Katzenstein Kuhn; Professor Dr. L.F. Meyer; Dr. E. Nassau; Dr. Edith Neustadt; Dr. Fritz Schiff; Dr. Tänzer; Dr. Günther Winkler (Nussbaum Archive, Leo Baeck Institute, New York, box 3, folder 12).

[1] Hygiene – Hahn; Psychology – Birnbaum; Character – Kronfeld; Dermatology – Pincus; Surgery (?) – Guttmann; Throat and Lungs – (No name written); Immunology – Isaak Frkf.; Orthopaedics – Veit-+Simon Frkf.; Obstetrics – Borchardt; Gynecology – Ashheim (?); Martin Weidenreich. Below the above list is still another, but with no roles assigned: Hollaender, Edelstein, Bernat, Loewenstein, Baeck, Moses (?) – unreadable, Czellitzer (?), Reich, Leschnitzer, Driesen Frkf (Nussbaum Archive, Leo Baeck Institute, New York, box 3, folder 12).
[2] June 4; quoted in Lipphardt, 2008, 269.
[3] Kühl, 1994, 78.

22. Author Max Brod (1884-1968) comments: "We petted the hounds of Hades, and they gnawed at their chains."[1]

### Jewish Resistance to Eugenics

23. Jewish-British pioneer in psychiatric genetics Aubrey Lewis (1904-1983), who subsequently accepts a Rockefeller Foundation Fellowship to study in Munich and Berlin, primarily in the laboratories of Ernst Rüdin, accuses Rüdin of being one of the main authors of the German sterilization law.[2]

## 1934

### Context

1. Nussbaum receives a menacing summons from the State Secret Police (the Gestapo) demanding by-laws, minutes, bank accounts, names of governors, related and affiliated organizations, publications, a membership list. The summons is dated October 15, and directs Nussbaum to appear at 9:00 am of the 17th with all requested documentation.[3] Nussbaum responds that he harbors no Marxist or communist sympathies, and that the Cooperative is non-political and has government licensing; he also provides his personal background, but no membership lists.

2. Within one generation, the average size of Anglo-Jewry has fallen from seven children to two, and the average age at marriage has increased by two and a half years for brides and by three years for grooms between 1904 and 1934.[4]

3. The German governmental office of German Jews (Reichsvertretung der deutschen Juden) awards a grant of 1,000 Reichsmarks to the Jewish Genetic Research and Eugenics Cooperative.[5] Given the political backdrop of the period, the grant is remarkable.

---

[1] Referring to the high regard that many German-speaking Zionists had for such racial ideologues as Heinrich Driesmans (1863-1927), Arthur de Gobineau, Houston Stewart Chamberlain, Ludwig Schemann (1852-1938), Ludwig Wilser (1850-1923), Ludwig Woltmann (1871-1907), Eugen Fischer (1874-1967), and Hans Günther. Doron, 1980, 421, after H. Meier-Cronemeyer, *Jügendbewegung*, I. Teil, Köln, Germania Judaica, 1969.

[2] Gottesman, 2005.

[3] Nussbaum Archive, Leo Baeck Institute, New York, box 3, folder 18.

[4] *Jewish Chronicle*, 1981.

[5] Reichsvertretung…, 1934.

4. Jewish-Lithuanian-American rabbi and theologist Mordecai M. Kaplan (1881-1983), founder of the Reconstructionist Movement: "What can exercise a more blighting effect upon all moral endeavor than the notion that there is no meaning or purpose to the world, and that is soulless in its mechanistic perfection...? We may accept without reservation the Darwinian conception, so long as we consider the divine impulsion or initiative as the origin of the process."[1]

5. Eugenicist Magnus Hirschfeld: "I protest against now being called a Jew and therefore despised and persecuted by Nazi swine. I am a German, a German citizen... born of German parents! And the same thing happened to me that happens to every newborn in all of Europe: they are stuffed in a straitjacket, christened or circumcised, and are to be raised in the belief system of those who rear them. Since my parents held to the Mosaic belief, I am preconceived as marked by the Mosaic stigma."[2]

6. Jewish geneticist Ursula Phillip (b. 1908), who left Germany in 1933, receives her Doctoral degree from the Kaiser Wilhelm Institute (KWI) for Biology in Berlin-Dahlem in the department of Richard Goldschmidt. (She becomes a Member of the British Eugenics Society in 1957 and a Fellow in 1977).[3]

7. Dr. Harry H. Laughlin of the Department of Eugenics of the Carnegie Institution of Washington makes recommendations on immigration to the Chamber of Commerce of the State of New York: "There is a movement now to make special legislative provisions for the Jews persecuted in Germany. If, as a result of persecution or expulsion by any foreign country, men of real hereditary capacity, sound in physical stamina and of outstanding personal qualities, honesty, decency, common sense, altruism, patriotism and initiative, can be found, they should, because of such qualities, and not because of persecution, win individual preference within our quotas and be welcomed as desirable human seed-stock of future American citizens. If any would-be immigrant cannot meet these standards, he should, of course, be excluded.... The Jews are no exception to rac-

---

[1] Kaplan, *Judaism as a Civilization*, 1934, 98; quoted in Cherry, 2003, 267.
[2] Quoted in Magnus Herzer, *Magnus Hirschfeld: Leben und Werk eines jüdischen, schwulen und sozialistischen Sexologen*, Frankfurt am Main, 1992, 25f; cited in Lipphardt, 2008, 201.
[3] Vogt.

es which are widely variable in family-stock quality within their own race.... High-grade Jews are welcome, and low-grade Jews must be excluded."[1]

8.  Concealing his ancestry, the Jewish Chairman of the German Association of Biologists Ernst Lehman preaches racial hygiene: "It is truly admirable what the biological will has accomplished since January 30, 1933 [occasion of a massive torchlight parade in Berlin to celebrate the appointment of Hitler as Chancellor of Germany].... This was the last chance to save Germany, which lies so close to the compact of settlement areas of Jews in the East, from becoming utterly Jewified.... We have freed our *Volk* from foreign races. And if we work on building up anew its racial composition, the next task, which arises inevitably from our basic biological knowledge, is to compensate within the ranks of our own *Volk* for the struggle for survival that operates freely in nature and thus to take the eugenic measure of sterilization."[2] At one point he states that "National Socialism is politically applied biology."[3] Even when his Jewish ancestry is exposed, he continues to promote these same ideas, but reverses himself after the war.

9.  Physician Walter Falk (Palestine): "The difference between Azhkenazi and Sephardi Jews is so great that biologically we can speak of two races."[4]

### Jewish Advocacy of Eugenics

10. I. Rubin (penname of psychologist and literary scholar Israel Rivka'i) writes an article entitled "The Ingathering of Exiles, Eugenically Considered' for *Moznayim*, the journal of the Jewish Writers' Association in Palestine," calling for "the creation of a *new corrected and perfected type of Hebrew*" on the basis of mixed marriages, not between Jews and non-Jews, but between Jews of different ethnic background: "It is an elementary eugenic truth, for example, that *mixed marriages* between nations and races have a strong eugenic value, that infusion of new blood can only heal and strengthen – and contrarily; if mating takes place only within the restricted

---

[1] *New York Times*, 1934.
[2] Deichman, 1996, 75-76.
[3] Deichman, 1996, 86.
[4] "Observation on the Development of Babies in Emek-Hayarden" (in Hebrew), *Harefu'a*, 393-400; citation on page 394, taken from Hirsch, 2009, 600.

confines of one nation, the result is nothing but degeneration and decline for this nation. From this perspective, if I may add in passing, the Nazis' racial doctrines, which lead a fierce battle against mixed marriages with non-Germans, non-Aryans, as though for the benefit of the German people, have nothing to do with the true eugenic doctrine. On the contrary: Hitler's 'racial doctrine' stands in absolute contrast to the 'doctrine of eugenics." Rivka'i proclaims the return of the Jews to the country of their ancestors to be a "great eugenic revolution in the life of the nation."[1]

11. Israel's national poet Chaim Bialik (1873-1934) declares at a press conference at the Hebrew University: "I too, like Hitler, believe in the power of the blood idea."[2]

12. Anonymous article in *Israelit*: "We had no need to learn from the National Socialists the significance of everything related to ancestry, heredity, and the character of a race or a people. Jewry's written and oral traditions harbor a wealth of material that obliges the religious Jew to take his place in full respect next to the newly awakened racialist, genetic, and eugenic scientific method, and in this sense from one's enemy."[3]

13. The *Palestine Post* enthusiastically reviews Enid Charles's "admirable book," *The Twilight of Parenthood: A Biological Study of the Decline of Population Growth*: "Mrs. Charles's work is a happy and convincing mixture of science and practice. She gives the reader in comprehensible form the results of the most recent technical studies in population problems, deals thoroughly with the eugenic side of family-restrictive tendencies, and, finally, 'with the advantage of being at the same time the mother of four children and a wage-earner in the academic profession,' she is able to turn to her concluding chapter – which has as its subject how, by the State subsidizing families, to keep up the population...."[4]

14. Lev (Leon) Davidovich Trotsky (né Bronstein, 1879-1940): "While the romantic numskulls of Nazi Germany are dreaming of restoring the old race of Europe's Dark Forest to its original purity, or rather

---

[1] Original article in Hebrew, 89-93; cited in Hirsch, 2009, 594.
[2] Bialik, 1934, 6.
[3] Lipphardt, 2006.
[4] *Palestine* Post, June 10.

its original filth, you Americans, after taking a firm grip on your economic machinery and your culture, will apply genuine scientific methods to the problem of eugenics. Within a century, out of your melting pot of races there will come a new breed of men – the first worthy of the name of Man."[1]

15. The physician Y. Rubin publishes in a periodical of the Hebrew Authors' Association a paper entitled "The Ingathering of the Exiles from a Eugenic Point of View." Rubin views "our life in the homeland, in its very essence" as "primarily a great and courageous national effort in the eugenic sense.... Anyone who does not recognize in the return of the sons to the land of their fathers a huge *eugenic* revolution in the life of the nation, does not discern the 'forest' from the 'trees'.... The essence is the sum total: *the production of a new Hebrew type*, restored and improved. *A psychobiological approach to the settlement of Palestine* – is a duty to us all!" (emphasis in the original).[2]

16. Jewish-German gynecologist and eugenicist Ludwig Fraenkel is forced to step down from his position of Director of the Women's Clinic of the University of Breslau and emigrates permanently (year approximate). London's *Jewish Chronicle* reports that Jewish physicians Weide and Schumann, Professors of Eugenics at the University of Dresden, have been dismissed from the positions.[3]

17. The *Palestine Post* publishes a rather positive article from "The Times Correspondent" on Nazi eugenicist Otto von Verschuer: "The giving of advice to those about to marry, far from being the monosyllabic affair once recommended by Punch, has *become* a complicated matter in National Socialist Germany; it is to include the medical examination, not only of the would-be happy pair but of their relatives.... This complicated process of gathering relatives for examination, however, would give way in time to a far simpler system, for all registry offices should be compelled to keep family records, and all public health offices should be obliged to create race and hygienic departments with eugenic card indexes. These should record vaccination, the results of school and all subsequent examinations, bodily characteristics and physical deficiencies, and

---

[1] Trotsky, 1934.
[2] *Moznayim*, 1(4), 89-93; quoted in Falk, 1998, 598.
[3] "Germany: Jew Murder Again," July 13, 18.

should be kept carefully up to date. Thus in time, as Baron von Ver-schner [sic] says, the task of the eugenics consultant would be great-ly simplified. He would no longer need to examine the whole fami-ly, but simply to take the names of the applicants, send for the cor-responding set of cards, and, from the information given on them, decide whether a marriage was permissible or not."[1]

18. Dr. Israel Rubin, an educator and literary critic, observing mixed marriages among various Jewish ethnic groups: "Here the 'ingather-ing of the exiles' in Palestine makes possible 'mixed marriages' not between Jews and non-Jews, but between Jews and other Jews.... Doesn't this in itself contain the hope of eugenic salvation to a great extent?"[2]

19. Jewish-Polish physician Gerson Lewin interprets "Talmudic hy-giene" as fundamentally serving eugenic purposes[3]

## 1935

### Context

1. Two laws designed by Adolf Hitler are approved by the National Socialist Party convention in Nürnberg on September 15. The citi-zenship law (Reichsbürgergesetz) strips Jews of German citizen-ship, designating them "subjects." The Law for the Protection of German Blood and German Honor (Gesetz zum Schutze des deut-schen Blutes und der deutschen Ehre) forbids marriage or sexual re-lations between Jews and persons of German or kindred blood.

2. German publisher and *SA-Gruppenführer* Julius Streicher (1885-1946): "The sperm of a man of a different race constitutes allogenic albumin. During intercourse the male sperm is absorbed completely or partially by the maternal compost and thus gains access to the blood. A single contact of a Jew with an Aryan woman suffices to poison the blood of the latter forever. Together with the allogenic albumin she also assimilates the allogenic soul. Even if she later marries an Aryan man, she will never again be able to bear pure Aryans, but only half-breeds in whose bosoms will dwell two souls,

---

[1] "Marriage and Heredity in Germany: Importance of Uncles and Aunts," Dec. 28, 1934, 9.
[2] Karpel, 2006, exact year not indicated; probably in *Moznayim 1*, 89-93.
[3] Lewin, 1934, 10; pointed out by Kamila Uzarczyk in Turda/Weindling, 289.

whose mixing will be physically apparent. For their part, her children will be hybrids, that is, ugly persons of unstable character and with a propensity for physical misery. This process is known as 'impregnation.'"[1] (Streicher was executed by occupational authorities after the war.)

3. American eugenicist Clarence G. Campbell in Berlin (after praising a number of non-German eugenicists): "It is from a synthesis of the work of all such men that the leader of the German nation, Adolf Hitler, ably supported by the Minister of Interior, Dr. [Wilhelm] Frick [1877-1946], and guided by the nation's anthropologists, its eugenicists, and its social philosophers, has been able to construct a comprehensive race policy of population development and improvement that promises to be epochal in racial history. It sets the pattern which other nations and other racial groups must follow, if they do not wish to fall behind in their racial quality, in their racial accomplishment, and in their prospect of survival."[2]

4. The Rockefeller Foundation withholds funding for genealogical demographic research at the Kaiser-Wilhelm-Institute of psychiatry in Munich.[3]

5. Jewish physician and eugenics proponent Arthur Kronfeld (1886-1941) emigrates from Germany to Switzerland.

### Jewish Advocacy of Eugenics

6. David Macht, a Baltimore pharmacologist and Yeshiva College Ph.D., pleads with Union of Orthodox Congregations of America members to deepen their appreciation for marriage, "called in Hebrew *Kiddushin*, or 'sanctification,'" for "the Biblical conception of matrimony and purity of marital life as denoted and connoted by the Hebrew term, *taharath ha-mishpahah*; and… a more conscientious and consistent devotion to *hinnukh*," which he defines as eugenics.[4]

---

[1] *Deutsche Volksgesundheit aus Blut und Boden*, Nuremberg, 3(1); translated from the French, Essner, 1995, 6.
[2] "The Biological Postulates of Population Study," *Bevölkerungsfragen: Bericht des Internationalen Kongresses für Bevölkerungswissenchaft,* Berlin, Aug. 1-Sept., ed. Hans Harmsen and Franz Lohse, Lehmann, Munich, 1936, 928, in translation quoted in Kühl, 1994, 34.
[3] Seidelman, 2001.
[4] Rothstein, 2000.

7. In Berlin Wilhelm Nussbaum finally provides the police with a list of founding members of the Jewish Genetic Research and Eugenics Cooperative, closes both the Cooperative[1] and his medical practice, and emigrates to the United States. Rabbi Leo Baeck writes a letter of recommendation, describing him as "one of our foremost authorities in the area of [genetic] patrimony and racial studies."[2] In an undated, untitled, and unsigned article contained in the archives of the Leo Baeck Institute Nusbaum debunks the concept of an Aryan race and concludes: "Despite this insight political hatred now rages against the non-Aryan thanks to the political propaganda. Instead of cooperation between different races, for example, despite the approval of the Reich Ministry, Jewish genetic research within the Society for Jewish Genetic Research was not allowed, and the Society had to cease its activities." (signature typed over with multiple x's)[3]

8. Nussbaum on extreme environmentalism: "Is it not a foolishness to insist on explaining man exclusively on the basis of his milieu? What good does it do, for example, when we use social funds to provide a home and minimal financial support for persons who are genetically incapable of taking care of this home?"[4]

9. Anthropologist and future President of American Eugenics Society Harry L. Shapiro upon return from a visit to Pitcairn Island, which is inhabited by descendants of the mutineers from the vessel The Bounty in 1789 and their Polynesian wives: "The Pitcairn Islanders show no ill effects of several generations of intermarriage. They are taller, and at least in some respects appear to be better developed physically, than either the English or the Polynesian races."[5]

10. Jewish-German eugenicist Franz Kallmann (1897-1965), the "father of psychiatric genetics," continues to work in the Deutsche Forschungsanstalt für Psychiatrie after the National Socialists come to power. Aware that he might be forced as a Jew to accept early retirement, he writes in September to German eugenicist Ernst Rüdin: "Since I will continue to feel myself to be a German and a Christian even if I spend the rest of my life working as a handyman among

---

[1] Nussbaum Archive, Leo Baeck Institute, New York, box 3, folder 19.
[2] Nov. 14, 1935, Nussbaum Archive, Leo Baeck Institute, New York.
[3] Nussbaum, untitled and undated.
[4] Nussbaum, undated.
[5] Pace, 1910.

negroes and Hindus, the only reasonable exits out of my current distress are those that offer the opportunity to work with minimal interference *within* the borders of my German Fatherland." Thanks to
Rüdin's strenuous efforts on his behalf, Kallmann is able to present
his research at the International Congress on Population: "The path
to a rapid and certain eradication of genetic predisposition to schizophrenia has been clearly defined. Early sterilization at onset of
fertility of all those whose illness has been engendered by heredity
and the exclusion [from the gene pool] of all heterozygotes and
those carriers who are manifestly handicapped are the crucial miles-
tones, while differential-diagnostic confirmation and reliable recognition of pre-psychotic and heterozygotic personality types is the
most urgent precondition. Data from our proband [the first affected
family member who seeks medical attention for a genetic disorder]
children and siblings on fertility- and genetic-burden data demonstrate unambiguously that the exclusion also of hetereozygotic carriers of schizophrenia in close blood relationship with schizophren-
ics is not only essential, but also feasible." Ironically, the Greater-
Berlin chapter of the National-Socialist Physicians' League takes
exception to Kallmann's views... and also to his Jewishness. Rüdin
responds: "On the whole I want to say that we had here a conflict of
interests and principles, in as much as I was committed to utilizing
Kallmann's research in the interest of science and especially on behalf of the sterilization law, which this research strongly supports,
but on the other hand it was a Jew who had produced these results
and placed them on our agenda."[1]

11.  The Palestine *Post*, edited by Gershon Agronsky, publishes an un-
critical review of Charles Chamberlain Hurst's (1870-1947) *Heredity and the Ascent of Man* (Cambridge University Press), eliciting
two letters from readers:

- "Reference to the review shows Dr. Hurst to state that eugenic
principles will be applied, if ever, only by that nation whose
leaders will awaken to the national advantage of breeding a su-
per-race capable of inheriting the earth. Whilst deprecating the
increasing percentage of mediocrities produced by democracy,
together with its mediocre leaders of narrow outlook, Dr. Hurst
goes on to admit that a dictatorship, such as that in Germany, in
its attempt to raise race levels misapplied the possibilities of

---

[1] Mildenberger, 2002, 188-191.

race improvement by emphasizing physical at the expense of mental qualities. Otherwise stated, the idea is that while a democratic society, in giving birth to leaders possess of limited vision and small coercive powers, is not likely to apply the science for the improvement of the race yet a dictatorial form of society by its very reverence for a rule of force, cannot fail but to misapply the science for the improvement of the race.... Without the voluntary participation of women who must bring the heaviest sacrifice, there is no solution to the eugenic problem. It is not enough that there might be a number of women in Palestine who would accept the decision of Eugenic Experts as to who shall be the fathers of their children. For the masses of mediocrities will go on multiplying easily and cheerfully even though confronted with the superior specimens of humanity exhibited to them by our pioneering women eugenicists." (S. Broyde, Kiryat Schmuel, Jerusalem.)

- "Is not the discussion of race improvement by producing less mediocrities merely empty talk so long as the intellectuals, the non-mediocrities are not free from the trait of credulousness? There is hardly any hope for race improvement in Palestine or elsewhere unless and until credulity is replaced by critical judgment. Let us nevertheless entertain the expectation that someday humanity will wake up to the necessity for rejecting humbug which has kept men in darkness from the beginning till the present day. Then, and then only, will race improvement begin without any assistance whatsoever from the theory of eugenics or genetics." (Leah Broyde, Jerusalem).[1]

12. Mr. A. R. Kaufman (1935, 1937) and Dr. Morris Siegel (1935, 1937) are indicated as members of the (British) Eugenics Society (renamed the Galton Institute in 1989).[2]

### 1936

### Context

1. At a December conference of the Soviet All-Union of Agricultural Sciences (VASKhNIL) a fierce debate breaks out between the geneticists and the Lysenkovites, led by Isai Izrailovich Prezent. The

---

[1] June 9, June 21, June 28.
[2] *Eugenics Review, Bulletin of the Eugenics Society, Biology and Society,* Eugenics Watch.

Russian plant selectionist Nikolai Vavilov (1887-1943) is arrested, but he writes a letter, published in *Izvestiia*, claiming that he is at liberty and that the arrest of a handful of criminals will in no way affect Soviet genetics.[1]

2.  The Prussian Secret Security Police issues an order to Jewish organizations in Berlin to use only German.[2]

3.  Jewish-German zoologist and proponent of eugenics Ernst Marcus (1893-1968): "If I… am expelled from the fatherland and made into a homeless beggar, I will see this as the greatest injustice and the deepest insult that could be inflicted on me. This I have not deserved, if people acted in accordance with the Prussian principle 'suum cuique.'"[3]

4.  Jewish-Russian historian Semyon Dubnov (1860-1941) writes that reading Darwin contributed to his loss of religious belief.[4]

5.  Jewish-German neurologist Albert Moll (1862-1939), co-founder of the Berlin Medical Society for Sexology and Eugenics (Die ärztliche Gesellschaft für Sexualwissenschaft und Eugenik), reverses his former position that homosexuality is innate rather than acquired, and thus welcomes the repressive measures of the National Socialist government, even though as a Jew he was stripped of his medical license in 1933.[5]

### Jewish Advocacy of Eugenics

6.  Architect of Jewish settlement in Palestine and eugenicist Arthur Ruppin in an article entitled "Selection of the Fittest": "It would naturally be desirable to have only 'racially pure' Jews entering Palestine, but a direct influence on the process by selecting those immigrants who most closely approach this racial type is not a practical possibility…. It would of course be preferable to have only strong

---

[1] *New York Times*, 1936; Zhuravsky, 1993: *Izvestiia*, Dec. 22, 1936.
[2] May 8, Nussbaum Archive, Leo Baeck Institute, New York.
[3] Deichman, 1996, 22.
[4] *Sefer ha-hayim*, Tel Aviv; cited by Dubin 1995, 94.
[5] "Albert Moll and Florence Tamagne, *A History of Homosexuality in Europe*, http://books.google.com/books?id=VV56dunio2EC&pg=PA211&lpg=PA211&dq=ei n+leben+als+arzt+der+seele&source=bl&ots=A-fDRXLemQ&sig=-9CHpxiY5SVexUmbGYHPnxfV39U&hl=en&ei=gm2kSYLlDpDUnQf40OGcBQ&s a=X&oi=book_result&resnum=4&ct=result, 211.

and healthy persons come to settle in Palestine, so that we would be assured a strong and healthy succeeding generation. Unfortunately this greatly desired objective cannot be implemented with such generalized simplicity."[1]

7.  The first professor of zoology in the Hebrew University Fritz Shimon (Frederick Simon) Bodenheimer (1897-1959): "It may be assumed with certainty that elements of the Jewish race which we call the "Sephardi" did not participate in the great mixture with the Slavic and Tartar nations that profoundly affected the Ashkenazi during the Khazar times and in the days of [Bogdan] Chmielnicki. On the other hand, we do not find among the Ashkenazi the mixture with the Black types, at least not to the extent found among the Sephardi. The Semitic type is most prominent among the Yemenites.... Zionism aspires directly, though not originally, to the unification of all the elements prevalent today among the Jewish race, in order to form a new, harmonious Jewish type."[2]

8.  Dr. Boris Shapiro is among the attendees at the Third International Congress of Eugenics in New York.[3]

### Jewish Eugenics under Attack

9.  Jewish-American geneticist Herman J. Muller writes a letter to Stalin suggesting the creation of a eugenic state: "In view of the immediately impending rise of discussion on matters relating to genetics it is important that the position of Soviet genetics on this subject should soon be clear. It should have its own standpoint, the positive, Bolshevik standpoint, to set against the so-called 'Race Purification' and perverted 'Eugenics' doctrines of the National Socialists and their allies on the one hand and against the 'laissez faire' and 'go slow' doctrines of the despairing liberals on the other hand. Most liberals take an attitude of practical hopelessness and impotence with regard to human biological evolution, declaring that little or nothing can be done. This is in line with their political individualism and hopelessness." Stalin rejects the proposal in favor of Ly-

---

[1] Arthur Ruppin, "Selection of the Fittest," *Three Decades of Palestine: Speeches and Papers on the Upbuilding of the Jewish National Home,* Tel-Aviv, 78-79; cited in Morris-Reich, 2006a, 8.

[2] Falk, 2007, 153.

[3] *Eugenics Review, Bulletin of the Eugenics Society, Biology and Society,* Eugenics Watch.

senkoism, and Muller hurriedly leaves the U.S.S.R.[1] It is possible
that Stalin's attack on genetics may have been triggered by the letter.

10. Solomon Grigorievich Levit's (1894-1938) pro-eugenics Medical
   Genetics Institute in Moscow is attacked in the newspaper *Komsomolskaia Pravda* as "the scalawagary of fascist and fascist-imitating
   racist scientists."[2]

11. Three Jewish-Russian geneticists – all opponents of Trofim Lysenko (1898-1976) – are arrested: Levit, Israèl I. Agol (1886-1936),
   and Max Levin (b. 1885). Agol is executed the very day that Muller
   leaves; even the translator of Muller's book is reportedly shot.[3]
   Walter Landauer (1896-1978), a German socialist geneticist and eugenicist who emigrated to the United States, writes a letter to German biologist Julius Schaxel (1887-1943), who fled to the Soviet
   Union, about the fate of Vavilov, Agol, and Levit. Schaxel responds
   that Agol is a criminal who has been properly arrested and that the
   Soviet people are opposed to a eugenics based on the principles of
   animal breeding.[4] Schaxel's pro-Soviet, anti-eugenics posturing
   fails to save him. He and his wife are both arrested, and he dies under unclear circumstances.[5]

12. Biochemist and Lysenko supporter Sergei Stepanovich Perov
   (1889-1967): "Levin has been arrested, Levit as well. Serebrovsky
   … has yet to distance himself from menshevizing idealism."[6] (The
   Mensheviks had been an almost exclusively Jewish political party,
   and the term had become a code word for "Jew.")

13. Jewish-Russian geneticist Aleksandr Serebrovsky publicly repents
   his pro-eugenics stance, stating that his 1929 article, "which recently was properly classified in *Izvestiia* as 'counterrevolutionary rav-
   ings,' represented a hideous 'leftist deviation,' an entire chain of the
   crudest political, antiscientific, anti-Marxist mistakes' linked to the
   ideology of Menshevizing idealism…. Created by the bourgeoisie,
   the class 'science' of eugenics serves in the capitalist countries as a

---

[1] Muller, 1936.
[2] MTs, 1936.
[3] Adams, 1990, 197.
[4] http://www.ihst.ru/projects/sohist/document/an/218.htm
[5] Deichman, 1996, 23.
[6] Zhuravsky, 1993: Spornye voprosy genetiki i selektsii. 1937. Moscow, 322.

device intended to defraud and oppress the exploited classes and the colonial peoples." Serebrovsky survives, but two of his brothers are arrested, he is expelled from the Communist Party, and the already typeset galleys of two of his books are destroyed.[1]

14. Jewish-Austrian feminist and eugenicist Bertha Pappenheim (1859-1936) dies after harsh interrogation by the Gestapo about an anti-Hitler remark.[2]

15. Biologist Richard Goldschmidt (1878-1958) is dismissed from his position as Director of the Kaiser Wilhelm Institute for Biology and emigrates to the United States.[3]

## 1937
## Context

1. The International Federation of Eugenics Societies holds a conference in Paris.[4]

2. Max Levien, a Jewish-German professor of the history and philosophy of science in Moscow and an advocate of a universalist eugenics, perishes in the Soviet purges.[5] Another victim is Aleksandr Ivanovich Muralov (1886-1937), a Jewish supporter of Trofim Lysenko and a senior Soviet official.[6]

3. Founding member of the American Eugenics Society Frederick Osborn (1889-1981) in a letter to Franz Boas describes attempts of reform eugenicists to develop a "sound program that will eliminate all of the old class and race biases of eugenics."[7]

4. *New York Times*: "The report on immigration control distributed recently to all members of the Chamber of Commerce of the State of New York was sharply attacked yesterday by Bernard S. Deutsch, president of the Board of Aldermen, and Rabbi Stephen S. Wise, at a conference of 1,500 members of Jewish labor, fraternal

---

[1] Pchelov, 2006.
[2] ORTHOMOM, 2005.
[3] Deichman, 1996, 20.
[4] Adams, 1990, 110-111.
[5] http://de.wikipedia.org/wiki/Spezial:Suche?search=adolf+basler&go=Artikel, accessed January 22, 2009.
[6] Zhuravsky, 1993.
[7] Kühl, 1994, 81; dated Oct. 11, 1937.

and other organizations at the Hotel Edison. ¶ The immigration control report made various recommendations 'for the information of the chamber' based on a study prepared for the chamber's special immigration committee by Dr. Harry H. Laughlin of the department of eugenics of the Carnegie Institute of Washington. Dr. Laughlin had pointed out that 'there is a movement now to make special legislative provision for the Jews persecuted in Germany' and asserted that if any would-be immigrant could not meet certain standards he should be excluded. ¶ 'Dr. Laughlin's 'purification of race theory,' Mr. Deutsch said in opening the conference, 'is as dangerous and as spurious as the purified Aryan race theories advanced by the Nazis, to which it bears suspicious resemblance. ¶ His singling out of the Jews for mention as a particular race group to be barred from general admission to the United States, despite the condescending tribute to so-called 'superior' Jews, is a knavish, deliberate slur upon the whole Jewish people, which differs only from the Nazi brand in that it is couched in more polite language... 'I think it is a filthy thing,' Rabbi Wise declared, 'for the Chamber of Commerce of the State of New York to have given that [report] out to the public without giving it to the members to act on. It was an attempt to move the people of the State and of the nation to form a premature judgment on the findings of one probably amateur eugenicist.'"[1]

5.    The first two directors of the Kaiser Wilhelm Institute for Brain Research (Institut für Hirnforschung) in Berlin, eugenicists Oskar Vogt (1870–1959) and Cécile Vogt-Munier (1875– 1962) are dismissed from their jobs.[2]

6.    Some members and officers of the American Eugenics Society (renamed Society for the Study of Social Biology in 1973): community leader Sidney Borg; sociologist David Victor Glass, Ph.D. (1937, 1957, 1977); professor Adolf Meyer.[3]

7.    A German propaganda image depicts Jews as the bastard sons of Asian and negroid racial groups.[4]

---

[1] "State Chamber Assailed by Jews," 1937.
[2] Müller-Hill, 2006; Weiss, 2005.
[3] *Eugenical News, Eugenics Quarterly, Social Biology*, Web site "Eugenics Watch."
[4] http://www.ushmm.org/uia-cgi/uia_doc/query/53?uf=uia_XanwLE

## 1938

### Context

1. 1,002,406 Arabs and 401,557 Jews are recorded as residing in Palestine.[1]

2. Jewish doctors are decertified in Germany.

3. In Buffalo New York, the average completed family size of professional Jews is 2.9, in contrast to 3.2 for businessmen, 3.5 for artisans, and 3.7 for peddlers.[2]

4. German Jews seeking refuge in the Soviet Union are depressed and terrified as they observe Russians being swallowed up in the purges.

5. British feminist and eugenicist Eleanor Rathbone (1872-1946) denounces the Munich Accords and pressures the parliament to grant entry to dissident Germans, Austrians and Jews.

6. German biologist and eugenicist Otto von Verschuer (1896-1969) pens an essay entitled "The Racial Biology of Jews," the research subsidized by Germany's National Socialist government.[3]

7. First Ashkenazi chief rabbi of the British Mandate for Palestine Abraham Isaac Kook (1865-1935): "The theory of evolution (*hitpattehut*) is increasingly conquering the world at this time, and, more so than all other philosophical theories, conforms to the kabbalistic secrets of the world. Evolution, which proceeds on a path of ascendancy, provides an optimistic foundation for the world. How is it possible to despair when we see that everything evolves and ascends? When we penetrate the inner meaning of ascending evolution, we find in it the divine element shining with absolute brilliance. It is precisely the *Ein Sof* [God's infinite light] *in actu* which manages to bring to realization that which is *Ein Sof in potentia.*"[4]

---

[1] Gilbert, 2007, 149.

[2] Uriah Z. Engelman, "A Study of Size of Families in the Jewish Population of Buffalo," University of Buffalo Series, XVI, Nov., 195-210; cited in Goldscheider, 1967, 203.

[3] One of nearly fifty articles, published in six volumes, under the title *Forschungen zur Judenfrage* (Studies on the Jewish Question), Hamburg.

[4] Kook, *Orot ha-Qodeš*, Jerusalem, 1938, II, 537; quoted in Cherry, 2003, 252-253.

## Jewish Advocacy of Eugenics

8.  Physician Arie Kochinsky in *Harefuah*, journal of the Hebrew Med-
    ical Association: "The contemporary Land of Israel is ready to sa-
    crifice anything for the sake of its youth. For the sake of a sound
    and far-sighted public control of the development of the young, and
    for a healthy population policy, we should use means for racial im-
    provement in order to prevent excessive proliferation of the mental-
    ly deranged and the worthless who are also unfit for social life."[1]

9.  Jewish-German author Käthe Rosenthal in *Jewish School Didactics*
    presents Jewish religious law as corresponding to the principles of
    eugenics.[2]

10. Sociologist David Victor Glass (1911-1978) defines "differential
    fertility" as one of the primary aims of the new field of demogra-
    phy.[3]

11. Prominent Jewish-Russian-American psychiatric researcher and
    eugenicist Aaron Rosanoff (1878-1943): "In all probability, even
    under ideal social and economic conditions, there will still be many
    cases of antisocial personalities, in which rehabilitation or a tolera-
    ble social adjustment will prove unattainable. For such cases, *per-
    manent* segregation and, probably, sterilization will have to be ar-
    ranged, for the double purpose of protection of society and eugenic
    effect. [F]or mental disorders which are determined solely by here-
    ditary factors the only prevention is through eugenics."[4]

12. Jewish-Polish neurologist and eugenicist Henryk Higier (1866-
    1942) praises German sterilization legislation of 1933 (Gesetz zur
    Verhütung erbkranken Nachwuchses): "The Germans pay little at-
    tention to environmental influences and stress hereditary factors as
    the source of diseases. Therefore, according to German law, it is le-

---

[1] 14, 1938, 223-226; cited in Lewit, 2003.
[2] Lippstadt, 2006.
[3] D. V. Glass and C. P. Becker, *Populatiion and Fertility*, London, Population Inves-
tiigation Committee, 1938, pg 50; cited in Oakley, 1992, 165.
[4] *Manual of Psychiatry and Mental Hygiene*, 642, 749-750; quoted by Jay Joseph,
*The Gene Illusion*, 25-26,
http://books.google.com/books?id=OyDQlKwRpfwC&pg=PA25&lpg=PA25&dq=aa
ron+rosanoff+eugenics&source=web&ots=4UXm_gmHs-
&sig=u0KX7Fn93BvKF1znIn-
VU01j938&hl=en&sa=X&oi=book_result&resnum=2&ct=result#PPA25,M1.

gitimate to sterilize some individuals if medical knowledge and experience justify the prediction that their offspring will suffer from severe physical and psychological disorders."[1]

13. Jewish-Austrian neurologist Walther Birkmayer (1910-1996), a member of the National Socialist German Workers Party since 1932 and a member of the *Schutzstaffel* (SS) since 1936: "In all cases where we see a danger for our people and are compelled to demand sterilization, this in no way contradicts the healing tradition, for to be a doctor means not only to promote the health of the individual as much as possible, but also not to lose from sight the interests of the community. In the post-war period we have all clearly witnessed how the broad masses found only pleasure in the degenerate, the unjust, the filthy, and the fallen. History has oft experienced such signs of decay, which have been overcome only thanks to a renewal of the blood and an influx [of new blood] from without. To our people was granted to give birth to a Genius who instinctively recognized and demanded that only the hereditary health of the people [*das Volk*] offers salvation from this degeneration. And we as fanatical youth must exterminate all that is sickly, impure, and degenerate in our people, in order that it over the generations might be empowered to fulfill its historic destiny."[2] A year later Birkmayer's "non Aryan background" is exposed and he is expelled from the SS by Reichsführer Heinrich Himmler (1900-1945), and also fired from his lectureship, although he retains his position in the University of Vienna Racial-Biological Institute. He serves as chief physician at the brain-injury hospital in Vienna until the end of the year, and in 1960 pioneers the use of the drug Levodopa, which is still the only truly effective treatment for Parkinson's disease.[3] [4]

14. Physician Arie Kochinsky argues in the journal *Harefuah* that the findings of a census of the mentally ill in Palestine should serve primarily as "a basis for methods to improve the race."[5]

---

[1] "Psycho-higiena społecna a sterylizacya eugeniczna," Medizina Społecna 11, 1-2, 3; cited by Kamila Uzarczyk in Turda/Weindling, 2007, 294.
[2] Neugebauer, 1998, 131.
[3] Hilchey, 2008.
[4] Carlson/Riederer/Stern, 1997.
[5] Feldman, 2009.

## Jewish Eugenics under Attack

15. Jewish physician and eugenicist Arthur Kronfeld (1886-1941) and his wife take Soviet citizenship.[1]

16. Gynecologist and eugenicist Bernard Aschner (1883-1960) emigrates from Vienna to the United States after the German-Austrian *Anschluss* (exact year unknown, possibly 1939.)[2]

17. Jewish-Russian eugenicist Solomon Levit is executed.[3]

18. Attacked in the press, Jewish professor of law and co-editor of the *Russian Eugenics Journal* Pavel Isaakovich (Isaevich) Liublinsky (1882-1938) supposedly dies as the result of a "fall" from the platform of the Leningrad commuter train that he had taken for 25 years.[4]

19. Yakov Arkad'evich Yakovlev (né Èpshtein, 1896-1938) and Mikhail Aleksandrovich Chernov (1891-1938), Jewish-Russian supporters of Lysenko, perish in the purges.[5]

20. Magnus Hirschfeld (1868-1935) (posthumous publication, from exile): "If a serious effort is to be made to breed a race of Nietzschean supermen and superwomen, the Race Offices should be promptly transformed into Marriage Advisory Boards, guided by hygienic and eugenist principles widely different from those upon which the present crude attempts at racist selection are based."[6]

21. Felix Abraham (1901-1938), assistant to Magnus Hirschfeld and possibly the first surgeon to perform a sex-change operation (1931), commits suicide either in Switzerland or possibly Florence.[7] Karl Giese (1898-1938), Hirschfeld's assistant, who remained in Berlin until the closing of the Institute for Sexual Science in 1933 and then failed to obtain a visa for a "safe" country, commits suicide in Czechoslovakia when Hitler's army occupies the country.

---

[1] "Kronfeld, Arthur."
[2] *British Medical Journal*, July 2, 1960, 73.
[3] Pchelov, 2006.
[4] Baranovsky, 2005.
[5] Zhuravsky, 1993.
[6] M. Hirschfeld, 1938, *Racism*, London; Kratz, 1980.
[7] "Abraham, Felix."

## Jewish Resistance to Eugenics

22. A petition authored by Isai Izrailovich Prezent and signed by other Soviet biologists terms eugenics a "black-hundreds delirium... that forms the basis of the racial theories of fascism, bestial chauvinism, and vicious misanthropy."[1]

## 1939

### Context

1. During the period 1914-1939 the Jewish population of the globe is estimated to have risen from about 13,500,000 to 16,500,000. The rate of growth is smaller than in the preceding period because of the reduction in natural increase caused by the spread of birth control among the Jews in Europe and America.[2]

2. In a letter to philosopher John Dewey (1859-1952) eugenics opponent Franz Boas (1858-1942) advocates "the subordination of the state to the interests of the individual."

3. On September 16 the journal *Nature* publishes a joint statement issued by America's and Britain's most prominent biologists. Some of them are Nobel Prize laureates, and H. J. Muller and Arthur G. Steinberg (1912-2006) are Jewish.[3] The document is widely referred to as the "Eugenics Manifesto." The authors explicitly decry antagonism between races and theories according to which certain good or bad genes are the monopoly of certain peoples.[4]

4. In September Hitler issues a secret order initiating a national euthanasia program intended to free up as many as 800,000 hospital beds for expected war casualties. The action is frequently confused with eugenics, even though it was targeted at institutionalized persons, and German eugenicists vehemently attacked euthanasia proposals.[5]

---

[1] Pchelov, 2006.
[2] Schmelz/DellaPergola, 2007.
[3] Jenkins, 2007, 1011.
[4] http://whatwemaybe.org/
[5] Aktion "T4"/"Wilde Euthanasie" (1939-1945); Aussage des "T4"-Leiters Viktor Brack: "Nutzlose Esser" 1946); Aus: DOC-NO426, in GSTA, Rep. 335, Fall 1, Nr. 202, Bl. 11; quoted in Kaiser *et al.*, 1992, 250.

5. Pavel Postyshev (1887-1939), a Jewish supporter of Lysenko and a senior Soviet official, is executed.[1]

### Jewish Advocacy of Eugenics

6. April 27: Rabbi Louis I. Newman (1893-1972) of Congregation Rodeph Sholom, New York, delivers a lecture entitled "The Modern Jew Looks at Programs of Human Betterment" at a conference on "The Relation of Eugenics and Church," sponsored by the American Eugenics Society. Among the other 140 attendees are Rabbi Sidney T. Goldstein and Rabbi Stephen A. Wise. Altogether six rabbis attend, and general support is expressed for eugenics.[2]

7. Physician Morris Siegel publishes his two-volume *Population, Race, and Eugenics*, the first volume advocating positive eugenics, and the second devoted to negative eugenics.

8. Jewish physician William Moses Feldman writes an article for the British almanac *Medical Leaves*: "Ancient Jewish Eugenics":

   • "The Greeks and Romans applied to the human race the method of the animal breeder; that is to say, they eliminated any deformed or weakly newborn infant by killing it…. The ancient Hebrews, on the other hand, infused a humanitarian spirit into their system, and by tempering their eugenics with mercy, and combining judicious selective mating with intelligent antenatal and postnatal care, they succeeded in rearing a race, not indeed of supermen, but one which is probably the most virile that ever lived…."

   • "It is true that infanticide, as a crime punishable by death, applied only to full-term babies, and, in theory, the killing of a premature baby before it reached its thirtieth day of postnatal life, i.e., before one could be quite sure of its viability, was punished by a fine only. In practice, however, even premature babies were adequately protected, because every baby was considered to have been born at full term unless convincing evidence to the contrary was produced (Yebamoth, 37a)."

---

[1] Zhuravsky, 1993.

[2] http://www.dnalc.org/ddnalc/ben/index.html?id=1686; Bozeman, 2004, 425-426, 429.

- "In agreement with the Hippocratic view, the Rabbis were of the opinion that before the fortieth day there was no animation in the foetus, which was therefore considered merely as a "bladder of water" (Niddah, 30a; K'ritboth, 7b, Rashi). Induction of abortion, therefore, was not a criminal offense."

- "The saying 'like father like son' was the guiding principle of the rabbis (Erubin, 70b) in matters related to match-making."

- "'Two imbeciles should not marry.'" (Yebamoth 112b.)

- "Indeed, according to a famous Talmudic commentator, it is permissible for a woman to be sterilized if she is likely to bear children who are going to be tainted with physical or mental disease. (Luria, on Yebamoth, iv, 44).[1]

9.  Michael Berman and Dr. Helen Rosenau are indicated as members of the (British) Eugenics Society.[2]

### Jewish Eugenics under Attack

10. An article appears in *Pravda* entitled "Phony Scientists Should Find No Place in the Academy of Sciences" attacking the Jewish-Russian biologist and eugenics advocate Lev Solomonovich Berg (1876-1950), who had been nominated for membership in the Academy of Sciences, and Russian geneticist Nikolai Konstantinovich Koltsov (1872-1940): "One can easily become convinced as to the total identicalness of the eugenic views of Professor Koltsov and those of contemporary fascist scientists."[3] Koltsov dies of a heart attack the following year, and his wife commits suicide.[4]

11. Since 1933, thirty biologists have been dismissed from German universities as non-Aryans or because they are married to non-Aryans.[5]

12. The Gestapo arrests 183 professors at the University of Kraków, among them 23 life scientists.[6]

---

[1] Feldman, 1939.
[2] *Eugenics Review, Bulletin of the Eugenics Society, Biology and Society,* Eugenics Watch.
[3] Bakh/Keller/Koshtoiants/Komarov, 1939.
[4] Adams, 1990, 198
[5] Deichman, 1996, 15..
[6] Deichman, 1996, 47.

13. Jewish-German anthropologist and eugenicist Wilhelm Poll emigrates from Germany to Sweden and dies shortly thereafter. Ironically, the judophobe Otto Verschuer acknowledges in an article in *Science* that Poll was the first researcher in Germany to follow up on Galton's ideas on twins studies.[1]

14. Jewish eugenicist of Australian-British origin Norman Haire writing to British eugenicist Havelock Ellis (1859-1939): "I have been very busy with medical refugees. It is very depressing to see all our old friends and acquaintances, who took part in international congresses, in distress, and to hear every week that some more of them have committed suicide."[2]

## 1940

## Context

1. American eugenicist Lothrop Stoddard (1883-1950) after a visit to Germany: "Without attempting to appraise the highly controversial racial doctrine, it is fair to say that Nazi Germany's eugenic program is the most ambitious and far-reaching experiment in eugenics ever attempted by any nation."[3]

2. From a note to curator of scientific institutes, Vienna: "Until the revolution of 1938, the leadership and organization of the research institute was largely in Jewish hands. The Jews were removed after the revolution and in so far as funds of the biological research institute… were being used for the personal work of Jews, Aryans were paid with these funds."[4]

3. Harvard sociologist Carle C. Zimmerman: "Any normal 2,000 babies born tomorrow will consist of 1,060 males and 940 females. Of the 940 females about 700 will grow up and marry and somewhat more than 600 will have children. Each must bear between three and four children on the average to sustain a population, presupposing greater losses some years than normal times. This immediately rules out the two-child family ideal…. When a country has set sail

---

[1] No. 90(2326), July 28; cited in Braund/Sutton, 2008, 28-29.
[2] Crozier, 2001, 308.
[3] Kühl, 1994, 53.
[4] Deichman, 1996, 18.

on a course of action, it must carry through a population policy to support it or go to its doom."[1]

4. A proposal is presented in late 1940 on behalf of the Irgun Zvai Leumi (National Military Organization) in Palestine to two German diplomats in Lebanon, offering to actively take part in the war on Germany's side. The NMO describes itself as "closely related to the totalitarian movements of Europe in its ideology and structure":
   a. "Common interests could exist between the establishment of a new order in Europe in conformity with the German concept, and the true national aspirations of the Jewish people as they are embodied by the NMO.
   b. "Cooperation between the new Germany and a renewed folkish-national Hebraium would be possible and,
   c. "The establishment of the historic Jewish state on a national and totalitarian basis, bound by a treaty with the German Reich, would be in the interest of a maintained and strengthened future German position of power in the Near East."[2]

5. The average number of children born to foreign-born Jewish women aged 45 and over declines to 3.8 from 7.2 in 1910.[3]

6. *New York Times*: "The Eugenics Publishing Company announced yesterday that it was distributing its customary year-end bonus of an extra week's salary to each employee."[4]

7. The estate of Jewish-German-American philanthropist James Loeb withholds funding for eugenics research at the Kaiser Wilhelm Institute of Psychiatry in Munich, and German Director Ernst Rüdin, desperate for support, turns to the SS for funding[5]

---

[1] *New York Times*, 1940a.
[2] Brenner, 2002, 301-303. At the time Lebanon was run by Germany's Vichy ally. The document was deposited in the German embassy in Turkey, where it was found after the war. Brenner quotes it after David Yisraeli, *The Palestine Problem in German Politics: 1889-1945*, Bar Ilan University, Israel, 1974, 315-317.
[3] DellaPergola, 2005, 108; citing W. H. Grabill, C,V, Kiser, and P. K. Whelpton in *The Fertility of American Women*, J. Wiley, 1958.
[4] *New York Times*, 1940b, Dec. 12..
[5] Seidelman, 2001.

## Jewish Advocacy of Eugenics

8. Zionist Arthur Ruppin publishes (in Hebrew) *The Jews' Wars for Survival*. Historian Amos Morris-Reich comments: "Even in this book Ruppin continues to put his faith in the cornerstones of Nazi race theory."[1]

9. Morris Siegel's *Population, Race, and Eugenics* is reviewed by H. R. Crosland of the University of Oregon: "The reviewer is wholly sympathetic to the author and to his point of view. Various eugenic schemes should be introduced and should receive wide application to increase the population of able persons and to reduce the population of the unfit."[2]

10. Tufts College professor of medicine Hyman Morrison: "Anthropologically we are not a pure race, for assimilation, active and passive, has been going on at all times. It is just as certain, however, that the Jewish strain has always been dominant in our people.... The Jews, a 'chosen people' through natural selection.... At the beginning of the twentieth century the total Jewish population was 12,000,000 and at present it is estimated between sixteen and seventeen millions...."[3]

11. Jewish-German physician and eugenicist Adolf (Israel after 1938) Gottstein (1857-1941), who chose not to emigrate: "One of the chief tasks of racial hygiene is the preservation of genetic patrimony for posterity. As far as a continuation of our current reproductive relationships is concerned, as can be read in all solid texts on racial hygiene, the predictions are quite *unfavorable* [stress in original]. The line of reasoning runs as follows: Those components of a hearty people which are less favorable have for a considerable length of time been reproducing themselves with greater vigor than those who are at an upper level or those who are moving upward.... Moreover, social welfare expenses at the cost of those who are creative are excessive, so that many survive who formerly would have been wiped out by natural selection. The inevitable consequence must be a progressing physical and cultural degeneration."[4]

---

[1] Morris-Reich, 2006a, 13.
[2] Crosland, 1940.
[3] Morrison, 1940.
[4] Koppitz/Labisch, 1999, 236.

## 1941

### Context

1. Relations between German racial hygienists and American eugenicists break off totally when America enters World War II.

2. Dismissed from his professorship, the Jewish-Austrian zoologist Heinrich Joseph (1875-1941) commits suicide, together with his wife.[1]

3. Jewish-German eugenicist Arthur Kronfeld and his wife take poison in Moscow, possibly fearing an imminent German victory.[2]

4. Eugen Fischer (1874-1967), one of the founders of the Kaiser Wilhelm Institute for Anthropology, Human Genetics, and Eugenics, gives a talk in German-occupied Paris in which he notes remarkable Jewish achievements, but goes on to make a statement supposedly uncharacteristic of him: "Their ethical tendency and all the activities by Jewish Bolsheviks lay bare such a monstrous mentality that we are no longer able to speak of inferiority but a species different from our own."[3]

5. July 31 order of German National-Socialist military and political leader Hermann Göring (1893-1946): "To the Chief of the Security Police and the SD, SS *Gruppenführer* [Reinhard] Heydrich [1904-1942], Berlin: In completion of the task which was entrusted to you in the Edict dated January 24, 1939, of solving the Jewish question by means of emigration or evacuation in the most convenient way possible, given the present conditions, I herewith charge you with making all necessary preparations with regard to organizational, practical and financial aspects for an overall solution (*Gesamtlösung*) of the Jewish question in the German sphere of influence in Europe. ¶ Insofar as the competencies of other central organizations are affected, these are to be involved. ¶ I further charge you with submitting to me promptly an overall plan of the preliminary organizational, practical and financial measures for the imple-

---

[1] Deichman, 1996, 19.
[2] "Arhur Kronfeld."
[3] Eugen Fischer, "Le problem de la race et la legislation raciale allemande," *Cahiers de l'Institut allemand*, 106; cited in English translation by Weiss, 2006, 72-73.

mentation of the intended final solution (*Endlösung*) of the Jewish question."[1]

## Jewish Advocacy of Eugenics

6.  The *Universal Jewish Encyclopedia* contains a strongly pro-eugenics article by Rabbi Max Reichler (1885-1957): "The rabbis of old, like the eugenists of today, measured the success of a marriage by the number and quality of its offspring. In their judgment the main objects of marriage were the reproduction of the human race, and the augmentation of the favored stock (Tur Eben Haezar 25).... The attempt to limit the increase of undesirable progeny resulted in three kinds of prohibitions: 1) against the marriage of congenital defectives; 2) against the marriage of personal defectives; 3) against consanguineous marriages."[2]

7.  Gynecologists Frances I. Seymour (1900-1954) and her husband Alfred Koerner (b. 1897) claim that a survey of physicians showed 3,649 babies born thanks to artificial insemination.[3]

8.  Nine months before the Japanese attack on Pearl Harbor, New Jersey businessman Theodore Newton Kaufman (1910?-?) self-publishes *Germany Must Perish!*: "There remains then but one mode of ridding the world forces of Germanism – and that is to stem the source from which issue those war-lusted souls, by preventing the people of Germany from ever again reproducing their kind. This modern method, known to science as Eugenic Sterilization, is at once practical, humane and thorough. Sterilization has become a byword of science, as the best means of ridding the human race of its misfits: the degenerate, the insane, the hereditary criminal.... The population of Germany, excluding conquered and annexed territories, is about 70,000,000, almost equally divided between male and female. To achieve the purpose of German extinction it would be necessary to only sterilize some 48,000,000 – a figure which excludes, because of their limited power to procreate, males over 60 years of age, and females over 45. Concerning the males subject to sterilization the army groups, as organized units, would be the easiest and quickest to deal with. Taking 20,000 surgeons as an arbitrary number and on the assumption that each

---

[1] Arad *et al.*, 1999, 233.
[2] Reichler, 1941, 192.
[3] Daniels/Golden, 2004.

will perform a minimum of 25 operations daily, it would take no more than one month, at the maximum, to complete their sterilization. Naturally the more doctors available, and many more than the 20,000 we mention would be available considering all the nations to be drawn upon, the less time would be required. The balance of the male civilian population of Germany could be treated within three months. Inasmuch as sterilization of women needs somewhat more time, it may be computed that the entire female population of Germany could be sterilized within a period of three years or less. Complete sterilization of both sexes, and not only one, is to be considered necessary in view of the present German doctrine that so much as one drop of true German blood constitutes a German."

9.   At Goebbels' direction the German press plays up Kaufman's call for genocide. A front page article about the book in the Berlin daily *Der Angriff*, July 23, 1941, appears under headlines calling it a "Diabolical Plan for the Extermination of the German People" and a work of "Old Testament Hatred." Extracts also appeared, for example, in the nationally circulated weekly paper *Das Reich*, August 3, 1941.

## 1942

### Context

1.   The Eugenics Publishing Company leases the entire fifteenth floor of 308 West 35th St. in Manhattan, comprising 16,500 feet, doubling its previous space. This fact refutes the claim that the eugenics movement had withered away in the 1930s. The *New York Times* notice further states that the same brokers also leased space to the Jewish Social Service.[1]

2.   A small questionnaire study shows that Jews are significantly more favorably disposed toward the use of contraception than are Christians.[2]

### Jewish Advocacy of Eugenics

3.   Physician Arie Kochinsky (Tel Aviv) in *Harefuah*, journal of the Hebrew Medical Association: "Nowadays, we see as a vital duty the determination of the qualitative structure of the population. We

---

[1] *New York Times*, 1942.
[2] Sappenfeld, Burt R. 1942.

want to grasp the social and biological structure of society from the qualitative point of view."[1]

4.   A. J. Jaffe of the U.S. Bureau of the Census reviews Henry Pratt Fairchild's 1939 *People, the Quantity and Quality of Population*: "The author has prepared an inspired popular presentation of the present knowledge and unsolved problems in the field of population. This book is aimed at the intelligent layman and as such is useful to the college student as supplementary reading material rather than as a textbook. Included in the content are discussions of Malthus, the factors in population growth, migration, optimum population, and eugenics. The author makes use of the word 'larithmics' to refer to the quantitative aspects of population, as contrasted with 'eugenics,' referring to the qualitative aspects."[2]

## 1943
## Context

1.   Johns Hopkins University medical historian Henry Ernest Sigerest (1891-1957): "I think it would be a great mistake to identify eugenic sterilization solely with the Nazi ideology and to dismiss the problem simply because we dislike the present German regime and its methods[3].... The problem is serious and acute, and we shall be forced to pay attention to it sooner or later."

2.   Iowa Senator (D) Guy M. Gillette (1879-1973) on resistance by Zionist leaders to a resolution of twelve Senators to rescue German Jews, because it did not specify relocation to Palestine: "These people used every effort, every means at their disposal, to block the resolution.... [They] tried to defeat it by offering an amendment, insisting on an amendment to it that would raid the question, the controversial question of Zionism,... or anything that might stop and block the action that we were seeking." Gillette also quotes a comment by a colleague the day the Senate was to vote on the measure: "I wish these damned Jews would make up their minds what they want. I could not get inside the committee room without being buttonholed out here in the corridor by representatives who said that

---

[1] Second National Convention of Neurologists and Psychiatrists in Eretz Israel, April 17-18; quoted in Lewit, 2003.

[2] Jaffe, 1942.

[3] *Civilization and Disease*, 106-7; quoted in Ludmerer, 1972, 105, and also Kühl, 1994, 105.

the Jewish people of America did not want the passage of the resolution."[1]

3.    At the request of David Ben-Gurion, eugenicist Joseph Meir formulates the "One Million Plan" – the medical program for absorption of one million immigrants to *Eretz-Israel*.[2]

4.    From a memo of Heinrich Himmler (1900-1945), head of the *Schutzstaffel* (SS), to Martin Bormann (1900-1945), head of the Chancellery, on racial background checks to reveal quarter-Jews: "I consider such checks to be absolutely indispensable perhaps even for *Mischlinge* of more distant degrees. Let this remain *entre nous*, but here we should resort to a process analogous to the breeding of plants and animals. The offspring of such mixed families should be subject to racial control by independent institutions and sterilized in cases of racial inferiority so as to prevent them from passing on their heredity. Perhaps you could let me know your opinion on this question?"[3]

### Jewish Advocacy of Eugenics

5.    Obstetrician and vice president of the American Eugenics Society Alan Guttmacher (1898-1974) expresses doubt in the artificial insemination figures claimed by Seymour and Koerner.[4]

6.    Feb. 22: Seventy-nine-year-old Zionist leader, peace activist, philosophical universalist, and co-founder of the eugenics-oriented Bureau for Jewish Statistics Alfred Nossig is accused of being a German agent and is executed by the Jewish Fighting Organization (Żydowska Organizacia Bojowa) in the Warsaw Ghetto. Israeli scholar Shmuel Almog comments: "Even if the execution of Alfred Nossig by the Jewish Fighting Organization can be characterized as a miscarriage of justice, it did contain a measure of poetic justice of which the Underground members could not have been oblivious. The man's life style was diametrically opposed to theirs and signified in many ways a negation of their own values. Both his real biography, and the legends spun around him, could serve as a sort of

---

[1] Wyman, 1984, 200.
[2] Joseph Meir, 1943; cited in Dvora Hacohen, *From Fantasy to Reality: Ben-Gurion's Plan for Mass Immigration*, Jerusalem, 1995, 264-265 (in Hebrew).
[3] Translated from the French: Essner, 1995, 27.
[4] Daniels/Golden, 2004.

litmus paper by which to distinguish between his decaying world and theirs... There is some irony in the fact that Nossig was a man of peace, whose ideas were based on political realism and a sense of compromise.... It is significant that he lost his life, having tried to negotiate with the authorities and to reach an accommodation with the powers that be, under circumstances that no longer warranted such approach. His was a generation that still spoke the language of nineteenth-century Europe, a language of old-world manners and values. It is doubtful whether he was capable of grasping the extent of the changes that took place. It is further questionable whether or not he realized what the Germans now stood for. In a sense, the old man had been living on borrowed time long before he was put to death."[1]

7.    Dr. Julius Isaac is a member of the (British) Eugenics Society, 1943, 1957.[2]

# 1944
## Context

1.    Hans Przibram (1874-1944), a Jewish-Austrian biologist and mentor of Lamarckian eugenicist Paul Kammerer, fails to obtain an American visa from Amsterdam and is deported to Theresianstadt, where he dies of exhaustion. His wife commits suicide the following day.[3]

2.    Jewish-German eugenics proponent Hans Goslar is arrested in Amsterdam and taken first to the transit camp Westerbork and then to Bergen-Belsen.[4]

3.    Dec. 17: Jewish Soviet writer and journalist Ilya Ehrenburg (1891-1967) makes an historic claim – six million Jewish victims of the Holocaust – and also asserts universal German complicity: "In the seized countries and regions the Germans killed all the Jews – the old men and the babies. Ask any German prisoner of war why his countrymen destroyed six million innocent people, and he will re-

---

[1] Almog, 1983, 28-29.
[2] *Eugenics Review, Bulletin of the Eugenics Society, Biology and Society,* Eugenics Watch.
[3] Deichman, 1996, 18-19
[4] Kramer, 2003.

spond: 'They're Jews. They have black hair (or red). Their blood is different.'"[1]

4.  *Palestine Post*: British physician J. A. Fraser Roberts reports that "tests of 3,361 children of Gaza [?] support the long held theory by experts in eugenics that children conceived in the winter months are more intelligent than those conceived in the summer."[2]

5.  *Palestine Post*: "Nazi eugenics are outlined by the official S.S. paper *Das Schwarze Korps*. 'If our parents' generation (it writes) had not fallen prey to the madness of the Malthusian doctrine of population restriction we might have been able to mobilize half a million more men. Do you realize how many machine gun crews and divisions this would mean? Do you know what effect this would have on our armaments? Because our parents speculated on material security, a great material insecurity has come on us. How silly it is of some parents to say: what is the good of having a lot of children if new wars will only kill them again?'"[3]

### Jewish Advocacy of Eugenics

6.  Jewish-German philosopher Hannah Arendt (1906-1975): "Of greater importance for Eichmann [1906-1962] were the emissaries from Palestine, who would approach the Gestapo and the S.S. on their own initiative, without taking orders from either the German Zionists or the Jewish Agency for Palestine. They came in order to enlist help for the illegal immigration of Jews into British-ruled Palestine, and both the Gestapo and the S.S. were helpful.... According to the story by Jon and David Kimche... they were not interested in rescue operations. 'That was not their job.' They wanted to select 'suitable material'.... They were probably among the first Jews to talk openly about mutual interests and were certainly the first to be given permission 'to pick young Jewish pioneers' from among the Jews in the concentration camps. Of course they were unaware of the sinister implications of this deal, which still lay in the future; but they too somehow believed that if it was a question of selecting Jews for survival, the Jews should do the selecting themselves. It was this fundamental error in judgment that eventually led to a situation in which the non-selected majority of Jews inevitably found

---

[1] Ehrenburg, 1994.
[2] May 7, 1944.
[3] Jan. 19.

themselves confronted with two enemies – the Nazi authorities and the Jewish authorities."[1]

7. Psychiatrist Kurt Levinstein (b. 1877) delivers a lecture at a Tel Aviv conference: "A person in whom hereditary mental illness has not been prevented or cured presents just as great a danger to the race as a regular patient, at the height of his suffering ... Eugenic prophylaxis is the only prophylaxis and the ideal prophylaxis for hereditary illnesses."[2]

## 1945

### Context

1. *Palestine Post*: "I am sure that Professor Ernest Rudin never so much as killed a fly in his 74 years. I am also sure that he is one of the most evil men in Germany.... Today Prof. Rudin will tell you that he has always been purely the scientist who has never been interested in politics and never took part in it; that as Professor of race hygiene at Munich University, he had to expound various racial laws.... Yes, he is disarmingly candid, and if you press him on the question of ultimate responsibility for the murder of millions of Jews and Slavs and other non-Nordics he will tell you again and again he had nothing to do with it, that he was always against it, and that the 'Nazis shamefully abused my name and my ideas in committing such atrocities.'"[3]

2. A pupil of Franz Boas, the anthropologist Margaret Mead (1901-1978) in a 1962 article proposes a "limitation on freedom" of research and, recalling specifically the period 1933-1945, calls – reluctantly – for self-censorship: "It is, perhaps, in the realm of genetics, constitutional differences, and learned differences in national character that the greatest confusion has arisen and that anthropologists have shown the least capacity for scientific detachment. Any suggestion that one group of individuals may be innately or experientially superior to any other is likely to meet with intense resistance. So, also, are the mildest eugenic proposals, such as the suggestion that if artificial insemination is to be used, care should be

---

[1] Arendt, 1963, 56; citing Jon and David Kimche, *The Secret Roads: The "Illegal" Migration of a People 1938-48*, London, 1954.
[2] Feldman, 2009.
[3] Bernstein, 1945, 5.

taken to choose donors from what appears to be superior and healthy stock. The resistance of anthropologists to eugenic proposals is partly reactive to the suggestion by biologists that it is possible to breed for such culturally determined types of behavior as cooperativeness, for example. Some of us felt during the Hitler period that it was important, not to highlight studies of innate traits, but to focus instead on the extent to which all normal human beings could learn any human culture. A somewhat similar ethical issue is raised by all questions about the popularization of scientific research.... We have to consider the effects of our materials if they are presented to a lay audience, and to discriminate between those materials that should be published in technical journals and those that may appropriately be presented for a wider audience."[1]

3.  Jewish eugenics proponent Hans Goslar dies from hunger and illness in the Bergen-Belsen concentration camp.[2]

4.  Jewish-German eugenicist Arthur Czellitzer perishes in the Sobibor concentration camp (date uncertain, possibly earlier).[3]

5.  Columbia University Press issues Ashley Montagu's (né Israel Ehrenberg, 1905-1999) *Man's Most Dangerous Myth: The Fallacy of Race*. The book signals a radical shift of Jewish public ideology in the aftermath of World War II, reversing former racialist orientation.

## 1945-1947

1.  In Great Britain the Jewish fertility rate is 11.6 per 1,000, compared to 16.8 for the total population.[4]

---

[1] Mead, 1962, 3, 10.
[2] Kramer, 2003.
[3] Kratz, 1980, quoting Andreas Pretzel.
[4] Hannah Neustatter, "Demographic and Other Statsical Aspects of Anglo-Jewry," in Maurice Freedman (ed.), *A Minority in Britain*, 1955, 82; cited in Goldscheider, 1967, 200.

## 1946

### Context

1. Fred Blair publishes *The Ashes of Six Million Jews*, issued by The People's Bookshop in Milwaukee, repeating the figure asserted by Soviet writer Ilya Ehrenburg in December 1944.

2. The Yiddish Scientific Institute publishes Max Weinreich's *Hitler's Professors: The Part of Scholarship in Germany's Crimes against the Jewish* People, New York.

3. Large-scale Jewish emigration to Palestine begins.

4. Jean-Paul Sartre: "In effect, the Jew is to another Jew the only man with whom he can say 'we.'"[1]

### Jewish Advocacy of Eugenics

5. Dr. Moses Jung, a son of the Chief Minister of the London Federation of Synagogues, offers a comprehensive course in "Family Living and Marriage" at the Young Israel Institute of Jewish Studies in New York. Among its topics are eugenics and family hygiene.[2]

6. Saul Rosenzweig of Western State Psychiatric Institute and Clinic in Pittsburgh, Pennsylvania: "As a psychodiagnostic art, clinical psychology derives, on the one hand, from the fields of evolutionary eugenics (Galton) and educational measurement (Binet), and, on the other, from those influences, including psychoanalysis and psychosomatic medicine, which may be summarized under the term *psychodynamics*.... As clinical psychology matures in a systematic psychodynamic setting and ceases to be a group of loosely organized techniques applied in mechanical fashion, the psychologist may expect to make an increasingly greater contribution to the study of the mental patient and achieve a greater degree of acceptance by his colleagues."[3]

7. Jewish-Russian-American physician Herman Harold Rubin (b. 1891), member of the American Eugenics Society, issues a new edi-

---

[1] Sartre, 1948, 101.
[2] *Jewish Chronicle*, Aug. 20, 6.
[3] Rosenzweig, 1946, 94, 100.

tion of his *Eugenics and Sex Harmony*, the previous edition having gone through 12 printings since 1933:

- "From dogs to kings – from mice to elephants – blood always tells. People of fine, clean blood, living in hygienic surroundings, will rear fine, healthy, properly nourished children. People of degenerated or deteriorated blood, living in unwholesome, unsanitary surroundings, will produce scrub children, degenerated children.

- "Will the Negro Disappear?... The quadroon is a composite, made by mixing together Caucasian and mulatto blood. In Negro social life the quadroon is considered better than the mulatto. Socially superior to the quadroon is the octoroon, the offspring of a Caucasian and a quadroon.... The offspring of the Caucasian and the octoroon will, perhaps, result in an apparently Nordic type.

- "Recognizing Polygamy Officially... If the white race is to replenish its horrible wastage, or if the women of Europe are to be accorded the right to exercise functions implanted in them by a fecund Nature, may it not be that some modification of the present monogamic relationship be established?... For the very factors that have liberalized centuries' [sic] old thought may sweep away barriers erected by church and social code and declare for a freedom of conduct – in channels eminently sane and scientifically correct – never before known on this earth. All of which will contribute to general happiness, greater courtesy, and a tolerance still sadly needed in most of our still bigoted and stupidly reactionary communities.

- "There is a method, developed some years ago, by Dr. Harry C. Sharpe, and used by him in the Indiana Reformatory, that offers an ideal solution to the problem of propogation [sic] by the unfit. It merely means that the man or woman, by an exceedingly simple operation in the case of the man, and by a relatively simple operation in the case of the woman, is rendered sterile.... If these principles were made universal, and if all who are defective, imbecile, or even of the low-grade moron type, were prevented by this means from propagating their species, we could, within a decade, raise the physical, moral, and intellectual level of the race in a most important way.

- "That we may have smoothed the road just a little for those who are to come, and helped make the world better for our having lived; that we have stilled a child's cry, dried a woman's tears, lifted the burden of sorrow from some stricken soul, and backed the faith in himself of one who fought unselfishly for the right – these are *real* achievements." [1]

## 1947

## Context

1. Stalin writes in a document, later declassified, that he considers "Michurinist" biology (Lamarckism) to be the only scientifically legitimate position.[2]

2. The term "genetic counseling" is coined.

3. The Central Conference of American Rabbis (Reform Movement) adopts a proposal made by the Committee on Mixed Marriage and Intermarriage: "With regard to infants, the declaration of the parents to raise them as Jews shall be deemed sufficient for conversion. This could apply, for example, to adopted children…. If the parents therefore will make a declaration to the rabbi that it is their intention to raise the child as a Jew, the child may, for the sake of impressive formality, be recorded in the Cradle-Roll of the religious school and thus be considered converted."[3]

4. British geneticist Eliot Slater (1904-1983, married to the sister of Boris Pasternak):

   - "It is sometimes debated whether it is, or is not, desirable for intermarriage between various races to occur. It is, of course true that much further knowledge is required on the results of mixed breeding before any final judgment can be made. Nevertheless, if racial intermarriage never occurred at all, the races of mankind would inevitably evolve in the course of time into separate species with interspecific sterility, and the foundations be laid of a more disastrous disunity of mankind even that which obtains today.

---

[1] Rubin, 1946, 24, 139, 156-157, 281-282, 546.*
[2] Babkov, 1998.
[3] Committee on Patrilineal Descent, 1947.

- " …it seems likely that the frequency of marriage with non-Jews has increased among [British] Jews in recent years, and is now of the order of one in eight marriages. Such a rate of intermarriage would, if maintained for a number of generations, assimilate the Jewish and non-Jewish populations. As this point was approached, the differences between Jews and non-Jews which, it is suggested, provide the potentiality for anti-Semitism would disappear."[1]

5.  Austria is the first country to criminalize "Holocaust revisionism."

- Other countries to follow suit are Belgium (1947, 1992), Spain (1971, 1995), Germany (1985, 1992, 2002, 2005), Israel (1986), Switzerland (1995), Portugal (1997?), Poland (1998), France (1990), Liechtenstein (2000), the Czech Republic (2001), Romania (2002, 2005), and The European Union (2007).[2]

- Among those eventually imprisoned are the British historian David Irving (b. 1938, arrested while driving through Austria in 2006) and German-Canadian publisher Ernst Zündel (b. 1939, arrested in the United States in 2003, deported to Canada and imprisoned in that country, redeported to Germany, where he was sentenced to five years in 2007).

- A *Wikipedia* article lists the following persons as "Notable Holocaust Deniers": "Mahmoud Ahmadinejad, Jean-Marie Le Pen, Mahmoud Abbas (Abbas' Moscow PHd [sic] involved denial, he has since partially retracted his position, Abdel Aziz al-Rantissi, Mohammed Mahdi Akef, Harry Elmer Barnes, Arthur R. Butz, Wendy Campbell, Thies Christophersen, Doug Collins (journalist), Günter Deckert, Léon Degrelle, David Duke, François Duprat, Robert Faurisson, Bobby Fischer, Roger Garaudy, Hutton Gibson, Jürgen Graf, Nick Griffin, Richard E. Harwood, Michael Hoffman II, Gerd Honsik, David Irving, James Keegstra, Fred A. Leuchter, Norman Lowell, Carlo Mattogno, Carl O. Nordling, Roeland Raes, Siegfried Verbeke, Dariusz Ratajczak, Ahmed Rami, Paul Rassinier, Otto-Ernst Remer, Michele Renouf, Manfred Roeder, Germar Rudolf, Bernhard Schaub, Israel Shamir, Gerald L. K. Smith, Wilhelm

---

[1] Slater, 1947, 19, 21.
[2] http://en.wikipedia.org/wiki/Laws_against_Holocaust_denial, accessed March 29, 2009.

Stäglich, Fredrick Töben, John Tyndall, Richard Verrall, Udo Walendy, Richard Williamson, Ernst Zündel, R. J. Rushdoony, Bela Ewald Althans (German-language article)." Separate *Wikipedia* articles are devoted to each of them.[1]

## 1948

### Context

1. The *American Jewish Year Book* revises its 1948 estimate of the world Jewish population from 11,373,350 to 11,303,350. The 1938-1939 estimate had been 15,300,000, making for a drop of roughly four million persons over that period.[2]

2. Jewish families seem to be relatively unaffected by the "baby boom."[3]

3. In a limited survey of parents of Jewish college students, college-educated Jews are found to have smaller families than do those with only a grammar-school education.[4]

4. The Universal Declaration of Human Rights is introduced. "Men and women of full age, without limitation due to race, nationality or religion, have a right to marry and found a family."

5. The State of Israel is created, setting the stage for a so-called "ingathering" of Jews.

6. Geneticist Iosif Abramovich Rapoport (1912-1990) defends the theory of genetic mutation at an August conference of the Soviet All-Union of Agricultural Sciences (VASKhNIL) and is attacked by persons accusing him of being a disciple of the American geneticist Thomas Hunt Morgan (1866-1945), the German evolutionary theor-

---

[1] http://en.wikipedia.org/wiki/Holocaust_revisionism#Notable_Holocaust_deniers, accessed March 29, 2009.

[2] 1938-1939:
http://www.ajcarchives.org/AJC_DATA/Files/1938_1939_7_Statistics.pdf; 1948: http://www.ajcarchives.org/AJC_DATA/Files/1948_1949_18_Statistics.pdf; 1950: http://www.ajcarchives.org/AJC_DATA/Files/1950_7_WJP.pdf.

[3] Liebman Hersch, "Jewish Population Trends in Europe," *Jewish People: Past and Present*, II, 11, Table 10, cited in Goldscheider, 1967, 200.

[4] Myer Greenburg, "The Reproductive Rate of the Families of Jewish Students at the University of Maryland," *Jewish Social Studies*, X, July, 230; cited in Goldscheider, 1967, 203.

ist August Weismann (1834-1914), and the Austrian pea experimen-
ter (the 'father of genetics') Gregor Mendel (1822-1884), and not
following the precepts of the Russian selectionist Ivan Vladimiro-
vich Michurin (1855-1935):

- Arno Arutyunovich Babadzhanian (1921-1983), Director of the
  Institute of Genetics of the Armenian Academy of Sciences:
  "Comrades!... Dr. Rapoport says: 'Soviet geneticists are not an-
  ti-Darwinists.' What do our Morganists intend in making such a
  declaration?... Who if not Morgan and the Morganists considers
  Darwinism to be a system for speculating on questions of evolution?... It is impossible to conjure up a greater anti-Darwinist
  than Weismann?... And aren't Rapoport's words on mutations
  and modifications not Weismannism?..."

- Rapoport: "But there are useful mutations, even many of them.
  Why do you close both eyes to them?"

- Babadzhanian: "First of all, these are useful mutations in a useless object." [Applause]... The Mendelists are not just enemies
  of proven successes, but also are potential enemies of future
  successes." [Applause]...

- Ivan Evdokimovich Glushchenko, also from the Institute of
  Genetics of the Armenian Academy of Sciences: "The Morganists maintain that the so-called chromosome theory has supposedly provided a basis for understanding heredity... but in reality they are merely attempting to conceal the true essence of
  Morganist views.... Laying a foundation for racism and eugenics – this is what captures the attention of modern Morganic genetics.... This is whom and what Morganic-Mendelian genetics
  serves...."

- V. A. Shaumian, specialist on cattle: "There is no such thing as
  non-political science. That has been proven long ago. Therefore
  all Comrade Rapoport's urgings about the coexistence of Michurinist and reactionary biology are in vain...."

- K. Iu. Kostriukov of the Kiev Medical Institute: "Comrade Rapoport has spoken as a true Morganist.... He has fallen captive
  to a hostile theory... For him the gene is dressed up in new, sty-
  lish clothing, biochemical clothing... Be honest, Comrade Rapoport!... Despite your claims that the gene is a physical particle, Comrade Rapoport, the gene is a pure fiction. The elec-

tronic microscope won't save you. You can peer at whatever tiny particles you like in an electron microscope, but they will be only pieces of a chromosome. You won't see a gene, because there is no such thing... Gene science is a false theory that is holding back the development of science."

• Plant geneticist Nikolai Vasilievich Turbin (1912-1998): "The time has come to put an end to reactionary Morganism's unrestrained propaganda in the ranks of biologists and agrobiologists; we have to create conditions essential for the development and spread of Michurinist genetics and Soviet creative Darwinism. This is the demand of our Soviet life, which is so severe and merciless with regard to the scrawny miscarriages of metaphysical thought...." [Extended applause]

• Rapoport refuses to recant and is forced off the stage and subsequently fired from his position.[1]

## Jewish Advocacy of Eugenics

7.  Jewish-French physician and eugenicist Isidore Simon (d. 1985) founds *Revue d'histoire de la médecine hébraïque.*[2]

8.  Jewish eugenicist William Nussbaum presents a paper "Anthropological Studies on German Jews (1933-1934)" at a Brussels conference, maintaining that Jews have mixed into the German population and cannot be considered a separate population.[3]

9.  The *Palestine Post* enthusiastically reviews *Man in the Modern World*, by biologist Julian Huxley (1887-1975): "For readers unacquainted with the writings of this famous scientist and head of UNESCO, *Man in the Modern World* will be a fascinating introduction. Huxley's style has that vividness and lucidity which brings even the most difficult scientific problems within the layman's understanding without becoming superficial. The selection illustrates all the facets of Huxley's brilliant mind. Their themes range from those drawn from his special domain – biology – like 'Eugenics and Society' and 'Race in Europe' – to treaties on 'Philosophy of a World at War'.... It seems noteworthy that the author has not eliminated from the

---

[1] Vystuplenie na sessii VASKhNIL, 1948, Aug. 2, http://www.rapoport-genetika.ru/course/institute/?id=68, accessed Dec. 29, 2007.
[2] *Bulletin of the History of Medicine* 78.3 (2004) 702-703.
[3] Simonson, 2006.

book some criticism of the Soviet attitude towards modern eugenics, though he has moved further in the direction of the Communist attitude to science since he first published the essay in question."[1]

10. Some members, fellows, and officers of the (British) Eugenics Society (renamed the Galton Institute in 1989): Dr. Philip M. Bloom (1948, 1957, 1977) (possibly Dr. Philip M. Bloomfield; British social reformer and feminist Eva Marian Hubback (née Spielman, 1886-1949); Dr. M. Schachter.[2]

### Jewish Rejection of Eugenics

11. A loyal follower of Franz Boas, anthropologist Melville Jean Herskovits (1895-1963) advocates cultural relativism in *Man and His Works* (later revised as *Cultural Anthropology*).

### Jewish Eugenicists under Attack

12. In Russia Jewish geneticist and eugenicist Vladimir Pavlovich Èfroimson (1908-1989) is arrested and sentenced to a forced-labor camp.[3]

## 1949

### Context

1. Canadian data indicate an urban Jewish fertility rate lower than than of the non-Jewish fertility rate. [4]

2. Russian Soviet biologist Ivan Evdokimovich Glushchenko (1907-1987) in an article in *Pravda* ridicules American geneticists in general and Jewish-American eugenicist Herman Muller specifically for exaggerating problems of "so-called overpopulation." Glushchenko describes American geneticists as agents of capitalism fixated on eugenics and sterilization, which is "a weapon in the hands of American reactionaries intent on pursuing political purposes."[5]

---

[1] Signed C.Z.K., April 16, 1948.
[2] *Eugenics Review, Bulletin of the Eugenics Society, Biology and Society,* Eugenics Watch.
[3] Vergasov. Undated.
[4] Nathan Goldberg, "The Jewish Population in Canada," *Jewish People: Past and Present*, II, 35-39; cited in Goldscheider, 1967, 200.
[5] Salisbury, 1949.

3. Mapai member of the Knesset Eliyahu Carmeli: "I'm not willing to accept a single Arab. I want the State of Israel to be entirely Jewish, the descendants of Abraham, Isaac and Jacob."[1]

## Jewish Advocacy of Eugenics

4. Jewish-French psychiatrist Isidore Simon: "As early as the Biblical and Talmudic period Jewish scholars utilized notions of heredity and eugenics, not solely in the interests of the individual, but also of the entire nation.... The Bible contains genealogical tables so minutely drawn up that 'they would gladden the heart of the most ardent modern eugenicist'"[2][3]

5. Jewish-German-American geneticist and eugenicist Curt Stern: "Eugenic thinking has always emphasized the well-being of mankind, even though much eugenic counseling was based on inadequate knowledge and has been harmful. In the future more knowledge will be gathered and will aid wise planning. Then genetic and eugenic counseling will become the foundation of human genetic engineering. Although eugenic problems are not as urgent as the pessimists believed, their ultimate importance can hardly be overestimated."[4] "To state that reproductive selection against severe physical and mental abnormalities will reduce the number of affected from one generation to the next by only a few percent does not alter the fact that these few percent may mean tens of thousands of unfortunate individuals who, if never born, will be saved untold sorrow."[5]

6. Israeli Minister of Finance Eliezer Kaplan (1891-1952) argues for "regulation of immigration in qualitative and quantitative terms." [6]

7. Nathan Isaacs (Officer of the order of the British Empire) and Martin Zander are indicated as members of the (British) Eugenic Society.[7]

---

[1] Segev/Weinstein, 1998, 47.

[2] Feldman, 1939, 35

[3] Simon, 1949, 54; reverse translation of Feldman quote.

[4] Stern, 1949a, 208.

[5] Stern, 1949b, 538.

[6] Shvarts *et al.*, 2005, 13.

[7] *Eugenics Review, Bulletin of the Eugenics Society, Biology and Society,* Eugenics Watch.

## 1950s-Early 1960s

1.  Nurit Kirsh of the Cohn Institute for the History and Philosophy of Science of Tel Aviv University about the effects of Zionist ideology on research into human population genetics in Israel and the manipulation of differences during this period: "In general, the Israeli researchers seemed careful not to formulate conclusions that would contradict the accepted Zionist narrative and indeed endeavored to reach conclusions that supported it.... As a result, they tended to provide intrapopulation explanations of differences between 'communities' and avoided speculating about the prevalence of intermarriage.... The Israeli researchers' publications never mention the eugenic and racial aspects of their research; nevertheless, they tried to use different terms and different criteria from those of German bioracial science and eugenics."[1]

## 1950

## Context

1.  Prominent Jewish-Russian cytologist Vladimir Yakovlevich Aleksandrov (1906-1995): "If one takes into account the enormous loss of professionals who had been fired from their jobs and also the fact that those who managed to keeps their positions were psychologically twisted by fear and greed, by 1950-1951 our [Soviet] biology seemed hopelessly doomed to total degradation."[2]

2.  Pablo V. Gejman and Ann Weilbaecher in the *Israeli Journal of Psychiatry and Related Sciences* explain the excessive reliance in the 1950s and 1960s on psychological counseling over medication in cases of schizophrenia as a negative reaction to the eugenics movement.[3]

3.  Sir Cyril Lodowic Burt (1883-1971), whose surname appears in eleven different databases of Jewish names,[4] publishes the first results of his studies of identical twins raised separately.

### Jewish Advocacy of Eugenics

---

[1] Kirsh, 2003.
[2] Aleksandrov, 1993.
[3] Gejman/Weilbaecher, 2002, 229.
[4] http://www.avotaynu.com/csi/csi-result.html?page=next.

4.  Eugenicist Chaim Sheba (Shiber) (1908-1971), having fled Austria
    in 1933, replaces eugenicist Joseph Meir as Israel's Minister of
    Health. Like Meir, Sheba argues for genetic vigilance in questions
    of immigration so as to protect the gene pool. Sheba is said to exer-
    cise an "unprecedented influence" on David Ben-Gurion.[1]

5.  Yitzak Rafael (1914-1999), Director of the Immigration Department
    of the Jewish Agency, issues a ruling requiring a pre-immigration
    medical examination, but the directive is largely ignored, and is
    made obsolete within months by the Law of Return, which is passed
    unanimously by the Knesset.[2]

6.  Jewish-American gynecologists Bernard Aschner (1883-1960) and
    Frances I. Seymour (1901-1954) lecture at the American Museum
    of Natural History's Division for Education in Heredity and Eugen-
    ics.[3]

7.  Jewish-American eugenicist Herman J. Muller delivers the presi-
    dential address to the American Society of Human Genetics.[4]

8.  Speaking to the Anglo-Jewish Association in London, geneticist and
    eugenicist Redcliffe Salaman describes Jews as an endogamous
    family rather than as a race.[5]

### Resistance to Eugenics

9.  Prolific writer Ashley Montagu (né Israel Ehrenberg) authors the
    United Nations Statement on Race, asserting that "scientific evi-
    dence indicates that the range of mental capacities in all ethnic
    groups is much the same" and rejecting predictions of degeneration.
    As a young man, Ehrenberg studied anthropology and psychology
    at the University of London with Charles Spearman and Karl Pear-
    son, leading figures in the eugenics movement.

---

[1] Shvarts *et al.*, 2005, 24-25.
[2] Shvarts *et al.*, 2005, 23.
[3] *New York Times,* 1950.
[4] Kallmann, 1952, 238.
[5] Endelman, 2004, 82.

## 1951

### Context

1.  The average size of Jewish families in Canada decreases from 3.6 in 1941 to 3.2, as opposed to a drop of 3.9 to 3.7 for non-Jewish births during the same period.[1]

2.  Despite a brief Jewish 'baby boom' in the early postwar years, from the 1950s on a renewed decline in Jewish birth rates ensues in Europe, America, and other Western countries. Though there has been persistent natural population growth in Israel, changes in the overall size of the Jewish world population have been, in the words of demographers Usiel Oscar Schmelz and Sergio DellaPergola, "rather limited."[2]

### Jewish Advocacy of Eugenics

3.  Ben-Gurion does not oppose internal deliberations in the Israeli government and Knesset on the subject of medical selection and even remarks that Israel faces "an immigration... different not only quantitatively but also qualitatively from previous immigration."[3]

4.  From an article by German-American eugenicist William (Wilhelm) Nussbaum: "A sociological type has been welded together by historical fate and monotheism. Always embedded in a cultural milieu between Acadian-Sumerian, Babylonian, and Egyptian cultures, this type had already come together by the time the Babylonian exodus and the return to Palestine took place. Even the clash with Jewish culture proved incapable of diluting the sociological Jewish type."

5.  From a response to Nussbaum's article by Saul Bernstein, editor of *Orthodox Jewish Life*: "An excellent article [reflecting] a great deal of study and learning... It seems to me... that even though it has been patently demonstrated that there is no such thing as a Jewish 'race' in the true anthropological sense, our truly unique history must have tended to produce a distinctive Jewish type or types. If that is so, then I think it would be worthwhile to indicate more or

---

[1] Louis Rosenberg, "The Demography of the Jewish Community in Canada," *The Jewish Journal of Sociology, I, Dec., 1959, 217-233*; cited in Goldscheider, 1967, 199.

[2] Schmelz/DellaPergola, 2007.

[3] Shvarts *et al.*, 2005, 16.

less precisely in what way our national experience may have affected the Jewish stock so as to differentiate it from other human families – always bearing in mind that such distinctions do not mark a fundamental racial differentiation from the rest of humanity."[1]

6.  The Israeli Law of Return granting all Jews the right to immigrate is restricted on medical grounds, partly of a eugenic nature, narrowing the gateway for new immigrants: declaring oneself a Jew no longer suffices; applicants must now prove their Jewishness by birth and genealogy, and they can be rejected for medical cause.[2][3]

## 1952

### Jewish Advocacy of Eugenics

1.  When Ben-Gurion offers a one-time grant of 100 lirots to any Israeli mother of ten or more, physicians protest on eugenic grounds, among them the highly respected and influential Joseph Meir: "We have no interest in the tenth or even the seventh child of the poor Mizrahi families. . . . We must pray for the second child of the families of the intelligentsia."[4]

2.  German-Israeli geneticist Elisabeth Wexler Goldschmidt (1912-1970) introduces two courses on eugenics at the Hebrew University. Later the courses are retained, but their name was changed from 'eugenics' to 'genetics.'[5]

3.  Delivering the presidential address to the American Society of Human Genetics, whose members he refers to as "a band of *friendship* emboldened by *singleness of purpose,*" Jewish eugenicist Franz Josef Kallmann (1897-1965) speaks of "the traditional fact that competent physicians have been unified by *the idea of a common cause* (emphasis in original).... If a broad moral platform is adopted for the discharge of the social and professional responsibilities of our discipline, it will not only be helpful in rallying a multitude of specialized groups around a common cause, but it will serve to vitalize the general appeal of our cause. The prospect of taking an active

---

[1] Nussbaum, 1951.

[2] Shvarts *et al.*, 2005, 23.

[3] Shhvarts *et al.*, 2005, 27.

[4] Joseph Meir, "Increasing Birth Rates or Increasing Fertility Rates?" *Eytanim*, 3-4 (1952), 76-77; quoted in Stoler-Liss, 2003, 114.

[5] Kirsh, 2004, 80.

part in such a promotional scheme may not seem attractive to a scientist, who is glued to a microscope or is brooding over the structural essence of life. In view of the growing complexity of human societies, however, it is apparent that our chance of inducing the employment of sound biological principles in future population and public health policies will depend on the efficacy of our joint promotional plans and endeavors."[1] Kallmann, who is considered the founder of psychiatric genetics, is a member of the American Eugenics Society.[2]

## Mixed Feelings

4. Medical editor Arnold Sorsby in *Jewish Chronicle*: "Dr. [Carlos Paton] Blacker, who has been Secretary of the Eugenic Society for the past 20 years, has done perhaps more than anyone else to turn the Eugenic Society that Galton created into a civilized institution. In his able and remarkable study [*Eugenics: Galton and After*] he does well to stress the scientific and humanitarian aspects of Galton's activities, for it is by these that he will be remembered. What is more questionable is whether his advocacy of one tiny remnant of Galton's programme is indeed necessary. This, however, does not detract from the value of this book as an exceptionally clear exposition of present-day Eugenics, and the fundamental science of genetics and population studies on which it is based. Bearing in mind the unsoundess of Galton's teaching, the Jewish reader will find little comfort in Galton's enthusiasm for some aspects of Jewish life. It is merely a variant of – 'some of my best friends are Jews' – and just as profound."[3]

5. Gynecologist and former Director of the American Eugenics Society Alan F. Guttmacher (1898-1974): "At Johns Hopkins, where I served for many years, we put into effect the Para Eight rule: any woman could be sterilized who had had eight births. When I went to Mt. Sinai in New York in 1952 and took over the obstetric-gynecology service, I found that sterilization for the private patient was relatively easy to acquire, but sterilization for the ward patient was unobtainable. In order to equalize sterilization on the two services I put into effect a rule which I am now not very proud of, but

---

[1] Kallmann, 1952, 243.
[2] *Eugenical News, Eugenics Quarterly, Social Biology*, Web site "Eugenics Watch."
[3] Sorsby, 1952.

in 1952 believe it or not, it was a radical rule. Any woman with her sixth living child irrespective of age could be sterilized, also any woman age thirty to thirty-five with five living children and any woman age thirty or more with her fourth living child. This became known as the law from Mr. Sinai. For far too long it was rather slavishly followed in American medicine. I regret to report that several hospitals still [1973] follow the law from Mt. Sinai. To be sure it has been modified in many institutions."[1]

## 1953

### Jewish Advocacy of Eugenics

1. Frida Laski, widow of Jewish-British eugenicist Harold Laski, is recorded as a member of the (British) Eugenics Society.[2]

### The Continuing Soviet Assault on Genetics and Eugenics

2. The *Great Soviet Encyclopedia* publishes a strongly pro-nurture, anti-nature article approved by censors for publication during the so-called "anti-Cosmopolitan campaign" (directed against the Jews).[3]

3. At an ideology-driven congress of Soviet scientists in which the lead-off presentation is delivered by Lamarckian Trofim Lysenko (1898-1976), Academician Ivan Evdokimovich Glushchenko accuses geneticist Iosif Abramovich Rapoport of promoting eugenics and calls "bourgeois genetics" a "false science."[4] Rapoport writes a letter to Nikita Khrushchev asking to meet to discuss the situation but receives no response.

---

[1] Robitscher, 1973, 55. Guttmacher's article is entitled "General Remarks on Medical Aspects of Male and Female Sterilization" (52-60).
[2] *Eugenics Review, Bulletin of the Eugenics Society, Biology and Society,* Eugenics Watch.
[3] Vvedenskij, 1953.
[4] Vsesojuznaja Akademija Sel'skokhozjaistvennykh Nauk, 1953.

## 1954

### Context

1. Bloom Syndrome is identified by the New York dermatologist David Bloom. Carrier testing successfully identifies 95-97% of Jewish carriers.[1]

2. Even though the Chief Rabbinate of Israel rules that "the sect of the Bene Israel in India is of the seed of the House of Israel without any doubt," several Israeli rabbis refuse to marry Bene Israel to other Jews after Baghdadi Jews who had resettled in India denounce intermarriage with those whom they consider to belong to an inferior caste, claiming this will lead to *mamzerut* (illegitimacy). After a series of sit-down strikes and hunger strikes, the Jewish Agency deports 337 individuals, although some are later permitted to return.[2]

### Jewish Advocacy of Eugenics

3. Contradicting the claim that the eugenics movement died out in the 1930s, Director of the Social Hygiene Division of the N.Y. Tuberculosis and Social Hygiene Division Jacob A. Goldberg notes a strengthening of eugenics instruction on a national level and calls for even greater effort in this direction: "Ten years ago it was still necessary to attempt to convince many educators that eugenics and family life education were essential obligations of the educational system. While some progress has been made on the college level in the interim, much still remains to be done insofar as implementation in the secondary schools is concerned.... There is a wide area of service open to those interested in eugenics. Perhaps one field that warrants immediate cultivation is broadening the base of those who should participate in the movement, if such it may be called. If we are to advance family life education, eugenics, and the technical aspects of heredity coupled with environmental factors, it may be necessary to go outside of the laboratory and the experimental animal cages; in part, at least, lay aside our computing machines, and move more effectively into the community."[3]

---

[1] James L. German III, "Jewish Genetic Diseases," *Jewish Genetic Disorders*, Sept., 2005, 68.
[2] Bar-Giora, 2007, 338.
[3] Goldberg, 1954, 39, 46.

4.  Biochemist Hermann Lehmann, Ph.D., is recorded as a member of the (British) Eugenics Society (1954, 1957, 1977, 1969-1971).[1]

5.  Jacob A. Goldberg and Alan F. Guttmacher are among those who either contribute articles to the *Eugenics Quarterly* or whose work is reviewed or advertised there.

## 1955

### Context

1.  The "Growth of American Families" study indicates an average size of Jewish families of 1.7, as opposed to 2.1 for Catholics and Protestants. Furthermore, Jews expect significantly fewer children (2.4) than either Catholics (3.4) or Protestants (2.9).[2]

2.  The play *The Diary of Anne Frank*, written by two non-Jewish authors, Albert and Frances (Goodrich) Hackett, is staged in Philadelphia with great success, making the book the canonical Holocaust text, translated into more than 55 languages, selling over 24 million copies world wide.[3]

3.  In the Soviet Union Jewish biologist Khilia Faivelovich Kushner publishes a pro-Lysenko text on animal genetics.[4]

4.  Jewish eugenicist Vladimir Pavlovich Èfroimson is released from a Soviet forced-labor camp.[5]

### Jewish Advocacy of Eugenics

5.  Dr. Abraham Stone of the Margaret Sanger Bureau: "The premarital consultation can... play an important role in preventive eugenics.... It is essential, however, that the physician who would undertake to guide young people in the eugenic aspects of their marriage should himself have a thorough understanding of genetics and of

---

[1] *Eugenics Review, Bulletin of the Eugenics Society, Biology and Society,* Eugenics Watch.
[2] Freedman/Whelpton/Campbell, "Differential Fertility among Native-White Couples in Indianapolis," XXI, July, 226-271; cited in Goldscheider, 1967, 198.
[3] Cole, 2000, 23.
[4] Kushner, 1955.
[5] Vergasov. Undated.

the heredity of human disease. Not many physicians are today well equipped to give competent genetic advice."[1]

6.   The Alfred A. Knopf publishing house issues *Cultural Anthropology* by anthropologist Melville Jean Herskovits (1895-1963), an abridged edition of *Man and His Works* (1948). Herskovits is a student of Franz Boas, and he attempts to reconcile Boas's cultural relativism with the still predominant school of physical anthropology. Herskovits notes that endogamy (inbreeding) results in lower variability.[2]

7.   Geneticist Franz Josef Kallmann (1955, 1957) and Ruth E. Weiss are recorded as members of the (British) Eugenics Society.[3]

8.   Some persons who contribute to the *Eugenics Quarterly* or whose work is reviewed or advertised there: H. O. Goodman, Melville Herskovits, A. J. Jaffe, Franz Kallmann, Harry L. Shapiro, Arthur G. Steinberg, Curt Stern, Abraham Stone.

### 1956

### Jewish Advocacy of Eugenics

1.   Physician Harry L. Shapiro of the American Museum of Natural History is elected President of the American Eugenics Society.[4]

2.   Lithuanian-American rabbi and theologist Mordecai M. Kaplan (1881-1983), founder of the Reconstructionist Movement: "[Darwinism] holds forth the promise of man's evolution into a much higher type of being than he is now."[5]

3.   Demographer Erwin S. Solomon, writing in the *Eugenics Quarterly*, cites figures indicating less differential fertility by IQ among Jews than among Christians.[6]

---

[1] Stone, 1955, 52.
[2] Herskovits, 1955, 70.
[3] *Eugenics Review, Bulletin of the Eugenics Society, Biology and Society,* Eugenics Watch.
[4] http://www.eugenics-watch.com/aeugensoc/aeback.html, accessed May 12, 2008.
[5] "Revelation of God in Nature: A piyut for the First Benediction of the Evening Prayer," *The Reconstructionist Answers*, New York, 1956: Reconstructionist Press; quoted in Cherry, 2003, 272-273.
[6] Solomon, 1956.

4.   A number of Jews are listed as members and officers of the American Eugenics Society.[1]

## 1957

### Context

1.   Swiss Jews are shown to have a lower fertility rate than the total population.[2]

2.   The U.S. Bureau of the Census shows a Jewish intermarriage rate of 7.2%, compared to 9% for Protestants and 21% for Catholics.[3]

3.   The *American Journal of Human Genetics* publishes a study of Jewish fingerprint patterns[4] – a project first proposed to the German police by Jewish-German eugenicist Wilhelm Nussbaum in 1921.

4.   According to the U.S. Census, Jews in the United States did not participate in the postwar 'baby boom' as much as Roman Catholics and Protestants.[5]

### Jewish Advocacy of Eugenics

5.   Jewish-German-American eugenicist and Berkeley professor of zoology Curt Stern: "Eugenics, the planning toward rational improvement of the genetic improvement of human populations, has been regarded as a utopian dream, and a dangerous one in addition.

---

[1] Professor Harry L. Bauer, professor of medicine and genetics Stanley Gartner (1956, 1974, 1975), zoologist Harold O. Goodman (1956, 1974, 1983-1985), Samuel Gottfried, professor of genetics Lissy Feingold Jarvik (1956, 1974, 1989), genetic psychiatrist Arnold R. Kaplan (1971-1972), professor of medicine Herbert Spencer Kupperman (1956, 1974), biometrician professor Howard Levene, physician Lena Levine, professor of zoology Max Levitan, professor of psychology Irving Lorge, Dr. Robert M. Stecher, life sciences scholar Ruth S. Stein, geneticist Arthur G. Steinberg, Ph.D., geneticist Curt Stern (1956, 1974), Dr. Harry Wallerstein. (Kaplan, *Questions Jews Ask: Reconstructionist Answers*, Reconstructionist Press, New York, 107; quoted in Cherry, 2003, 273.) Contributers to the *Eugenics Quarterly* and authors whose whose work is reviewed or advertised there include: K. Z. Altshuler, Alan Guttmacher, F. J. Kallmann, Amram Scheinfeld, Harry L. Shapiro, Sam Shapiro, Erwin S. Solomon, Arthur G. Steinberg, Curt Stern.
[2] Kurt B. Mayer, "Recent Demographic Developments in Switzerland," *Social Research*, XXIV, Summer, 350-351; cited in Goldscheider, 1967, 200.
[3] Goldstein/Goldscheider, 1966, 386.
[4] Sachs/Bat-Miriam, 1957.
[5] Rosenthal, 1957.

Many of our concerns have been recognized as being excessive or even unnecessary, and our eagerness to act precipitously has rightly decreased. But, while the attempt at improvement beyond the present average could well be relegated to the future, even the most determined opponents of positive eugenics have now realized with the fervor of converts that negative eugenics, the attempt to inhibit the increase of harmful genotypes, is an immediate necessity."[1]

6. Writing in the *British Journal of Sociology*, Gordon Rose presents British criminology as developing along eugenics lines, in opposition to the Lombrosian school and the biosocial theories of Havelock Ellis. Rose singles out eugenicists Francis Galton and Karl Pearson as the forerunners of the new thinking and praises Cyril Burt's "classic" *Young Delinquent*: "The study's great strength is the combination of clinical insight and case material with sophisticated statistical techniques; a superstructure built upon a strong foundation of wide academic knowledge."[2]

7. Some members, fellows, and officers of the (British) Eugenics Society (renamed the Galton Institute in 1989): Mr. P. J. Manasseh; Dr. and attorney Felix Ferdinand Tietze; Dr. J.W. Tietze; Leonard Wilensky.[3]

8. Persons who contribute to the *Eugenics Quarterly* or whose work is reviewed or advertised there: Kenneth Z. Altshuler, Paul Glick, Philip Levine, Franz J. Kallmann, Curt Stern.

## 1958

### Context

1. Israel supports a statement by geneticists attending the Tenth International Genetics Conference in Montreal condemning the Soviet refusal to permit attendance by those of its scientists who did not support Lysenko.[4]

---

[1] Stern, 1957, 748.
[2] Rose, 1958, 58.
[3] *Eugenics Review, Bulletin of the Eugenics Society, Biology and Society,* Eugenics Watch.
[4] Schmeck, 1958.

## Jewish Advocacy of Eugenics

2.  An advertisement in the *New York Times* announces the 36th edition Hannah and Abraham Stone's pro-eugenics *Marriage Manual*. Since its original publication in 1935, 350,000 copies of the book have been sold, and it has been translated into twelve languages.[1]

3.  Jewish-American eugenicist Joshua Lederberg (1925-2008) shares Nobel Prize in medicine for research on gene exchange in bacteria.

4.  Continued strong Jewish participation in eugenics publications.[2]

## 1959

### Context

1.  Physical anthropologist Gabriel Lasker (1912-2002): "The analytical-comparative method has largely replaced the racial typing of individuals. This is particularly true of the studies by those who attempt a genetic analysis of race differences."[3]

### Jewish Advocacy of Eugenics

2.  President of the American Eugenics Society Harry L. Shapiro delivers his *Presidential Address* at a joint dinner meeting of the American Society for Human Genetics and the American Eugenics Society, August 22, Montreal: "Now I need not stress that in the tradition of democracy the problem of guarding and even improving the genetic quality of our population must be solved by a program that has popular acceptance. Any solutions that smell of force or duress cannot be admitted. There is, therefore, a serious and heavy responsibility for those who believe in eugenics. They must bend every effort to encourage investigations of an impartial and scientific nature to determine what factors are affecting the quality of our population and whether or not the balance between it and culture is sound. And based on such researches they must develop a program of informa-

---

[1] Jan. 12, pg. BR34.

[2] Persons who contribute to the *Eugenics Quarterly* or whose work is reviewed or advertised there: Kenneth Z. Altshuler, Baruch S. Blumberg, Kurt Hirschhorn, Franz Kallmann, Arnold A. Kaplan, P. Levine, Amram Scheinfeld, William J. Schull, J. N. Spuhler.

[3] Lasker, 1959.

tion and action that will be acceptable to a free, democratic society. The task is difficult, but it is worthy of our best."[1]

3. Continued strong Jewish participation in the eugenics movement.[2]

## 1960

## Context

1. The U.S. Food and Drug Administration (FDA) approves the contraceptive pill.

2. The capture of Adolf Eichmann by the Israeli Secret Service and his subsequent trial in Jerusalem the following year stimulates popular interest in the slaughter of Jewish civilians during World War II, and the term 'Holocaust' begins to be used with regularity in the United States and Great Britain.[3] Holocaust historian Tim Cole comments: "The need for greater identification with the Diaspora Jew was heightened by the growth of 'Canaanism' amongst sections of Israeli youth during the 1950s. These 'Canaanites' rejected any links between contemporary Israel and the Diaspora Jews. Rather than identifying with the European Jews who had experienced the Holocaust, 'Canaanism' identified with Israeli Arabs." Attended by 376 journalists from fifty countries along with 166 Israeli journalists and observers, Eichmann's trial is filmed in its entirety by the U.S. film and television company Capital Cities. After the trial the Holocaust gains broad acceptance as a topic.[4]

3. Linguist Ron Kuzar: "The Canaanite movement proposed a radical alternative to Zionism.... Their views capitalized on a radicalization of intra-Zionist tendencies which were quite popular in the local Jewish community of the 1940s and 1950s, which idolized the

---

[1] Shapiro, 1959.

[2] Some members, fellows, and officers of the (British) Eugenics Society (renamed the Galton Institute in 1989): Dr. Kalman Freid; Mr. P. J. Manasseh; Dr. and attorney Felix Ferdinand Tietze; Dr. J.W. Tietze; Leonard Wilensky. (*Eugenics Review, Bulletin of the Eugenics Society, Biology and Society,* Eugenics Watch.) Persons who contribute to the *Eugenics Quarterly* or whose work is reviewed or advertised there: I. Lester Firschein, David Goldberg, Alan F. Guttmacher, Franz J. Kallmann, Richard Levins, Amram Scheinfeld, Sheldon J. Segal, Harry L. Shapiro, Robert M. Stecher, Curt Stern, W. F. Wertheim, Irving B. Wexler.

[3] Noted by Cole, 2000, 7.

[4] Cole, 2000, 57-58, 63, 67.

healthy, tall, tanned, down to earth, native *sabra* 'Jew born in Palestine/Israel' as the inverse image of the diasporic Jew.... Having its early roots in European extreme right-wing movements, notably Italian fascism, it exhibited an interesting blend of militarism and power politics towards the Arabs as an organized community on the one hand and a welcoming acceptance of them as individuals to be redeemed from medieval darkness on the other."[1]

4. Jewish geneticists successfully launch a bitter campaign to prevent an international genetics conference from taking place in Germany, changing venue to The Hague.[2]

### Jewish Advocacy of Eugenics

5. The "Growth of American Families" study continues to indicate that Jews expect and desire fewer children than do either Catholics or Protestants.[3] The U.S. Bureau of the Census indicates the quotient of actual births to Jewish women of childbearing age to be about three quarters as high as for the country as a whole.[4]

6. UNESCO publishes *The Jewish People*, by President of the American Eugenics Society Harry L. Shapiro.

7. Persons who contribute to the *Eugenics Quarterly* or whose work is reviewed or advertised there: A. Falek, Ronald Freedman, Paul C. Glick, David Goldberg, Kurt Hirschhorn, Alan F. Guttmacher, P. A. Jacobs, William Schull, Sheldon J. Segal, Harry L. Shapiro, Robert M. Stecher, Arthur Steinberg.

---

[1] Two Brief Introductions to Hebrew Canaanism,"
http://www.geocities.com/alabasters_archive/kuzar_intros.html, accessed January 26, 2009.
[2] Kirsh, 2004, 88-89.
[3] Freedman/Whelpton/Campbell, "Differential Fertility among Native-White Couples in Indianapolis," XXI, July 71-72, 90-91, 247-252, Tables 33, 46; cited in Goldscheider, 1967, 199.
[4] Glick, 1960, 38.

## 1961

## Context

1.  Dutch Jews are shown to have a lower fertility rate than the total Dutch population.[1]

### Jewish Advocacy of Eugenics

2.  President of the American Eugenics Society Harry L. Shapiro creates the Hall of the Biology of Man in the American Museum of Natural History. An editorial in the *New York Times* describes the Hall as "the newest thing of value in this city, [showing] through the most modern museum techniques the evolution and biology of this fellow we know so much about and yet so little."[2]

3.  Geneticist Jack B. Bresler is part of a team proposing in the *Eugenics Quarterly* to do a study of blood samples of Brown-Pembroke students to identify indicators of intelligence. The team cites as inspirational a 1924 statement by S. J. Holmes in his study of University of California students: "Of course the group studied represents a selected class differentiated in several aspects from the general population. But it is a very important class. It is a class which furnishes much of our intellectual leadership and its biological trend is a matter of some moment."[3]

4.  Herman J. Muller, winner of the Nobel Prize in Medicine for research on the genetic effects of radiation: "an ever wider over-all view will emerge, and a surer, greater over-all plan, or rather, series of plans. To create them and put them into effect will then enlist our willing efforts. And the very enjoyment of their fruits will bring us further forward in our great common endeavor: that of consciously controlling human evolution in the deeper interests of man himself."[4]

5.  Jewish-Austrian-American geneticist and physician Kurt Hirschhorn writes a favorable review in *Eugenics Quarterly* of Otmar Freiherr von Verschuer's twin-based textbook of human genetics,

---

[1] "Dutch Jewry: A Demographic Analysis," *The Jewish Journal of Sociology*, III, Dec., 195-243; cited in Goldscheider, 1967, 200.

[2] *New York Times*, 1990.

[3] Bresler *et al.*, 1961, 11.

[4] Muller, 1961, 649.

calling it a "valuable collection of rare morphological diseases" written "in the Fischer-Bauer-Lenz tradition."[1] The generally praiseful tone of the review is remarkable in light of Verschuer's virulent anti-Jewish views.

6.     Journalist and author Amram Scheinfeld (1897-1979): "But if we today are the product of chance, our descendants do not have to be…. Genetics and the related sciences have proved beyond question that we can guide, if not control, the destinies of those who follow us by selecting the units of biologic and social inheritance which we pass on to them."[2]

7.     Continued strong Jewish participation in the eugenics organizations and publications.[3]

## 1962

### Context

1.     The council of the Chief Rabbinate of Israel rules that marriage with members of the group Bene Israel (a Judaizing group which has immigrated from India) is permissible, but the rabbi registering the marriage is bound to investigate the maternal ancestry of every applicant so as to establish that there has been no intermixing with non-Jews – over *at least* three generations. The ruling, which is more rigorous even than Germany's 1935 Nuremberg racial-purity laws that extended back only two generations, is vehemently protested by the Bene Israel.[4]

2.     Louis B. Brinn in *Harofé Haiviri*, the Hebrew Medical Journal, discusses twelve hereditary diseases that appear with particular frequency among Ashkenazi Jews and mentions endogamy and genetic isolation as possible explanations. Writing in the *Eugenics Quarter-*

---

[1] Hirschhorn, 1960.

[2] Scheinfeld, 1961, 588.

[3] Robert Kuttner, Ph.D., and journalist and author Amram Scheinfeld are recorded as members of the (British) Eugenics Society. (*Eugenics Review, Bulletin of the Eugenics Society, Biology and Society,* Eugenics Watch.) Persons who contribute to the *Eugenics Quarterly* or whose work is reviewed or advertised there: Jack B. Bresler, Melvin Embep, Arthur Falek, I. Lester Firschein, Kurt Hirschhorn, Emmanuel Margolis, Gitta Meier, Amram Scheinfeld, William Schull, Sheldon J. Segal, Harry L. Shapiro, Hirsch Lazar Silverman.

[4] Bar-Giora, 2007, 338.

*ly*, University of Michigan geneticist R. H. Post rejects this hypothesis and points out that the effect of consanguineous marriage is eliminated by a single generation of random marriage. Post also notes that cousin marriages have not been frequent among Jews for the "last generation or two."[1]

### Jewish Advocacy of Eugenics

3.  Sociologist Erich Rosenthal studies Jewish fertility patterns on the basis of a nationwide sample survey conducted by the U.S. Bureau of the Census in 1957:

    -   Jewish fertility is as little as 73.6% of Roman Catholic and 79.5% of white Protestant fertility between 1943 and 1957.

    -   The fertility of Jewish women has lagged behind for at least two generations.

    -   At least part of the above differential is explained by the high levels of Jewish urbanization, education, and professionalization, all three factors being associated with low fertility.[2].

4.  Arthur Falek and Irving Isadore Gottesman are among the contributors to the *Eugenics Quarterly* or whose work is reviewed or advertised there.

### Jewish Egalitarianism

5.  Ashley Montagu: "If we are to succeed in clarifying the minds of those who think in terms of 'race' we must cease using the word, because by continuing to use it we sanction whatever meaning anyone chooses to bestow upon it, and because in the layman's mind the term refers to conditions which do not apply. There is no such thing as the kind of 'race' in which the layman believes, namely, that there exists an indissoluble association between mental and physical characters which make individual members of certain 'races' either inferior or superior to the members of certain other 'races.' The layman requires to have his thinking challenged on this subject. The term 'ethnic group' serves as such a challenge to

---

[1] Post, 1965b.
[2] Rosenthal, 1961.

thought and as a stimulus to rethink the foundations of one's beliefs."[1]

## 1963

## Context

1. A sample survey of the Jewish population of the Providence, Rhode Island, metropolitan area shows a clear inverse relationship between socioeconomic status and fertility among first-generation Jews, but other studies seem to indicate greater homogeneity and convergence in the fertility patterns of third-generation Jews.[2]

2. Israeli geneticist Arieh Szeinberg: "If future investigations fail to show occurrence of a similar abnormality [G6PD deficiency] among the local populations in European countries from which the Ashkenazim come, these findings will provide a marker demonstrating a common origin of the Ashkenazim and of the other Jewish communities. If on the other hand, a similar frequency of the gene pool is found among Poles, Russians, or Germans, the possibility of sporadic mutations in any part of the world will have to be considered." Israeli geneticist Nurit Kirsh attacks Szeinberg's logic as being politically influenced (by Zionism): "similarity between Jews and other Jews (however slight) is explained in terms of common origin, whereas similarity between Jews and non-Jews is explained in terms of sporadic mutations, with both speculations being based on missing data…. Again, every result has a possible explanation, except the possibility of intermarriage and conversion."[3]

## Jewish Advocacy of Eugenics

3. Harry L. Shapiro of the American Museum of Natural History steps down as President of the American Eugenics Society, having served six years.[4]

---

[1] Montagu, 1962, 926.
[2] Goldscheider, 1967, 202.
[3] Szeinberg's comment is taken from "G6PD Deficiency among Jews – Genetic and Anthropological Considerations," *The Genetics of Migrant and Isolate Populations*, ed. Elisabeth Goldschmidt, Williams and Wilkins, New York, 69-70, as quoted by Kirsh, 2003-2004, 79-80.
[4] http://www.eugenics-watch.com/aeugensoc/aeback.html, accessed May 12, 2008.

4.  The Ciba Foundation convenes a conference in London under the title "Man and His Future," at which three distinguished biologists and Nobel laureates (Herman Muller, Joshua Lederberg, and Francis Crick) all speak out strongly in favor of eugenics. Muller stresses the need to avoid genetic deterioration whereas Lederberg counsels waiting for forthcoming findings in molecular biology and cytology, which promise to be more efficacious than traditional selection methodologies. Geneticist Arthur Falek congratulates Ciba on having published the presentations.[1]

5.  Writing in the *Eugenics Quarterly*, geneticist Arnold R. Kaplan denies the existence of major differences in mental endowment between Caucasians and Negroes.[2]

6.  Persons who contribute to the *Eugenics Quarterly* or whose work is reviewed or advertised there: Arthur Falek, Arnold R. Kaplan, Mortimer Spiegelman, Arthur G. Steinberg, Anthony Zimmerman.

## 1964

### Context

1.  Point 2 (of 15) of the American Eugenics Party platform: "NO PERSECUTION. No race or stock is to be harshly treated. All Caucasian stocks (Germans, Jews, Italians, Poles, etc.) are to remain separate and free from persecution or abuse and must unite to ward off the non-Caucasian genetic threat."[3]

2.  *Look* magazine publishes an influential article entitled "The Vanishing American Jew."[4]

3.  J. J. Groen, Professor of Medicine at the Hadassah Hebrew University Hospital in Jerusalem: "Gaucher's disease is much more frequent among Jews than among non-Jews. Among the Jews it has been observed among the Ashkenazim only. This observation supports the hypothesis that the Jews were ethnically already heterogeneous before their Diaspora and that the present Ashkenazim are descendants of a group, which existed already as such in biblical

---

[1] Falek, 1965. The reviewed volume is edited by G. Wolstenholme and published by Little, Brown, and Co.

[2] Kaplan, 1963, 190.

[3] *American Eugenics Party Platform*, 1964.

[4] Morgan, 1964.

times. The present paper gives reasons for believing that the Ashkenazim are the continuation of the inhabitants of the kingdom of Judea, comprising the tribes of Juda and Benjamin, whereas the Sephardim and Oriental Jews can be considered as descendants of the inhabitants of the Kingdom of Israel, which comprised most of the other tribes."[1]

4.  The 1962 ruling of the Israeli chief rabbinate that any member of the group Bene Israel (a Judaizing group that has immigrated from India) wishing to marry a Jew must be investigated to prove that his parents, grandparents, and great-grandparents have not intermarried with non-Jews culminates in a strike, and the Sephardi Chief Rabbi Yitzhak Nissim (1896-1981) is burned in effigy. Finally Prime Minister Levi Eshkol issues a statement: "the government of Israel reiterates that it regards the community of the Bene Israel from India as Jews in every respect, without any restriction or distinction, equal in their rights to all other Jews in every matter, including matters of matrimony."[2]

## Jewish Advocacy of Eugenics

5.  Wayne State University Professor of Economics Samuel M. Levin appears not to be aware of later claims that eugenics had supposedly withered away in the 1930s: "Malthus found himself dealing with questions which may well be considered as coming within the purview of *modern eugenics* [emphasis added].... It is certain that his conceptions will continue to give light to people coming to grips with a number of difficult problems facing the society of our day."[3]

6.  The Academy of Political Science publishes *The Geography of Intellect* by Jewish-American scholars Nathaniel Weyl (1910-2005) and Stefan T. Possony (1913-1995), who make a persistent case for Jewish intellectual superiority.

7.  Continuing Jewish participation in eugenics organizations and publications.[4]

---

[1] Groen, 1964, 548-549.
[2] Bar-Giora, 2007, 338.
[3] Levin, 1964, 51, 54.
[4] Some members, fellows, and officers of the (British) Eugenics Society (renamed the Galton Institute in 1989): A. R. Kaplan, Ph.D.; geneticist Michael Lerner, Ph.D. (1964, 1977). (*Eugenics Review, Bulletin of the Eugenics Society, Biology and Socie-*

## Jewish Rejection of Eugenics

8.  The Free Press of Glencoe brings out Ashley Montagu's *The Concept of Race*, which attacks anthropologist Carleton Coon's thesis that human beings are the product of the independent but convergent evolution of five different subspecies or races. Montagu's view is supported in the *Eugenics Quarterly* by John F. Kantner of the Population Council, who refers to Coon's book with sarcasm but concedes he has not read it. Kantner maintains that the concept of subspecies has been a failure in dealing not only with *homo sapiens*, but with other species as well.[1]

## 1965

## Context

1.  Demographer Larry D. Barnett: "The major difference in desired family size appears to emerge between Catholics on the one hand and Protestants and Jews on the other."[2]

2.  Demographer Calvin Goldscheider in the *Jewish Journal of Sociology* notes an unambiguously inverse relationship of fertility to socioeconomic status among first-generation Jews, but hypothesizes that the differences may diminish for second- and third-generation Jews, but the data are too limited to draw definitive conclusions. At the same time he sums up previous studies of Jewish population patterns: "low fertility patterns characterize not only contemporary Jewish couples in the United States but have been documented to apply to Jews as early as 1880 in the United States and in many western countries for at least the last century."[3]

3.  Geneticist Arnold R. Kaplan writing in *Eugenics Quarterly* on schizophrenia: "There is little scientific controversy concerning expected eventual discovery of chemical and/or biological correlates with the psychoses. Otherwise, the phenomena could be regarded as supernatural or metaphysical entities."[4]

---

*ty,* Eugenics Watch.) Persons who contribute to the *Eugenics Quarterly* or whose work is reviewed or advertised there: Morris Fisbein, Aviva B. Kesselman, Samuel M. Levin, Harry L. Shapiro, Medora Steedman-Bass.

[1] Kantner, 1965.
[2] Barnett, 1965, 163.
[3] Goldscheider, 1965, 235.
[4] Kaplan, 1965, 132.

4. Letter to the editor of *Science* from Benjamin Ginsburg: "the pursuit of knowledge must be coordinated with the adjudged safety with which this knowledge, if attained, can be entrusted to human nature in its present state of moral development. Otherwise we put ourselves in the position of the small boy who blows himself up by playing with explosives." Geneticist Curt Stern: "Have we not had enough lessons concerning the dangers of the puristic ivory-tower attitude still dominant in many 'scientists' and 'humanists'? The best answer that scientists can make to those who would misapply such a technique is to join with others in showing how it can be applied constructively."[1]

5. University of Michigan geneticist R. H. Post: "Since Gentiles today are rapidly changing their ways of life in the direction of the traditional Jewish environment – with rapidly growing towns and cities, rapidly increasing protection of slightly or afflicted persons through modern public health and individual health facilities – it seems reasonable to consider the question of whether the frequencies of...genetic diseases – might become progressively greater in future generations."[2]

## Jewish Advocacy of Eugenics

6. Chief Rabbi (retired) of the British Commonwealth of Nations and 'father' of modern Jewish medical ethics Lord Immanuel Jakobovits: "the killing of an unborn child is not considered as murder punishable by death in Jewish law.... The Christian tradition disputing this view goes back to a mistranslation in the Septuagint.... In the Jewish view the viability of a child is not fully established until it has passed the thirtieth day of its life...."[3]

7. Citing Margaret Mead as an influential thinker, sociologist Hyman Rodman reflects the paradigm shift that occurred in intellectual thought in the middle of the twentieth century: "It is now widely accepted that the 'storm and stress' of adolescence is not universal, but depends on the way that adolescence is organized by society."[4]

---

[1] Ginzburg/Stern, 1965.
[2] Post, 1965b, 164.
[3] *Abortion and the Law*, D. T. Smith (ed.), 1965, 1967; reprinted in Rosner/Bleich, 2000, 141-143.
[4] Rodman, 1965, 450.

8. From the obituary of Jewish geneticist and eugenics proponent Franz J. Kallman in *Eugenics Quarterly*: "His contributions to the field of human genetics were established with his classic work on the genetics of schizophrenia. By 1936, his opposition to Nazi laws requiring compulsory sterilization of mentally ill patients made his position in his native Germany untenable. Coming to the United States, he founded almost single-handedly the discipline of psychiatric genetics in this country.... As members of his department and long-time co-workers, we share his loss with the scientific community."[1]

9. Journalist and author Amram Scheinfeld (1897-1979), a member of the (British) Eugenics Society: "A sensible eugenics program... would seek to replace the reckless or haphazard direction of human evolution with intelligent and carefully planned guidance. In this we must think not merely of ourselves, but of our descendants to come.... But where there is or will be the wisdom and discretion to chart a proper eugenic course is another matter."[2]

10. Jewish participation in eugenics publications continues.[3]

### Jewish Resistance to Eugenics

11. Jewish-American geneticist Richard Lewontin performs some elementary mathematics that is very revealing as to the thinking behind the anti-eugenics movement: "We hear a good deal about the question, 'How much genetic punishment can human populations absorb?' That's something we really would like to know. It might be so very great that it would never be of any eugenic concern, or it might not. We would like to know if it's at all possible with present human data to give a rough estimate of the total differential mortality available in the human species – or let's say in a Caucasian popu-

---

[1] Erlenmeyer-Kimling *et al.*, 1965.

[2] Scheinfeld, 1965, 714.

[3] Persons who contribute to the *Eugenics Quarterly* or whose work is reviewed or advertised there: Lauretta Bender, Bernard Berelson, B. Catz, Leon Jacob Cole, Arthur Falek, J. D. Finkelstein, B. Fish, Ronald Freedman, A. M. Gittelsohn, Paul C. Glick, E. Goldschmidt, Alan F. Gutmacher, Franz Kallmann, John F. Kantner, Arnold B. Kaplan, D. Klein, P. Kunstadter, Louis Levine, Max Levitan, R. Levins, Sarah Lewit, Richard Lewontin, S. Milham, Jr., Ashley Montagu, Melvin Moss, H. V. Muhsam, Edward Pohlman, Ina Samuels, Lee E. Schacht, William J. Schul, Harry L. Shapiro, S. E. Snyderman, Mortimer Spiegelman, Robert M. Stecher, Aurthur Steinberg, Gary A. Steiner.

lation, if you want to make it more exact – that would still enable a population to replace itself. In, other words, to put it very crudely, if a human female is capable of producing 200 fertilized eggs in her lifetime, can we dispense with 198 of them or so and leave just a couple left over? What is the right number?"[1]

## 1966

## Context

1.  An Israeli medical study of the epidemiology of mental disorders establishes a rate of 45.3 cases per 1,000 adults.[2]

### Jewish Advocacy of Eugenics

2.  Rabbi and leading expert on Jewish medical ethics Moses D. Tendler in an essay entitled "Population Control – The Jewish View": "man has been granted a junior partnership in the management of this world. Imbued with the spark of Divine Intelligence, man is permitted, even *required* [emphasis added], to use his partnership rights to regulate his own affairs, on condition that he does not violate the by-laws of this God-man relationship that are formulated in the Torah. What if the present projections prove to be more accurate than those made by Malthus? We are told that at the present rate of increase in world population, 300 million tons of *additional* grain annually will be needed by 1980. This is more grain than is now produced by all North America! What guidelines have been set down for our instruction in this yet hypothetical situation? The Jew as a world citizen is personally concerned with famine in India and China. However the Noachidic Laws which serve as Torah (instruction) for all humanity demand a proper sequence of actions. Before a Jew can support birth-control clinics in overpopulated areas of the world, he must insist that there be heroic efforts made to utilize fully the agricultural potential of the world."[3]

3.  Soviet geneticist Nikolai Iosifovich Shapiro (1906-1987) writes a warm and lengthy obituary in memory of his colleague Aleksandr Serebrovsky, but despite the relaxed intellectual climate avoids discussing the political pressures exercised on Serebrovsky.

---

[1] Post, 1965a, 60-61.

[2] Maoz *et al.*, 1967, 282-283.

[3] *Tradition*, fall, 1966; reprinted in Rosner/Bleich, 2000, 119.

4.   Geneticist Joshua Lederberg: "When it becomes possible, vegetative or clonal reproduction of many proven genotypes (*i.e.*, persons of undisputed value to society) will offer significant advantages over other methods of improving man's biological potential…. The chief hurdle to any radical approach like clonal reproduction is the requirement of human experimentation, but this can be greatly reduced by thorough trials in subhuman primates."

5.   Persons who contribute to the *Eugenics Quarterly* or whose work is reviewed or advertised there: Benson E. Ginzburg, David Goldberg, Calvin Goldscheider, Sidney Goldstein, Arnold G. Kaplan, Morton Kramer, Louis Levine, B. M. Mandelbrote, Edward Pohlman, Amram Scheinfeld, Harry L. Shapiro.

### 1967

### Context

1.   The film *Guess Who's Coming to Dinner*, written by William Rose and directed by Stanley Kramer, presents an appealing vision of black-white intermarriage even as Jews heatedly debate on how to staunch the increasing rate of Jew-gentile intermarriage. The U.S. Supreme Court rules that State bans on interracial marriage are unconstitutional. (Loving vs. Virginia).

2.   Astra Books brings out *Malthus and the Conduct of Life* by Samuel M. Levin, who presents Malthus in a favorable light, noting that Malthus recognized but was not alarmed by questions of population quality.

3.   Such differing Jewish historians as Jacob Neusner, Deborah Lipstadt, and Norman Finkelstein all note the significance of the Jewish victory in the Arab-Israeli war in establishing the Holocaust Memorial Movement.[1] Journalist Judith Miller argues: "… there was nothing *inherently* exploitative in the Jewish push for monuments, memorials, and public tributes to the period of their most intensive suffering. *But* the linkage of the Holocaust with campaigns to raise money and enhance support for the State of Israel marked the beginning of serious abuse and misuse of the Holocaust…. American Jews discovered that the Holocaust could be used as a weapon not

---

[1] See Cole, 2000, 9-12.

only for garnering sympathy at home, but also for insisting on unquestioning support for Israel abroad."[1]

## Jewish Advocacy of Eugenics

4.  Birmingham mental health specialist Pauline C. Shapiro writes that the way in which Parliament's 1933 Brock Report on childbearing by persons of subnormal mentality "has passed into oblivion is indicative of a habit of shunning insoluble problems, especially those that concern handicap and questions of its inheritance and control.... The advent of oral contraceptives brings hope to the large families of non-copers, that by thoughtful administration of new techniques such families may be reduced to a size with which they will be better able to cope. This should result not only in an increase in happiness of the families concerned but also in a reduction of delinquency, poverty and child neglect."[2]

5.  *The Bulletin of Atomic Scientists* and the journal *BioScience* publish an article by Jewish biologist Leonard Ornstein calling for a "vigorously pursued... conservative eugenics" intended, at the very least, to prevent genetic decline. Not only do the editors make no mention of the claim that eugenics supposedly was rejected by the scientific community decades earlier, they even stress that they are reprinting the article "because they felt it deserved the attention of a larger audience of biologists."[3]

6.  Rabbi Eliezer Yehudah Waldenberg allows abortion following amniocentesis during the first trimester if the fetus is determined to have Tay-Sachs Disease. "If there is a strong suspicion that the fetus will be born physically deformed and suffer greatly, one can allow abortion prior to forty days of conception and perhaps even up to three months of the pregnancy before the fetus begins to move." Waldenberg also allows termination of pregnancy for Tay-Sachs Disease up to the seventh month of pregnancy because the defect, the anguish, the shame, the physical and mental pain and suffering of the parents are inestimable.[4]

---

[1] Cole, 2000, 11.

[2] Shapiro, 1967, 257; referring to the "Brock Report," *Report on the Departmental Committee on Sterilization*, 1934, HMSO.

[3] Ornstein, 1967.

[4] Waldenberg KY. Responza Tzitz Eliezer, vol. 9, #51: 3, Jerusalem, 1967.

7.  Continuing Jewish participation in eugenics organizations and publications.[1]

## 1968

## Context

1.  Writing in *Eugenics Quarterly*, sociologist Erich Rosenthal notes that his 1961 estimate of American Jewish fertility as two-thirds that of American Protestants and Catholics has since "shown no upswing in the birth rate. On the contrary, there is some evidence that, like the total birth rate, the Jewish birth rate also has declined since then."[2]

2.  Tim Cole, former Paul Resnick Resident Scholar at the Center for Advanced Holocaust Studies at the United States Holocaust Memorial Museum: "by the end of the 1960s and early 1970s, the new-left broadened their attack into one over American involvement in Vietnam and the position of the Palestinians in Israel. The questioning in Germany became less about the 'Holocaust' than about the perceived faults of capitalism and the establishment."[3]

3.  Sociologist Erich Rosenthal notes in *Eugenics Quarterly* that intermarriages between Jews and non-Jews in Indiana have increased to 51.7% from 46.3% in 1960, indicating a crude average intermarriage rate of 48.8%. The criterion used for defining Jewishness is religion; presumably the intermarriage rate for non-religious Jews is higher still.[4]

4.  Rabbi and professor of Jewish law and ethics J. David Bleich (b. 1936) on German Talmudist Rabbi R. Jacob Emden (1697-1776): "Emden reasons that if the mother may destroy herself completely she may certainly destroy a part of her body. Hence he concludes

---

[1] Persons who contribute to the *Eugenics Quarterly* or whose work is reviewed or advertised there: K. Z. Altschuler, Marianne E. Bernstein, B. Cohen, R. Freedman, J. Lederburg, D. Goldberg, I. I. Gottesman, J. Hirsch, A. Kaplan, A. Katz, B. Malzberg, H.Muller, C. Stern, N. Weyl, S. Possoni, Melvin Zelnik. Some members, fellows, and officers of the (British) Eugenics Society (renamed the Galton Institute in 1989): Miss A. Jacob; Dr. E. Posner; professor of anthropology P. L. Workman (1967, 1977).

[2] Rosenthal, 1968, 287.

[3] Cole, 2000, 9.

[4] Rosenthal, 1968, 278.

there can be no prohibition against the destruction of a bastard fetus since its life is legally forfeit."[1]

## Jewish Advocacy of Eugenics

5.  J. David Bleich again: "There are a number of latter-day authorities who are explicit in their opinion that feticide is a rabbinic rather than a Biblical offense. Perhaps the most prominent of these is the renowned seventeenth century scholar, R. Aaron Koidonover, author of the famed commentary of *Sede Kodashim, Birkat ha-Zevah*.[2]

6.  Persons who contribute to the *Eugenics Quarterly* or whose work is reviewed or advertised there: William E. Feinberg, Joseph Felsenstein, Bertram Fleshler, H. Green, Bernard Greenburg, Arnold R. Kaplan, R. C. Lewontin, Samuel Levin, N. Mantel, Ashley Montagu, Edward Pohlman, Erich Rosenthal, J. Samuelson, Melvin Zelnik.

## Jewish Rejection of Eugenics

7.  Subsequent to the 1967 Arab-Israeli war, the Holocaust Memorial Movement is launched, with eugenics targeted as the ideology of genocide. So effective is the campaign that polls show that many more Americans can identify the Holocaust than Pearl Harbor or the atomic bombing of Japan.[3] Those who are familiar with the term "eugenics" begin to associate it with "Holocaust" and "racism."

## 1969

## Context

1.  Robert Sinsheimer (b.1920), a molecular biologist at the California Institute of Technology, publishes an article in *Engineering and Science* "favoring 'a new eugenics': "For the first time in all time, a living creature understands its origins and can undertake to design its future."[4]

---

[1] "Abortion in Halachic Literature," *Tradition*, Winter, 1968; reprinted in Rosner/Bleich, 2000, 179-180.
[2] "Abortion in Halachic Literature," *Tradition*, Winter, 1968; reprinted in Rosner/Bleich, 2000, 159.
[3] Finkelstein, 2000, 11.
[4] http://www.penguindust.com/lisa/journals/1999/991104.html.

2. British physicist and novelist C.P. Snow (1905-1980): "Is there something in the Jewish gene pool which produces talent on quite a different scale from, say, the Anglo-Saxon gene pool? I am prepared to believe that it may be so. Take any test of achievement you like – in any branch of science, mathematics, literature, public life. The Jewish performance has been not only disproportionate, but almost ridiculously disproportionate."[1]

### Jewish Advocacy of Eugenics

3. Geneticist Irving Isadore Gottesman (b. 1929) is elected an officer of the American Eugenics Society.

### Jewish Rejection of Eugenics

4. Radicalized by the Vietnam War, a predominantly Jewish group of Harvard and MIT students and faculty members forms the Marxist organization "Science for the People." They promote biological egalitarianism and attack scholars pursuing sociobiology or eugenics.

5. Prominent Orthodox rabbi Moshe Feinstein (1895-1986) advocates tearing out pages of science textbooks that contain references to evolution or other "matters of heresy."[2]

## 1970-2005

1. A natural population increase in Israel is balanced out by population decline within the Diaspora, producing zero Jewish population growth worldwide.[3]

2. American Jewish intermarriage rate for 1970s: 28% (in contrast to 17% prior to 1970.)[4]

## 1970

### Context

1. The Knesset defines the term 'Jew' as meaning "one who was born to a Jewish mother or who converted to Judaism." This is a partial

---

[1] Snow, 1969.

[2] *Igrot Moshe, Yoreh De'ah*, vol. 3, responsum 73 (New York, Noble Press, 323; cited by Rena Selya in Cantor/Swetlitz, 2006, 194.

[3] Schmelz/DellaPergola, 2007.

[4] *National Jewish Population Survey*, 2002.

victory for those demanding traditional religious criteria, but keeps the door open to those who didn't fit that definition; the amendment also grants the right of immigration to the child, grandchild, or spouse of a Jew.

2.  Professor of medicine Fred Rosner notes that rabbinic authorities are divided on whether or not to consider a child conceived by artificial insemination illegitimate (a *mamser*).[1]

3.  A translation into English of one of the first books to connect eugenics with the Holocaust appears, written by Bernhard Schreiber, whose father was allegedly a Luftwaffe officer and who received a grant from a large university: *The Men behind Hitler: A German Warning to the World*, possibly printed in Les Mureaux, France. The German original does not appear until 1972 in a self-publication by the author.[2]

### Jewish Advocacy of Eugenics

4.  Psychiatric geneticist and a director of the American Eugenics Society Irving Isadore Gottesman (b. 1929): "The essence of evolution is natural selection; the essence of eugenics is the replacement of 'natural' selection by conscious, premeditated, or artificial selection in the hope of speeding up the evolution of 'desirable' characteristics and the elimination of undesirable ones."[3]

### 1971

### Jewish Advocacy of Eugenics

1.  Genetic Laboratories, Inc. and Iatric Corporation announce plans to open frozen sperm banks in New York City, fulfilling the dream of Jewish-American Nobel Prize winner Herman J. Muller for a scientific human breeding program.[4]

2.  "Differential reproduction in individuals with mental and physical disorders," conference sponsored by the American Eugenics Society and the Bio-medical Division of the Population Council, held in

---

[1] "Artificial Insemination in Jewish Law," *Judaism*, fall, 1970; reprinted in Rosner/Bleich, 2000, 125-137.
[2] *Die Männer hinter Hitler: Eine deutsche Warnung an die Welt.*
[3] Quoted by Cavanaugh-O'Keefe, 1995.
[4] "From the Day of Deposit – A Lien on the Future," *New York Times*, Aug. 22, E7.

New York, Nov. 13-14, 1970, Authors: Irving I. Gottesman and L. Erlenmeyer-Kimling, place of publication: Chicago.

## 1972

### Context

1.  American historian Kenneth M. Ludmerer describes eugenicists who supported Hitler's race policies as a minority within the eugenics movement and decries such views as a "perversion of the true eugenic ideal as seen by well-meaning men deeply concerned about mankind's genetic future."[1]

### Jewish Advocacy of Eugenics

2.  Lippincott publishers bring out *Heredity in Humans* by Amram Scheinfeld, a member of the (British) Eugenics Society.

## 1973

### Context

1.  Abortion is legalized in the United States by judgment of the Supreme Court (*Roe v. Wade*). A poor woman is four times as likely to experience an unplanned pregnancy as a higher-income woman,[2] and thus is more likely to request an abortion. Although Jews constitute 2% of the population, women who identify themselves as Jewish account for only 1.3% of abortions.[3]

### Jewish Advocacy of Eugenics

2.  In response to an inquiry by the Association of Orthodox Jewish Scientists (AOJS) seeking guidance on Halachic parameters for Tay-Sachs screening, Rav Moshe Feinstein publishes a *teshuva (responsum)*, which reads, in part:

    *   ". . . It is advisable for one preparing to be married, to have himself tested. It is also proper to publicize the fact, via newspapers and other media, that such a test is available."

    *   "The largest community-based screening program for Jewish genetic diseases which was created and is maintained under Ha-

---

[1] Ludmerer, 1972, 117.
[2] Wind, 2006.
[3] http://www.abortionno.org/Resources/fastfacts.html.

lachic parameters is Chevra Dor Yeshorim Committee for Prevention of Jewish Genetic Diseases ('Dor Yeshorim').... A confirmed incompatible couple is strongly discouraged from getting married to each other. By not getting married, such couples are not faced with such difficult decisions as what to do with a fetus that is afflicted with a genetic disorder."[1]

- Geneticist Michael Sagi of the Hadassah University Hospital in Jerusalem comments: "The program is regarded by the Israeli geneticists, as well as by the religious community, as a very successful one. We would suggest that similar programs be offered in other communities in which marriages are prearranged and termination of pregnancy is not accepted."[2]

## 1974

### Context

1. U.S. Supreme Court Justice Arthur Goldberg (1908-1990) on the passage of the Jackson-Vanik Amendment denying the Soviet Union Most Favored Nation Status in trade: "The stain of Holocaust abandonment has finally been removed."[3]

### Jewish Advocacy of Eugenics

2. Mucolipidosis IV (ML IV) is first described, occurring mostly among Ashkenazi Jews. Children with ML IV appear normal at birth but develop signs of central nervous system deterioration during the first year of life. Sitting is delayed and most people with ML IV do not walk. Motor and mental retardation are usually mild to moderate, and are slowly progressive. Some patients may become more severely retarded in the second or third year of life. It may lead to blindness in later years. No effective treatment is available, but prenatal testing is available for carriers and parents of previous affected children.[4]

### Jewish Rejection of Eugenics

3. Princeton University Press brings out Leon Kamin's (b. 1928) *The Science and Politics of IQ*, in which Kamin denounces the late

---

[1] http://jewishgeneticscenter.org/rabbis/overview/, accessed May 15, 2008.
[2] Sagi, 1998, 427.
[3] Lazin, 2005, 91.
[4] "Jewish Genetic Diseases," *Jewish Genetic Disorders*, Sept., 2005, 69-70.

prominent English psychologist Cyril Lodowic Burt's (1883-1971) studies of separately raised monozygotic twins and expresses skepticism about Burt's data, as also does the psychologist Arthur Jensen (b. 1923), who – unlike Kamin – is known for his hereditarian position.[1] In 1966 Burt reported an IQ correlation of 0.77 among 53 pairs of identical twins whom he had studied. Accusations that Burt falsified his data attract huge publicity, but Burt's findings have since been replicated repeatedly, including Thomas Bouchard's study of 8,000 twin pairs, which came up with a correlation of 0.76 for identical twins reared separately and 0.87 for those reared together. In another study of adopted children, conducted by Sandra Scarr and Richard A. Weinberg, also at the University of Minnesota, the adoptees' IQ scores correlated significantly more positively with those of their biological than with those of their adoptive parents.[2]

4.  Jewish-American biologist and politically left opponent of Neo-Darwinism, sociobiology, and eugenics Richard Lewontin (b. 1929): "For Muller, human progress meant enriching the species for a few superior genotypes while for Dobzhansky it means increasing, or at least maintaining, genetic diversity. Neither view admits the possibility that genetic variation is irrelevant to the present and future structure of human institutions, that the unique feature of man's biological nature is that he is not constrained by it."[3] This startling statement comes from a professional geneticist who is a long-standing (1966-1977) member of the American Eugenics Society.

5.  The group Science for the People, informally led by Jonathon Beckwith of Harvard and Jonathan King of the Massachusetts Institute of Technology, comes out in opposition to genetic screening at the Boston Hospital for Women (affiliated with Harvard's medical school) for chromosomal aberrations, particularly for XXY or XYY patterns, because males identified as being XYY are stigmatized as carriers of the "criminal chromosome." Although the medical school faculty supports the research by a vote of approximately 200 to 30, one of the principal investigators shuts down the project, say-

---

[1] Arthur R. Jensen. "Scientific Fraud or False Accusations? The Case of Cyril Burt." In D. J. Miller and M. Hersen (eds.), *Research Fraud in the Behavioral and Biomedical Sciences*. New York: John Wiley, 1992.

[2] Wright, 1997, 63.

[3] Lewontin, 1974, 31.

ing he is worn down. Beckwith and King write: "...we feel that the major effort in approaching the issue of behavioral problems should be one of changing the social and psychological (inseparable) conditions which separate them. We consider the attempts to determine a genetic basis for anti-social behavior, a diversion with harmful effects."[1]

# 1975

## Context

1. Presenting much the same views as those of Kamin, a CBS news special *The IQ Myth* declares that not only are IQ tests relatively useless as measures of intelligence, but that they are biased as well, for "it's economic class that marks the main dividing line on IQ scores."[2]

2. The American Jewish Committee, the Anti-Defamation League, and the American Jewish Congress all file amicus briefs with the Supreme Court in opposition to Affirmative Action programs even as Jewish groups pursue an intense search for allies within the Afro-American community under the banner of civil rights.[3] (Bakke vs. Regents of University of California)

3. Hungarian-Jewish anthropologist Raphael Patai (1910-1996) and Jennifer Patai-Wing in *The Myth of the Jewish Race* maintain that Jews of any given area tend to resemble non-Jews of that area more than they do each other.

4. A highly controversial ruling by Chief Justice of Jerusalem's Rabbinic Court Rev Eliezer Y. Waldenberg: "There is no greater abomination than insemination with a foreign donor's seed."[4]

5. Ben Zion Bokser (1907-1984), Polish-American major Conservative rabbi of the Forest Hills Jewish Center: "The willful consignment of a defective child to extermination is an intolerable breach of man's duty to show reverence for life, to foster it, protect it, and perfect it, to the extent that the biological facts permit it. It is also a negation of the basic thrust operative in nature, to bring to birth life,

---

[1] Culliton, 1975, 1285.
[2] Snyderman/Rothman, 1986, 83.
[3] Finkelstein, 2000, 36-37.
[4] Responsa, *Tritz Eliezer*, vol. 9, section 51, part 4; quoted in Green, 1991, 4.

sometimes in a defective state, giving man the opportunity to render supportive care, to develop compassion and a sensitivity to suffering."[1]

## Jewish Advocacy of Eugenics

6.  Paris rabbi Dr. Elie Munk discusses "the effect of the [Jewish religious] laws on married life and sexual relations on eugenics…. This system of prophylactic hygiene on the widest possible scale has brilliantly proved itself in its application to the Jewish people. In spite of the fact that the reasons for many prohibitions are beyond the capacities of human logic, the historical fact remains that the so-called ritual laws have conferred on Israel, throughout the generations, an extraordinary vigour and power of physical resistance. In the midst of living conditions often characterized by the most intense misery and the extremes of privation, these laws have sufficed, because of their real sanitary and hygienic value, to form a chain of generations perfectly healthy in mind and body and secure against disease and death to a remarkable degree."[2]

7.  Chief Rabbi of the British Commonwealth of Nations, former Chief Rabbi of Ireland, and Spiritual Leader of New York's Fifth Avenue Synagogue Dr. Immanuel Jakobovits (1921-1999): "Jewish law certainly went very much further than any other in ancient and medieval times in cultivating the eugenic ideal by prudent legislation and counsel…. Of particular interest are some of the regulations in the chapter on the obligation "to endeavor to take a fitting wife" (E.H., ii). Special emphasis is placed on the choice of a partner equipped with the highest intellectual and moral virtues…. These and many similar provisions in Jewish law are clearly motivated by eugenic considerations for the moral excellence of the progeny. The Talmud recognizes the hereditary element in the determination of character and virtue when it counsels a man seeking worthy children to examine the brothers of his prospective wife, 'since most children take after their maternal uncles.'… Insane persons may not contract a valid marriage at all (E.H., xliv). While the marriages of the deaf-and-dumb are effective, albeit only rabbinically (ib., I), the rabbis refused to make any provision for the legislation of marriages with or between lunatics. The declared reason for this refusal is that such

---

[1] Bokser, 1975, 137-138.
[2] Jakobovits, 1959, xxi-xxii.

marriages could never be happy or peaceful and not, as has been suggested, 'because they would produce backward children'. But this law is obviously still of eugenic interest."[1]

## Jewish Rejection of Eugenics

8.  By a vote of 72 to 35, with 32 abstentions, the UN General Assembly "determines that Zionism is a form of racism and racial discrimination," essentially declaring the State of Israel to be illegitimate (Resolution 242). As a counterbalance Jewish groups increase funding to the Holocaust Memorial Movement, which in turn attacks the eugenics movement with increasing fury.

9.  Fayard Publishers in Paris brings out Marc Hillel and Clarissa Henry's *Au nom de la race*, initiating the association in books of eugenics and "Holocaust."

## 1976

## Jewish Advocacy of Eugenics

1.  Geneticist Irving I. Gottesman is elected a vice-president of the American Eugenics Society.

2.  A woman who was conceived thanks to artificial insemination conducted by gynecologist Frances I. Seymour, leader of the National Research Foundation for the Eugenic Alleviation of Sterility, Inc., writes in a *New York Times* article entitled "Report from a test-tube baby": "My parents' story made sense of my childhood memories of the lovely lady I knew as Dr. Seymour. Whenever I had visited her office, she always fussed and beamed like a surrogate relative. When I did well on the psychological and intelligence tests she arranged, she praised me lavishly.... Knowing about my A.I.D. origin did nothing to alter my feelings for my family. Instead, I felt grateful for the trouble they had taken to give me life. And they had given me a strong sense of roots, a rich and colorful heritage, a sense of being loved."[2]

3.  Random House brings out *The Thirteenth Tribe: The Khazar Empire and Its Heritage* by Jewish-Hungarian writer Arthur Koestler (1905-1983), who makes the case that Ashkenazi Jews are not des-

---

[1] Jakobovits, 1959, 154-156.
[2] Atallah, 1976, 155.

cended from the Israelites of antiquity, but from the Khazars, a Turkic people in the Caucasus who converted to Judaism in the 8th century and were later forced to move westwards. The thesis is not a new one, but it has previously been studiously ignored by Jewish scholars. Ashkenazi Jews are estimated as comprising approximately 82% of all Jews in the world.[1]

4.  Jewish-American eugenicist Nathaniel Weyl describes disease as a "eugenic force": "the ravages of disease are generally not aristocidal, but the opposite."[2]

## Jewish Rejection of Eugenics

5.  Hutchinson Publishers in London issues Marc Hillel's and Clarisa Henry's *Children of the SS*, and McGraw-Hill brings it out in America under the title *Of Pure Blood* on Heinrich Himmler's *Lebensborn* program: It is reviewed by Holocaust historian Lucy S. Dawidowicz: "Marc Hillel and Clarissa Henry, a husband-and-wife journalistic team, insist on telling Lebensborn's story as a piece of political pornography, as if it had in fact operated as a stud farm, as if all the babies delivered at its homes were illegitimate, and as if Lebensborn were primarily responsible for the kidnappings of the foreign children....Pandering to the pornographic appetites of today's reading public, Hillel and Henry offer salacious surmise and innuendo where they lack real evidence. Written in breathless tones of moral outrage, *Of Pure Blood* is a prurient exploitation of Nazism and its malevolent racial doctrines, intended to titillate while it informs."[3]

6.  Random House and *Knopf* bring out Alan Chase's *The Legacy of Malthus: The Social Costs of the New Scientific Racism*. Although pro-eugenics books still predominate, the stress by book publishers on an association of eugenics and racism can be said to have begun.

## 1977

## Context

1.  In Göttingen, Vanndenhoeck und Ruprecht Publishers bring out Kurt Nowak's (1942-2001) book on euthanasia and sterilization in

---

[1] Merkel, 1997.
[2] Weyl, 1976, 243. For a hypothetical mathematical of aristocide, see Glad, 1998.
[3] Dawidowicz, 1977, 43-44.

Hitler's Germany and the confrontation it caused with the Evangelical and Catholic Churches.[1]

2.  Jewish-Hungarian anthropologist Raphael Patai: "As far as the significance of the Jewish Nobel record goes, I leave it to the reader to draw his own conclusion from the fact that the Jews, who constitute less than a half a percent of mankind, have won more than 15 percent of the prizes generally recognized as the highest accolade of modern times."[2]

### Jewish Advocacy of Eugenics

3.  A study of Bloom's Syndrome in Israel: "An effort was made to identify all individuals with Bloom's syndrome living in Israel between September 1971 and September 1972. Each of the eight individuals located were Jewish and could readily be classified Ashkenazic. The frequency of the Bloom's syndrome gene in Ashkenazim was estimated to be 0.0042 (minimum), implying a heterozygote frequency greater than 1 in 120."[3]

4.  Writer Izzy Siev: "Social Eugenics will accomplish the following: It will reduce future welfare roles. It will reduce our prison population. It will reduce future crimes. It will consist of a simple three-point plan.... First, the Aid to Dependent Children will be phased out.... Second... Persons on Welfare who consent to sterilizations will receive a one thousand dollar cash bonus.... Third, all prison inmates will be give [sic] a three-year reduction in their prison sentences if they consent to sterilization."[4]

5.  Israel's penal code permits abortion in cases when the fetus may have a physical or mental birth defect.[5]

6.  Rabbi J. David Bleich indicates that the elimination of Tay-Sachs Disease is a goal to which all concerned individuals subscribe, but that, while the couple may quite properly be counseled with regard

---

[1] *"Euthanasie" und Sterilisierung im "Dritten Reich" : die Konfrontation der evangelischen und katholischen Kirche mit der Gesetz zur "Verhütung erbkranken Nachwuchses' u.d. "Euthanasie"-Aktion.*

[2] Patai, 1977, 342.

[3] German *et al.*, 1977.

[4] Siev, 1977, 14-15.

[5] Wikipedia, "Abortion in Israel," http://en.wikipedia.org/wiki/Abortion_in_Israel, accesses August 16, 2008.

to the risks of having a Tay-Sachs child, failure to bear natural children is not a *halachically* viable alternative. He further voices concern that if the fetus is found to have Tay-Sachs Disease by pre-natal testing, abortion may not be sanctioned in Jewish law. Bleich concludes that screening programs for the detection of carriers of Tay-Sachs Disease are certainly to be encouraged. He is critical of Rabbi Eliezer Yehudah Waldenberg and points out that the latter's permissive ruling on abortion for Tay-Sachs Disease is contrary to the decisions of other contemporary rabbinic scholars including Rabbi Moshe Feinstein.[1]

7.  Some members, fellows, and officers of the (British) Eugenics Society (renamed the Galton Institute in 1989): University of California biology professor D. Baer, Dr. Conrad van Emde Boas, T. J. David, Dr. J. A. Davidson, President of the Institute of Jewish Affairs Lord Arnold Abraham Goodman, Mrs. Nathan Isaacs, Ph.D., Mr. P. A. Vygodsky.[2]

### Jewish Rejection of Eugenics

8.  Jewish-American zoologist, egalitarian, and prominent opponent of eugenics Stephen Jay Gould reviewing reprint of Allen G. Roper's 1913 *Ancient Eugenics*: "Roper continually invokes Plato's myth of the metals (*Republic*, Book 3) to affirm the innateness of unequal worth and the layering of men into social classes as a reflection of nature. But he seems to forget that Plato, at least, had the decency to brand it as a lie manufactured in the interest of social stability."[3]

### 1978

### Context

1.  Geneticists Arthur Ernest Mourant, Ada C. Kopeć, and Kaziemira Domaniewska-Sobczak in *The Genetics of the Jews*: "Each major community as a whole bears some resemblance to the indigenous people of the region where it first developed and, within each community there is some relation between the compositions of the separate Jewish sub-communities (national, etc.) and those of the peoples among whom they have recently lived. Nearly all Jewish

---

[1] Cited in Rosner, 1998, 409.
[2] *Eugenics Review, Bulletin of the Eugenics Society, Biology and Society,* Eugenics Watch.
[3] Gould, 1977, 626.

communities show a substantial number of African Negroid marker genes, such as to imply a total Negro admixture of the order of 5 to 10 percent. These admixtures are readily explained by slavery and concubinage by Patai and Wing (1975).[1]

2.   NBC's nine-and-a-half-hour television miniseries *Holocaust* is watched more than 120 million times over four consecutive evenings. Tim Cole, former Paul Resnick Resident Scholar at the Center for Advanced Holocaust Studies at the United States Holocaust Memorial Museum, comments: "It was this TV show, above anything else, which turned the term [Holocaust] into a 'household' name in the United States.... The result of this interest in the United States has been the 'Americanisation of the Holocaust'.... The 'Holocaust' is now considerably less important in Europe where it physically took place than it is in America where it has been embraced as a statement of faith."[2]

### Jewish Advocacy of Eugenics

3.   Rabbi and professor of Jewish law and ethics J. David Bleich: "The Gemara, Yevamot 64b, declares that a man should not marry into an epileptic or leprous family, i.e. a family in which three members have suffered from these diseases. This ruling is obviously a eugenic measure designed to prevent the birth of defective children.... While natural misfortunes may not be avoidable, man does not have the right to act in a manner which will result in harm to others. It follows, *a fortiori*, that overt intervention in natural processes which might cause defects in the fetus would be viewed by Judaism with opprobrium.[3]

### Jewish Rejection of Eugenics

4.   Ira Levin's (1929-2007) 1976 novel, *The Boys from Brazil*, is made into a screenplay by Heywood Gould, starring James Mason, Gregory Peck, and Laurence Olivier (production budget $12,000,000).[4] Although it is a work of fantasy, it is associated in the public consciousness with eugenics.

---

[1] Mourant *et al.*, 1978, 57.
[2] Cole, 2000, 12-14.
[3] "Test-Tube Babies, *Tradition*, 1978; reprinted in Rosner/Bleich, 2000, 101.
[4] http://www.the-numbers.com/movies/1978/0BYFB.php.

## 1979

### Context

1.  In France a 'New Right' movement is promoted by *Figaro* magazine and the Club de l'Horloge, rejecting both Marxism and materialism and reaching back to European cultural and aristocratic traditions. The group asserts that racial genetic differences must be recognized and opposes "race-mixing," which they perceive as a source of "maximum genetic disorder" that can only lead to "an immense retrogradation" of European stocks. A *New York Times* article states that they are not pointedly anti-Jewish.[1]

2.  Over two-thirds of purported émigrés to Israel 'drop out' and choose other countries.[2]

3.  Droste Publishers in Düsseldorf brings out Reiner Pommerin's (b. 1943) book on the sterilization of the "Rheinland Bastards" in Germany: the children of African soldiers and German women during the post-World War I occupation.[3]

4.  Claude Vorilhon (b. 1946), leader of the "Raëlian" religious group advocating cloning, writes to Rabbi Israel Abou Khatzira, Rabbi Mordekhai Shaarabi, Rabbi Halberstamm, Rabbi Menachem Schneerson, Professor Leon Manitou Ashkenazi, Israeli President Menachem Begin, and Chief Rabbi Ouiadia Yossef asking for a facility in Jerusalem.

5.  Donor insemination is allowed in Israel. [4]

6.  NBC's nine-and-a-half-hour television miniseries *Holocaust* is watched by more than 14 million Germans.[5]

---

[1] Kandell, 1979.
[2] Gittelman, 1989, 163.
[3] *Sterilisierung der Rheinlandbastarde: Das Schicksal einer farbigen deutschen Minderheit 1918-1937.*
[4] Birenbaum-Carmeli *et al.*, 1992, 80.
[5] Cole, 2000, 13.

## 1980

### Context

1.  American Jewish intermarriage rate for 1980s: 41% (up from 28% in 1970s, and 13% prior to 1970).[1]

2.  An Israeli study of Cochin Jews, also called 'Black Jews,' who have a high frequency of consanguineous marriages, indicates that they show particular similarities with Yemenite Jews, and also with the indigenous populations of southern India." About 4,000 reside in Israel.[2]

### Jewish advocacy of Eugenics

3.  Historian Daniel Winkler: "Racism, class bias, and violation of reproductive freedom are not part of the core notion of eugenics.... It is most plausible to view eugenics as sharing moral dilemmas with much of public health, and the critical issues of distributive justice."[3]

### Rejection of Eugenics

4.  Religious-leftist Berlin psychologist Peter Kratz attacks the new cult of Jewish-German eugenicist Magnus Hirschfeld as an early defender of gay rights: "Many of them [Hirschfeld's followers] felt themselves to be part of the educated bourgeoisie, whose irrational mindset as a forerunner of National Socialism was analyzed so magnificently by Lucácz. Many had become atheists, so that they rejected being assigned to Jewry by the extreme right."[4]

## 1981

### Context

1.  Julian Simon (1932-1998) writes *The Ultimate Resource,* discounting concerns regarding overpopulation and resource exhaustion.[5]

---

[1] National Jewish Population Survey, 2002.
[2] Cohen *et al.*, 1980
[3] Winkler, 1998, 455.
[4] Kratz, 1980.
[5] Simon, 2001.

2.  Bastian Till's German-language study of German psychiatry appears, attempting to trace a trajectory leading from eugenics to euthanasia.[1]

## Jewish Advocacy of Eugenics

3.  Geneticist Irving I. Gottesman steps down as vice-president of the American Eugenics Society, having served five years and been a member since 1969.[2]

4.  Rabbi Moshe Hershler (1905-1997) advocates genetic engineering and gene therapy, raising the question of the permissibility (or lack thereof) of experimenting with gene therapy to try to save the life of a child with thalassemia or Tay-Sachs Disease if the unsuccessful outcome of the experimentation would be a shortening of the child's life. Hershler is of the opinion that gene therapy and genetic engineering may be prohibited because he who changes the [Divine] arrangement of creation is lacking faith [in the Creator], and he cites as support for his view the prohibition against mating diverse kinds of animals, sowing together diverse kinds of seeds, and wearing garments made of wool and linen (Leviticus 19:19).[3]

## 1982

### Context

1.  The Soviet Union radically curtails Jewish emigration.[4]

2.  Gallup poll: While 44 percent of all Americans believe the world was created more or less as it currently is within the last ten thousand years, and over 80 percent supported the inclusion of creationism in the public schools, 85 percent of Orthodox Jews support the teaching of the theory of evolution. According to the 1990 National Jewish Population Survey, less than a quarter of those who identify themselves as Jewish agree that "the Torah is the actual word of God."[5]

---

[1] *Von der Eugenik zur Euthanasie: Ein verdrängtes Kapitel aus der Geschichte der deutschen Psychiatrie*, Verlagsgemeinschaft Erl, Bad Wörishofen.

[2] http://www.eugenics-watch.com/aeugensoc/aeoff.html.

[3] Moshe Hershler, "Genetic Engineering in Jewish Law," *Halacha Urefua*, Chicago, vol. 2, 1981, 350-353; cited in Rosner, 1998, 411.

[4] Lazin, 2005, 129.

[5] Cited by Shai Cherry in Cantor/Swetlitz, 2006, 185.

3. Web site announcement: "The Center for Jewish Genetic Diseases at The Mount Sinai Medical Center in New York City is the first center in the world devoted to the study of diseases that affect Ashkenazi Jews. Established in 1982, the Center has a twofold mission: 1) to improve the diagnosis, treatment, and counseling of patients and their families suffering from Jewish genetic diseases and 2) to conduct intensive research to combat these inherited diseases.... During the Center's history, its researchers have been awarded over $50 million in research and training grants by the National Institutes of Health. The Center has become the focus for the training of young physicians and scientists in the care of patients and in laboratory research to improve the diagnosis and treatment of Jewish genetic disorders. In addition, we have established a master's degree program to train genetic counselors, which focuses primarily on the genetic and psychosocial counseling involved with the diagnosis, management, and treatment of patients and families with Jewish genetic diseases."[1]

## Rejection of Eugenics

4. A "Theoretical Letter Concerning the Moral Arguments against Genetic Engineering of the Human Germline Cells" resolves "that efforts to engineer specific genetic traits into the germline of the human species should not be attempted." The letter is signed largely by clergymen of various denominations, including Pat Robertson and Jerry Falwell, and also three rabbis. By contrast, twenty of the signatories are Catholic. The document is circulated and was evidently drafted by Jeremy Rifkin's Foundation on Economic Trends in Washington. D.C.

## 1983

### Context

1. In Frankfurt am Main S. Fischer Publishers brings out Ernst Klee's study of the destruction of "life not worth living" under National Socialism.[2]

2. Demographer Leo Davids in *Israel Social Science Research*: "Although a late starter compared to most European and North Ameri-

---

[1] http://www.mssm.edu/jewish_genetics/overview.shtml, accessed July 12, 2008.
[2] Klee, *Euthanasie im NS-Staat:Die Vernichtung lebensunwerten Lebens.*

can countries, Israel is now moving rapidly towards a late modern type of family situation, characterized by high levels of contraception, of female employment, of divorce, with declining fertility which may approach zero population growth in the not too distant future."[1]

3.  The Central Conference of American Rabbis (Reform Movement) reaffirms its 1948 resolution that a child is presumed to be Jewish if one parent is Jewish, as long as the parents and child identify with Judaism.[2] [3] The decision unleashes a firestorm of Orthodox condemnation, described by author David Landau: "For the Orthodox, 'marrying out' is indeed a disaster akin to a death. When it strikes an Orthodox family, parents and siblings of the offender will symbolically tear their clothes and sit on low stools like mourners, to receive the consolation of other relatives and friends. And even in families that are not inclined to grieve publicly in the traditional mode, the newlyweds will often be ostracized and solemnly 'cut out' of the father's or grandfather's will." Director of the World Jewish Congress Israel Singer suggests that "Within a few years, all the presidents of the Jewish federations around the country will be goyim" and complains that many Jewish philanthropic organizations are drawn from a "Waspish Jewish plutocracy" in which Orthodox membership is minimal, while the proportion of non-Orthodox converts and patrilineal descendants is steadily rising. Defending the Reform position, Rabbi Alexander Schindler (1925-2000) looking back years later, says he has no regrets whatsoever. "We have fifty thousand children of mixed marriages in our movement. In most cases, the father is the Jewish partner.... There are compelling sociological reasons for patrilinealism. While not condoning intermarriage, we welcome the partners in mixed marriages into our communities. Once a mixed marriage is a fact, you have two choices: to bring them in, or to sit shiva (that is, to mourn as for the dead)."[4]

---

[1] Davids, 1983, 37-38.
[2] Berck, 2006.
[3] Committee on Patrilineal Descent.
[4] Landau, 1993, 299-300 (years of Singer's and Schindler's statements are not indicated by Landau).

## Limited Jewish Advocacy of Eugenics

4.  Fred Rosner, Director of Medicine at the Long Island Jewish-Hillside Medical Center: "If the purpose of Tay-Sachs screening is to provide genetic counseling about reproduction and mating options, few will argue against screening. If the purpose, however, is to introduce couples at risk to the benefits of prenatal diagnosis by amniocentesis with the specific intent of recommending abortion of affected fetuses, a procedure that may be contrary to the religious dictates of the client, then screening should not be performed. The religious teachings of the Jewish people must be considered if co-operation from the rabbinate and compliance from the clients is to be obtained in any screening program."[1]

## 1984

## Context

1.  Israeli psychologists M. Nathan and R. Gutman continue the tradition of studying twins, first proposed by Francis Galton."[2]

2.  British Independent Television Enterprises produces *The Master Race*, scripted by Charles Foster and Julia Spark.

## Jewish Rejection of Eugenics

3.  The New *Yorker* magazine publishes in serial form *In the Name of Eugenics: Genetics and the Uses of Human Heredity* by Daniel Kevles.[3]

## 1985

## Context

1.  Knesset delegate Meir Kahane proposes a "Law to Prevent Assimilation between Jews and Non-Jews." It is criticized as a replay of the 1935 Nuremburg Laws forbidding miscegenation between Jews and non-Jews.[4]

---

[1] Rosner, 1983, 44.
[2] Nathan/Guttman, 1984.
[3] Kevles, 2003.
[4] Cole, 2000, 139-140; citing G. Cromer, "Negotiating the Meaning of the Holocaust: An Observation on the Debate about Kahanism In Israeli Society," *Holocaust and Genocide Studies 2*, No. 2, 1987, 290.

2.  Knopf brings out historian Daniel Kevles's *In the Name of Eugenics: Genetics and the Uses of Human Heredity.* Generally critical of the eugenics movement, it proves to be extremely influential and is widely cited. Kevles writes that racists and supporters of Hitler's Germany "constituted a rapidly diminishing minority, most of them isolated on the far political right."[1]

### Jewish Advocacy of Eugenics

3.  After losing four children to Tay-Sachs Disease, Orthodox Rabbi Joseph Eckstein founds Dor Yeshorim to screen for recessive genes that disproportionately affect Jews. If two persons intending to marry are discovered to be carriers, they are encouraged to find new partners. Testing is later expanded to encompass Bloom syndrome, Canavan disease, Cystic fibrosis, Familial dysautonomia, Fanconi anemia (type C), Gaucher disease, Glycogen storage disease (type 1), Mucolipidosis (type IV), and Niemann-Pick disease.

4.  Ethiopian Jews begin to arrive in Israel. Although they are at first called *falasha*, the term comes to be viewed as perjorative and is replaced by 'Beta Israel.' Although they physically differ strikingly from the other Jewish population of Israel and genetic tests show little or no Jewish connection, considerable effort is expended to prove a genetic relationship.[2]

5.  Physician and eugenicist Isidore Simon dies, and his *Revue d'histoire de la médecine hébraïque*, which has published some 600 articles since he founded it in 1948, ceases publication.[3]

### A Mixed Jewish Reaction to Eugenics

6.  Rabbi Moshe Feinstein is asked whether or not it is advisable for a boy or girl to be screened for Tay-Sachs Disease, and if it is proper, at what age the test should be performed: "...it is advisable for one preparing to be married, to have himself tested. It is also proper to publicize the fact, via newspapers and other media, that such a test is available. It is clear and certain that absolute secrecy must be maintained to prevent anyone from learning the result of such a test performed on another. The physician must not reveal these to any-

---

[1] Kevles, 1986, 347.
[2] See Kaplan, 1992; Quirin, 1992; Parfitt/Egorova, 2005.
[3] *Bulletin of the History of Medicine* 78.3 (2004) 702-703.

one...these tests must be performed in private, and, consequently, it is not proper to schedule these test in large groups as, for example, in Yeshivas, schools, or other similar situations." Nevertheless he condemns abortion for Tay-Sachs Disease and even questions the permissibility of the amniocentesis which proves the presence of a Tay-Sachs fetus, since amniocentesis is not without risk.[1]

## 1986

### Context

1.  Westdeutscher Verlag brings out Gisela Bock's study of forced sterilization and racial and feminine politics under National Socialism.[2]

### Jewish Advocacy of Eugenics

2.  An Israeli screening for carriers of Tay-Sachs eliminates the disease among the children of newly wed couples.[3]

### Jewish Rejection of Eugenics

3.  Historian Sheila Faith Weiss: "Both under the Kaiser and under the swastika, German race hygiene was a tool whereby population could be managed in the interest of power. By focusing attention on the logic of eugenics, the link between the seemingly disparate programs of Schallmayer's nonracist eugenics and Nazi race hygiene becomes painfully evident."[4]

4.  Jewish-American-Swedish eugenics opponent Elof Axel Carlson: "Perhaps the most important lesson for the idealistic geneticist who hopes to elevate humanity to direct our own evolution is to reflect on the repeated vulgarization of complex genetic studies, reduced to simplistic models of human health and behavior, by those with more fervor than professional competence. It may take many generations for us to compensate for the genetic consequences of our present breeding habits, but it would be folly, considering today's standards

---

[1] Moshe Feinstein, response: Even Haezer, Part 4, #10, Bnei Brak, 1986; cited in Rosner, 1998, 409.
[2] Bock, 1986.
[3] Rosner, 1998.
[4] Weiss, 1986, 46.

of cultural prejudices, to intentionally modify those habits as public policy in the name of eugenics."[1]

5.  Jewish-American political scientist Ira H. Carmen reviewing Daniel Kevles's *In the Name of Eugenics*: "All in all, Kevles has rendered a distinctively valuable contribution with his penetrating historical review. What we now need to complete the picture is a sequal [sic] entitled *In the Name of Euculture*, which would document with equal vigor and scholarship the careers and preachments of those who today and yesterday have argued that genetics counts for nothing in explaining hominid social and intellectual life, that our species therefore possesses cultural choices constrained only by environmental circumstance, and that contrary propositions should not be entertained even as hypotheses to be studied scientifically, else man will be tempted to deracinate his unique moral integrity."[2]

## 1987

### Context

1.  The paperback edition *The Flamingo's Smile: Reflections in Naural History* (W. W. Norton) by Eugenics foe Stephen Jay Gould appears with adulation from the writer David Quammen (b. 1948), who calls Gould "one of the sharpest and humane thinkers in the sciences."[3]

2.  Historian Sheila Faith Weiss: "the designation of the Jews as an unfit, surplus, and disposable group is not unrelated to the emphasis implicit in German race hygiene regarding 'valuable' and 'valueless' people. For the eugenicists, human beings were in some sense variables – objects easily managed or manipulated for some abstract 'good.' In one of humankind's most barbaric acts to date, there is more than a hint of where the desire to be rid of a 'valueless' population can lead."[4]

---

[1] Carlson, 1986, 532.
[2] Carmen, 1986.
[3] Back cover.
[4] Weiss, 1987, 195, 236.

## 1988

### Context

1. For the first time a large number of books appear that stress the connection of eugenics with Hitler's Germany, including Suhrkamp's *Rasse, Blut und Gene: Geschichte der Eugenik und Rassenhygiene in Deutschland* (Race, Blood, and Genes: A History of Eugenics and Racial Hygiene) by Peter Weingart, Jürgen Kroll, and Kurt Bayertz.

2. Expert on medical ethics of Jewish religious law Rabbi Moshe David Tendler: "As medical science advances, the number of diseases detectable will increase until conceivably most of them can be predetermined. Thus if a society has a eugenic concern then we should remove all defective genes, to prevent the birth of a defective child. Are we prepared to live with this knowledge? There are over 3,300 known genetic diseases."[1]

### Jewish Rejection of Eugenics

3. Anti-eugenics activist Barry Alan Mehler (b. 1947) defends his Ph.D. dissertation at the University of Illinois at Urbana-Champaign: *A History of the American Eugenics Society, 1921-1940.*

## 1989

### Context

1. Historian Peter Weingart presents the standard argument that eugenics withered away on its own rather than having been politically suppressed: "Race hygiene having fallen into political and moral disrepute in the wake of the collapse of the Fascist state and its crimes, and with no governmental demands on that science or ideological support for it, human genetics began to launch attempts at professionalization as a depoliticized science."[2]

2. After Soviet leader Mikhail Gorbachev (b. 1931) resolves to allow free emigration of Soviet Jews, it becomes evident that over 90 percent wish to resettle in the United States. Partly at the request of Israeli Prime Minister Yitzhak Shamir (b. 1915), the United States

---

[1] Tendler, 1988, 87.
[2] Weingart, 1989, 282.

institutes a quota limiting immigration to the United States to 40,000, leaving the would-be Americans with no option other than either to remain in the U.S.S.R. or to emigrate to Israel.[1]

3. *New York Times*: "At Auschwitz, it is inscribed in stone: four million people died in the Nazi camps. But Yehuda Bauer [b. 1926], one of the foremost historians of the Holocaust and a sworn enemy of those who deny its reality, says that the number of victims was less than half that. Why is Mr. Bauer, the Director of the Division of Holocaust Studies at the Hebrew University of Jerusalem's Institute of Contemporary Jewry, insisting that far fewer people, including far fewer Jews, died at Auschwitz than is commonly reported? 'A historian's first duty is to tell the truth,' Mr. Bauer said. And in this case the truth is horrible enough. Exaggerating the number of dead at Auschwitz, he said, "would only be grist for the mills of the deniers of the Holocaust.""[2] A commemorative plaque at the camp claiming that 4 million people were murdered there is removed. The current plaque reduces that number to 1.5 million.[3]

### Jewish Advocacy of Eugenics

4. David W. Weiss, Hebrew University/Hadassah Medical School, Jerusalem: "The diaries and interrogation records of many of the major German war criminals make it clear that sadism was not, in many instances, a motivating element. Neither was rabid Jew-hatred an invariable component in the makeup of these persons. The far more common denominator was a scientifically framed, dispassionate belief in 'racial hygiene' (*Rassenhygiene*). It also became apparent to this writer, while serving in an intelligence unit of the United States Army in Germany after the war, that not a few of the accused mass murderers were idealists in every sense of the word but that which would have idealism subsume a commitment to the supreme value of individual human life and being."[4]

5. Jewish-American historian and eugenicist Nathaniel Weyl (1910-2005) describes "at least a thousand years of the history of the Jews of Europe. That history has sometimes been considered as a vast experiment, in which status was based on intellect serving religion,

---

[1] Lazin, 2005.
[2] Steinfels, 1989.
[3] Rense, 2005.
[4] Weiss, 1989, 151, 153, 157-158.

in which the intellectuals were commanded not to be chaste, but to be fertile, in which the rich and successful sought brilliant rabbinical scholars as husbands for their daughters, and in which family prestige was measured by pedigrees of scholarship."[1]

## Jewish Rejection of Eugenics

6. May 17-19: The Center for Biomedical Ethics at the University of Minnesota convenes a conference on bioethics and the Holocaust. The report on the conference, written by bioethicist Arthur Caplan, associates eugenics with euthanasia, genocide, and criminal medical experimentation, as in testing exposure to fatally cold temperatures. Caplan and colleagues "all argued that the racist underpinnings of Nazi ideology were firmly rooted in the racial hygiene theories prominent in German biology during the 1920s and 30s – long before Hitler came to power."[2]

# 1990

## Context

1. American Jewish intermarriage rate for 1990s: 46% (up from 41% in 1980s, 28% in 1970s, and 13% prior to 1970)[3]

2. The Committee to Examine In Vitro Fertilization in Israel adopts personal liberty as its fundamental principle. In a minority decision Rabbi Dr. M. Halpern cites a Halachic judgment that in some situations the good of the child dictates that it not be born.[4]

3. Researchers at the Neurology Department of Hadassah University Hospital, Ein Kerem, find the genetic defect which causes Creutzfeld-Jacob Disease (CJD) – a very rare disorder similar to Alzheimer's but affecting middle-aged people and killing them within months. This may be the first time a gene responsible for a neurological disease has been discovered and the nature of its defect defined in Israel. When asked if an abortion should be recommended because of a disease the unborn child may get at fifty, Ruth Gabizon, one of the researchers responsible for the discovery, responds:

---

[1] Weyl, 1989, 136.
[2] Caplan, 1989.
[3] National Jewish Population Survey, 2002.
[4] Landau, 1996. 39-42.

"We have asked the Ministry of Health to establish a committee to look into all the ramifications and give us guidance."[1]

4. American historian Mark B. Adams in *The Wellborn Science* (Yale University Press) presents eugenics as a phenomenon of history, not a school of thought vying for attention in today's marketplace of ideas. Eugenics, he maintains, was not repressed, but was naturally abandoned when geneticists disassociated themselves from it.[2] The general tone of books on eugenics becomes shriller, as in Troy Duster's *Backdoor to Eugenics* (Routledge, London).

5. Rabbi Arnold Wolf: "It is a simple fact that in New Haven, the Jewish community of 22,000 spends about ten times as much money on the Holocaust memorial as it does on all the college students in New Haven. I think that is shocking.... The community is saying: 'We have money for Holocaust, and that's all'.... It seems to me the Holocaust is being sold...."[3]

## Jewish Advocacy of Eugenics

6. Demographer Michael S. Teitelbaum steps down as President of the American Eugenics Society, having served five years.[4]

7. Jewish-American biologist and eugenicist Robert Sinsheimer: "The moral traditions of Christianity, Judaism, Islam, Buddhism, seem unlikely to provide a common view. Evolution may well have found a new arena for Darwinian selection. As sentient organisms, each of us might like to be able to choose his or her genome, but given the arrow of time, this will (mostly) never be possible. We cannot cut free from our genetic moorings. We (collectively) can only ever choose between genomes determined by chance – as have served us thus far – and, some-day, genomes determined by the best intentions of our predecessors. Thus the human condition evolves."[5]

8. Eliezer Waldenberg (1915-2006) of the Supreme Rabbinical Court in Jerusalem rules that abnormality of the fetus is sufficient justifi-

---

[1] Levavi, 1990.

[2] Adams, 1990, 200.

[3] A. J. Wolf, "The Centrality of the Holocaust Is a Mistake," in M. Berenbaum, *After Tragedy and Triumph: Essays in Modern Jewish Thought and the American Experience*, New York, 1990, 44-45; quoted in Cole, 2000, 1.

[4] http://www.eugenics-watch.com/aeugensoc/aeback.html, accessed May 12, 2008.

[5] Sinsheimer, 1990.

cation for termination of pregnancy with the first trimester provided there is no fetal movement.[1]

## 1991

### Context

1.  Israel's pronatalist policy includes child support up to 18 years of age, rising significantly with the fourth child, paid maternity leave, assistance in housing for young couples, and subsidized day-care centers.[2]

2.  Religious representatives and physicians in Israel are reported to label women undergoing abortion "reproductive deviants."[3]

3.  Rabbi Mordechai Halperin in the *Journal of Medical Ethics* of the Dr. Falk Schlesinger Institute for Medical-Halachic Research, Israel: "If before his death the man did not explicitly or implicitly agree to have his semen removed after his death for his wife to bear his children, then it is strictly forbidden to do so and there is no ha-lachic dispensation for performing the procedure. Second, if he did give explicit or implicit consent to the procedure, then the matter may depend on the different opinions among the poseqim [jurists] and a qualified halachic authority must be consulted."[4]

4.  Israeli bioethicist Noam J. Zohar: "Below a certain (somewhat hazy) threshold, we should be prepared to grant that some definite changes in genetic makeup will be properly judged as unaffecting personal identity. The functional perspective can provide a sensible account of the idea... of minimal genetic change, leaving the emb-ryo's identity unaltered. Modifications of genetic makeup judged in this perspective to be minor – will be, if beneficial, properly con-ceived as 'genetic therapy.'"[5]

5.  Zohar on surrogacy: "...if we allow artificial insemination from a donor in cases of male infertility, than (sic) we must put our minds to the parallel dilemma of a couple which is childless due to the woman's infertility. If bearing children is not only a right but also a

---

[1] Ziz-Eliezer, IX, No.51, ch.3; cited in Bleich, 1977, 113.
[2] Birenbaum-Carmeli *et al.*, 1992, 79.
[3] Birenbaum-Carmeli *et al.*, 1992, 79.
[4] Halperin, 1991, 28.
[5] Zohar, 1991, 275, 287.

duty, the implication of prohibiting a surrogate arrangement is forcing men into the tragic choice of foregoing children and failing to perform the *mitzvah*, on the one hand, and divorce, on the other. Moreover, forbidding or voiding surrogacy contracts flouts not only the autonomy of the married man, but also the autonomy of the woman contracting to bear the child."[1]

## Jewish Advocacy of Eugenics

6. In Israel a state-sponsored program of genetic counseling is introduced for the Bedouin minority, which suffers from a high prevalence of recessive genetic diseases, including thalassemia and congenital hearing loss, due to its traditional practice of polygamous cousin marriage.[2]

## 1992

## Context

1. At the Second International Congress of Yemenite Jewish Studies, held under the auspices of the Institute of Semitic Studies and the Committee for Jewish Studies, Princeton University, Israeli scholar Dr. M.A. Weingarten claims that Yemenite Jews are genetically close to Yemeni Muslims and that they are quite distant from other Jewish groups. The audience is described as "outraged."[3]

2. Israel's "Single Parents Law" entitles single parents to a special education grant and priority in vocational training programs and day-care programs. Another pronatalist measures is the non-funding of contraceptive means.

3. Book publishers finally begin promoting the association of eugenics with Hitler's Germany and racism in a major fashion. Some of the titles appearing this year include:

   • Arthur L. Caplan, *When Medicine Went Mad: Bioethics and the Holocaust*, Humana Press.

   • Elazar Barkan, *The Retreat of Scientific Racism: Changing Concepts of Race in Britain and the United States between the World Wars*, Cambridge University Press.

---

[1] Zohar, 1991, 18-19.
[2] Raz/Atar, 2004.
[3] Parfitt/Egorova, 2005, 204.

- Pauline Mazumdar, *Eugenics, Human Genetics, and Human Failings: The Eugenics Society, Its Sources and Its Critics in Britain*, Routledge.

- Christoph Beck, *Sozialdarwinismus, Rassenhygiene, Zwangssterilisation und* Vernichtung *"lebensunwerten" Lebens: Eine Bibliographie zum Umgang mit behinderten Menschen im "Dritten Reich"* [i.e. sterilization and murder of the handicapped], Psychiatrie-Verlag.

- Ulrike Schulz, *Gene mene muh, raus musst du: Eugenik, von der Rassenhygiene zu den Gen- und* Reproduktionstechnologien, AG SPAK.

- Jochen-Christoph Kaiser, Kurt Nowak, and Michael Schwartz, *Eugenik, Sterilisation, "Euthanasie": Eine Dokumentation*, Buchverlag Union.

- Franco Rest, *Das kontrollierte Töten* [Controlled Murder]*: Lebensethik gegen Euthanasie und Eugenik*, G. Mohn.

- Peter Propping, *Wissenschaft auf Irrwegen* [Paths of Madness]*: Biologismus, Rassenhygiene, Eugenik*, Bouvier.

- Gabriel Regine, Bärbel Maul, and Peter Sandner, *Informations- und Arbeitsmaterialien für den Unterricht zum Thema "Euthanasie"-Verbrechen* [the "crime" of euthanasia] *im Nationalsozialismus.*

- Veslemøy Kjendsli, *Kinder der Schande* [Children of Shame]*: Ein "Lebensborn-Mädchen" auf der Suche nach ihrer Vergangenheit*, Luchterhand.

- Daniel Kevles, *Controlling the Genetic Arsenal*, manuscript.

- Caroline Moorehead, *Himmler's Children*, Newspaper Publishing plc.

- Stefan Kühl, *The Nazi Connection: The American Eugenics Movement and the Racial Policies of German National Socialism*, M.A. thesis, Johns Hopkins University.

- Martin Arthur Elks, *Visual Rhetoric: Photographs of the Feeble-Minded during the Eugenics Era 1900-1930*, Ph.D. thesis, Syracuse University.

## Jewish Advocacy of Eugenics

4. A cumulative total of nearly one million young Jewish adults have been tested throughout the world for Tay-Sachs Disease, revealing more than 36,000 heterozygotes and 1,056 couples whose children are at risk. Of 469 fetuses diagnosed as affected, 451 (96%) are aborted.[1]

## 1993

### Context

1. The Bioinformatics Genome Center at the Weizmann Institute of Science in Rehovot and the National Laboratory for the Genetics of Israel Populations at Tel Aviv University are founded.[2]

2. The American Museum of Natural History radically updates the Hall of the Biology of Man, originally conceived by and built under the supervision of President of the American Eugenics Society Harry L. Shapiro, renaming it the Hall of Human Biology and Evolution.

### Jewish Advocacy of Eugenics

3. Canavan Disease, an autosomal recessive illness with a carrier frequency of 1 in 38 Ashkenazi Jews (1 in 5,000 active sufferers), is traced to a genetic defect, leading to a detection rate of almost 99% in this population.[3]

4. Chief Rabbi of the British Commonwealth of Nations, former Chief Rabbi of Ireland, and Spiritual Leader of New York's Fifth Avenue Synagogue Dr. Immanuel Jakobovits urges that if the techniques became available, then scientists should help eradicate the "abnormality of homosexuality." Gay groups are outraged.[4]

---

[1] Kaback M. *et al.*, "Tay-Sachs Disease: Carrier Screening, Prenatal Diagnosis, and the Molecular Era," *Journal of the Medical Association*, 270, 2307-2315; cited in Levin, 1999, 208.

[2] Segal, 1998, 23.

[3] "Jewish Genetic Diseases," *Jewish Genetic Disorders*, Sept., 2005, 68.

[4] Kossoff, 1993.

5.    Attitudes among Geneticists regarding Eugenic Policies[1]:

| The Statement (agree with) | Israel | Germany | USA |
|---|---|---|---|
| Before marriage, responsible people should know whether they or their prospective partner carries a genetic disorder that could be transmitted to their children. | 73% | 23% | 44% |
| A woman should have prenatal diagnosis if medically indicated by her age and family history. | 68% | 34% | 38% |
| It is not fair to a child to bring it into the world with a serious genetic disorder. | 68% | 18% | 40% |
| It is socially irresponsible knowingly to bring an infant with a serious genetic disorder into the world in an era of prenatal diagnosis. | 68% | 8% | 26% |
| The existence of people with severe disabilities makes society more rich and varied. | 10% | 38% | 24% |
| *Number of geneticists responding* | 23 | 255 | 1084 |

**Conflicting Jewish Views**

6.    Upon learning that geneticists Jerry Hall and Robert Stillman have achieved the first cloning of human embryos, anti-interventionist Jeremy Rifkin declares "This is the dawn of the eugenics era." Painting a dark picture of "standardized human beings produced in whatever quantity you want, in an assembly-line procedure," Rifkin organizes protests outside George Washington University and other reproductive-research institutions. By contrast, Arthur Caplan, director of the Center for Bioethics at the University of Minnesota, considers the cloning of human embryos medically appropriate under certain circumstances, as, for example, when a woman knows she is about to become sterile, either because of chemotherapy or

---

[1] Numbers extracted by Israeli scholar Yael Hashiloni-Dolev from data supplied to her by American Jewish-American bioethicist Dorothy C. Wertz (1938-2003) on basis of study done by Wertz and John C. Fletcher: "Geneticists Approach Ethics: A Survey in 37 Nations," *Social Science, Ethics and Law*, Shriver Center, 1993-1995; cited in Hashiloni-Dolev, 2007, xiv.

through exposure to toxic substances, and clones an embryo for future use; or when a couple knows that their children may inherit hemophilia or cystic fibrosis.[1]

7. Geneticist Jon Beckwith: "Today, dealing with the concerns about the social consequences of the new genetics and the Human Genome Project is being relegated, for the most part, to ethicists, social scientists, lawyers, and other non-scientists. Yet, those involved in the science have a key role to play and a responsibility to ensure that progress in their field is not used to harm rather than benefit people. This role calls for more knowledge of history and less hubris."[2]

## 1994

### Context

1. The National Laboratory for the Genetics of Israeli Populations is initiated by the Israel Academy of Sciences and Humanities as a national repository for human cell lines. The laboratory's Web site stresses the large ethnic variance between Israeli populations,[3] but does not discuss the political implications of this variance, although it does show the "Related Web site" of the Israel Ministry of Foreign Affairs.[4]

2. A Report of the Israeli Ministry of Justice shows that Israeli physicians support confidentiality of adults involved in donor assisted conception over the child's right to genetic identity.[5]

3. Oncologist Elliott Perlin: "Jewish law is permissive with respect to diagnostic techniques and genetic therapy if they can be performed with acceptable risk and will benefit humankind. However, Jewish and secular ethicists have not yet defined the limits of genetic therapy."[6]

4. The Bnei Menashe ('Children of Menasseh,' Hebrew בני מנשה, na) ethnic group from India's North-Eastern border states of Manipur

---

[1] Elmer-DeWitt/Bjerklie, 1993.
[2] Beckwith, 1993.
[3] http://nlgip.tau.ac.il, accessed May 12, 2008.
[4] http://www.mfa.gov.il/MFA, accessed May 12, 2008.
[5] Landau, 1998.
[6] Perlin, 1994, 335-336.

and Mizoram who claim descent from one of the Lost Tribes of Israel, begin arriving in Israel under the Law of Return. Linguistically, they are Tibeto-Burmans and belong to the Mizo, Kuki and Chin peoples (not to be confused with the so-called 'Bene Israel,' also from India.).[1]

5. Steven Spielberg's (b. 1946) *Schindler's List* is awarded seven Oscars at the Academy Awards Ceremony and is watched by 25 million Americans at movie theaters and 65 million on television.[2] Holocaust historian Tim Cole: "Spielberg's story does not stand alone. It is merely the most successful of the 'at least 40 films and 35 books that relate stories of Christian rescue of Jews which have appeared in the last decade."[3]

6. The publication of *The Bell Curve* by Charles Murray and Jewish-American psychologist Richard Herrnstein rekindles the debate about race and intelligence: "Jews – specifically, Ashkenazi Jews of European origins – test higher than any other ethnic group. The literature indicate that Jews in America and Britain have an overall IQ mean somewhere between a half and a full standard deviation above the mean, with the source of the difference concentrated in the verbal component."[4] Fifty-two prominent scholars and scientists, some of whom are Jewish, sign a collective statement in the *Wall Street Journal* describing the book as accurately reflecting the views of 'mainstream science,' as opposed to egalitarian views disseminated in the popular media.[5]

7. V. B. Penchaszadeh of the Department of Pediatrics, Beth Israel Medical Center, New York: "Worldwide, genetic diseases affect no less than 5% of all newborns. Most are caused by altered genes transmitted at conception, while a lesser share are due to chromosomal abnormalities – quantitative imbalances in the genetic material leading to various disorders including Down syndrome."[6]

---

[1] http://en.wikipedia.org/wiki/Bnei_Menashe, accessed January 25, 2009.
[2] Cole, 2000, 73-74.
[3] Cole, 2000, 89; citing A. H. Rosenfeld, *Thinking about the Holocaust; After a Half Century*,
[4] Herrnstein/Murray, 1994, 275.
[5] Gottfredson *et al.*, 1994.
[6] Penchaszadeh, 1994.

8.  Legal expert Dena S. Davis: "Because there is a lack of progressive Jewish materials on bioethics and a host of traditionalist writings, the non-Jewish scholarly world has tended to focus on the latter, if only by default. The prestigious *Encyclopedia of Bioethics*, for example, includes seven articles on Jewish topics, all by traditionalist scholars. This situation is exacerbated by the fact that Reform writers (and 'maverick' traditional writers) tend to identify themselves as such and to make clear to what extent they agree with or differ from the traditional sources they quote. Orthodox writers tend to publish books with titles such as *Jewish Bioethics*, which talk about 'the' Jewish view on various questions without giving the reader a clue that there might be other Jewish perspectives."[1]

9.  Evolutionary psychologist Kevin MacDonald: "At a fundamental level, a closed group evolutionary strategy for behavior within a larger human society, as proposed here for Judaism, may be viewed as pseudospeciation: Creation of a closed group evolutionary strategy results in a gene pool that becomes significantly segregated from the gene pool of the surrounding society. Within the strategizing group, there is increasing specialization so that the group is able to become extremely adept at occupying a specific type of niche that is commonly available in human societies. If the strategizing group then undergoes a diaspora and therefore lives among a wide range of human societies, members of the strategizing group, like conspecifics in the natural world, will have greater genetic ties with the dispersed members of their ingroup than with the other members of the society in which they live. Moreover, the within-group genetic commonality predisposes strategizing group members to relatively high levels of within-group altruism and cooperation, while the genetic barrier between the strategizing group and the surrounding society facilitates instrumental behavior directed toward the surrounding society. Moreover, the strategizing group is able to protect itself against freeloading individuals by instituting powerful social controls and belief systems so that a significant level of altruism is maintained within the strategizing group and cheaters who compromise group interests are punished."[2]

---

[1] Davis, 1994.
[2] MacDonald, 1994, 19-20.

## Jewish Advocacy of Eugenics

10. A prenatal diagnosis program using amniocentesis and chorionic villus sampling is offered free of charge to all pregnant Israeli women 37 and older; Jewish women are much more accepting of the procedure than are non-Jewish women.[1]

11. Oncologist Elliott Perlin: "The question might be asked why germline gene insertion is necessary. Is not somatic-cell therapy capable of correcting all genetic diseases? Probably not. How can one insert genes into the appropriate cells of brain tissue, liver tissue, etc.? Also, it is only through germline gene therapy that one could eliminate a defective gene from the family line."[2]

## Jewish Rejection of Eugenics

12. Anti-eugenics activist Barry Mehler in the magazine *Reform Judaism* writes disparagingly of studies of identical twins with their "legacy of Dr. Josef Mengele's twin experiments at Auschwitz." He is responded to by geneticist Irving I. Gottesman, who accuses Mehler of "character assassination" and comments on his conjuring up of Mengele: "This McCarthyistic tactic is especially offensive to the many Jewish scientists now engaged in the battle against ignorance of the causes of human suffering. As a Jew whose grandmother, five uncles and aunts, and some thirty more relatives were murdered in Auschwitz, and as a behavioral geneticist who has used the methods of twins, adoptees, and families... I am appalled by the blanket indictments and guilt-by-loose-association leveled at contemporary researchers." Mehler beats a retreat, writing that his article was merely "an exposé of... the eugenics-oriented Pioneer Fund.... It is most regrettable that Jewish scientists have become beneficiaries of the Pioneer Fund; it is sad and ironic that a Jew would use the victims of the Holocaust to uncritically defend twins research.... We must be vigilant. As Jews, we know all too well the danger of complacency in the face of evil."[3]

13. Germanist and historian Sander L. Gilman reviewing Kevin MacDonald's *A People That Shall Dwell Alone*: "Given the discussion of *The Bell Curve* and the question of Jewish superior intelligence,

---

[1] Davidov *et al.*, 1994.
[2] Perlin, 1994, 338.
[3] Mehler, 1994.

it is of little wonder that the sociobiologists would eventually come to the hoary chestnut. The manipulation of the idea of Jewish superior intelligence, as the other pole of the bell curve, can be seen in Kevin MacDonald's book (and its projected sequel on the sociobiology of anti-Semitism). MacDonald generally follows Nathaniel Weyl's 'eugenic' argument [*The Geography of American Achievement*]. However, MacDonald does call his own approach an 'evolutionary' one rather than a 'eugenic' one. His publishers are less subtle and sell his book under the advertisement 'Jewish Eugenics.'"[1]

## 1995
## Context

1. Political scientist and journalist Walter Truett Anderson (b. 1930): "it's not hard to imagine ugly scenarios connected with attempts to make it go away: self-appointed censors of scientific research, or police swooping down on the Jews of 'Dor Yeshorim' and telling them they must go back to having children with Tay-Sachs. Eugenics – whether we call it that or not – is here to stay. It is basically information which, once out in the world, tends to increase and circulate whether government approves or not. Eugenics becomes another one of the ever-growing class of things that ordinary people will learn about and fold into their lives, another set of choices they will make as the real arbiters, more than ever before, of their destinies."[2]

2. University of Manchester political scientist Hillel Steiner: "Future persons have no rights against present ones, the latter are fully at liberty to modify their own offspring's genetic endowments and those of their further descendants. And since they are self-owners, no one is at liberty to force others to do so or to prevent them from doing so. Nor therefore can mating or sterilization be permissibly reinforced."[3]

3. David Heyd, professor of philosophy at the Hebrew University of Jerusalem: "From society's point of view the prevention of homosexuality may be considered in either positive or negative terms. On the one hand, homosexuality could reduce fertility in a society which is badly in need of population growth. On the other hand, active steps to reduce the number of homosexuals may be seen as a

---

[1] Gilman, 1995a, 198.
[2] Anderson, *et al.*, 1995.
[3] 133, 138.

threat to the status and prospects of the existing homosexual community, which deserves respect from society."[1]

4.   A series of polls show that roughly 95% of Americans have heard of the term "Holocaust" and 85% claim to know what it means.[2]

5.   A psychological study from Tel Aviv University: "consanguineous couples who received pre-marital genetic counselling had fewer children, estimated their genetic risk as lower but its subjective significance as higher, and perceived genetic disorders as more severe."[3]

### Jewish Advocacy of Eugenics

6.   Rabbi David Moses Feldman in an article on eugenics and Judaism:

- The Old-Testament laws against consanguinity and incest have a rationale in eugenics.

- The Talmud counsels that a wife be chosen prudently, with regard to her intellectual and moral virtues, and also her maternal uncles (Bava Batra 110a).

- The Mishnah speaks of the qualities that a father bequeaths to his son: "looks, strength, riches, and length of years" (Eduyot, II, 9).

- Rabbi Judah the Pious advised against marriage with a niece because of potential genetic consequences.[4]

7.   Executive Director of Fairness and Accuracy in Media Dan Stein: "Immigration has created an altered sense of our vulnerability to outside forces in controlling our destiny and passing on to future generations a nation with the same qualities as those we inherited from our ancestors."[5]

---

[1] Heyd, 1995, 296.
[2] T. W. Smith, "The Polls – A Review: The Holocaust Denial Controversy," *Public Opinion Quarterly 59*, No. 2, 1995, 118; cited in Cole, 2000, 7.
[3] Shiloh *et al.*, 1995.
[4] Feldman, 1995.
[5] Quoted in Nelkin/Michaels, 1998, 53.

## 1996

## Context

1.  Surrogacy is legalized in Israel and paid for by the State.[1] As pointed out by David A. Frenkel of Ben-Gurion University, the law encourages contractual surrogacy with "gestational carriers."[2] Jewish religious law does not delegitimize the children of unmarried women, thus making it possible to combine Jewish legal principles with modern legal practices. In vitro fertilization and embryo transfer are preferred by some rabbis as a form of fertility treatment that does not violate the literal Halachic precepts against adultery.[3]

2.  Orren Alperstein Gelblum, mother of a child with Canavan disease, a rare, fatal neurological degeneration that strikes children, mainly those of Eastern European Jewish ancestry: "I do believe that people should have the information and should know realistically what it is they're facing. It's a very private decision that people make who discover that they're carriers. I can tell them the very complex, difficult experience it could be to have a Canavan child, but the decision of what to do is a very private one. If I were in the position of being pregnant and I knew I was carrying another Canavan child, I would choose not to have another Canavan child. But that doesn't diminish the experience that I had with my child who passed away."[4] Researchers think that one out of 40 Ashkenazis may carry the defective gene. Both parents must be carriers to pass it on. A child conceived by two carriers has a 50% chance of becoming a carrier and a 25% chance of becoming mentally and physically incapacitated.[5]

3.  Rabbi and physician Mordechai Halperin: "The Mishna (*Yadayin*, 4,3) emphasizes that only *prohibitive*, strict decisions require juridical substantiation while permissibility or leniency needs no supportive precedent. The absence of a prohibitive substantiation is to be equated with halachic permissibility. This implies that any technological innovation is permissible unless there is a halachic reason for prohibiting it. If in the *broad range of halachic sources* no reason is

---

[1] Kahn, 2000, 140.
[2] Frenkel, 2001, 606.
[3] Kahn, 2000, 74.
[4] Wahrman, 2002, 108.
[5] Lipsyte, 1996.

found for their prohibition, Jewish law permits the use of such technologies."[1]

## Jewish Advocacy of Eugenics

4.  At a panel discussion of the 7th International Conference on Judaism and Contemporary Medicine, Rabbi and Ph.D. Moshe Tendler comments: "We've been wary of eugenics since the Hitlerian era. Hitler used eugenics to destroy our people – bad eugenics, false eugenics, but eugenics never the less. We're now doing with our testing programs, we are confirming indeed, you have bad genes. What follows therefrom could very well be pleading with you, don't you pollute the gene pool. Don't have children. Followed by legislation to make sure you don't have children, which is the horror of eugenics when it becomes legislated, as indeed history has shown it does so. Once we enter into a screening program, the next question will be so what does public policy say about one percent of an Ashkenazi woman population that carries this bad gene, about the Jewish people, their responsibility to world society not to have children so that that gene dies out."

5.  Nobel Prize winner and geneticist Joshua Lederberg (1925-2008) responds: "I was provoked by Dr. Tendler's remarks. I was so exorcised about the issues of eugenics… that I introduced a counter program that I called euphenics. My belief and hope – firm belief – was that genetic analysis is not going to stop at testing, it's not going to stop at diagnosis. And just as was implied here, that a still deeper examination of the effects of these genes will result in direct remedies, not merely diagnosis, not merely of the people that might be affected by it that is affecting the phenotype, hence euphenics, as against efforts to go after the genotype, the eugenic program. I'm quite confident we will see that. We will see that with Alzheimer's. We'll see that with breast cancer just along the lines we just indicated. So there is that ray of hope for the future."[2]

6.  The Knesset enacts the *Agreements to Carry Embryos Law 5756-1996*. Professor of Business Administration David A. Frenkel of the Ben-Gurion University of the Negev, Beer-Sheva, comments: "There is a danger of 'commodification' of children. Abusing wom-

---

[1] Halperin, 1996.
[2] Lederberg *et al.*, 1996.

en of low-socio-economic status as breeding machines may be another outcome. No clear responsibility is imposed on the 'intended parents' before the child's birth. Splitting motherhood is another social problem that has to be dealt with. So far the sperm of the husband from the 'intended parents' has to be used, but further steps may follow. It is not certain that a policy of 'positive eugenics' will not develop."[1]

## 1997

## Context

1.  The Chief Rabbi of Petah Tikvah orders his employees not to validate marriages for members of the group Bene Israel, which has immigrated to Israel from India.[2]

2.  Simon and Schuster publishes Alan Dershowitz's *The Vanishing American Jew*: "...American Jews – as a *people* – have never been in greater danger of disappearing through assimilation, intermarriage, and low birth rates."[3]

3.  Conservative Rabbi Joel Meyers on a ruling by the Committee on Jewish Law and Standards of the Rabbinical Assembly: "The sole position is that the religious status of the child follows that of the gestational mother in cases involving surrogacy and in all other cases."[4]

4.  In Israel surrogacy is legalized for married women and is paid for by the State.[5] Jewish religious law does not delegitimize the children of unmarried women, thus making it possible to combine Jewish legal principles with modern legal practices. In vitro fertilization and embryo transfer are preferred by some rabbis as a form of fertility treatment that does not violate the literal Halachic precepts against adultery.[6]

---

[1] Frenkel, 2001, 605.
[2] Parfitt/Egorova, 2005, 207.
[3] Dershowitz, 1997, 1.
[4] Berck, 2007.
[5] Kahn, 2000, 140.
[6] Kahn, 2000, 74.

## Jewish Advocacy of Eugenics

5. Israel Penal Law on "Interruption of Pregnancy (312-321) acknowledges an embryopathic indication as a just cause for abortion *throughout* pregnancy and states that termination is allowed in case "the newborn is likely to have a mental or physical defect."[1]

6. Medical ethicist and Rabbi Louis Waldman at Knesseth Israel in Far Rockaway, New York: "We believe that you can conquer nature, master nature, even manipulate nature for the ultimate welfare and benefit of mankind."[2]

7. May 26-29: a conference is held in Jerusalem and Tel Aviv with the title "Eugenic Thought and Practice: A Reappraisal towards the End of the Twentieth Century," proceedings published in *Science in Context*.[3] The editors of *SiC* write that some of the advocates of eugenics were well motivated, but naively utopian, and go on to complain that "To label a policy 'eugenics' became *ipso facto* to condemn it.... We aimed to confront rather than simply dismiss, the ethical questions. In these ways, we hope to contribute to a literature that is increasing not just in size but in sophistication." The impressive sponsorship of the conference gives evidence of a new, less shrill ideological tone: it is convened by the Cohn Institute for the History and Philosophy of Science and Ideas at Tel Aviv University, the Edelstein Center for the History and Philosophy of Science, Technology, and Medicine at the Hebrew University of Jerusalem, and the Van Leer Jerusalem Institute. It is also supported by the Israel Academy of Science and Humanities in cooperation with the Division for Development and Public Relations of the Hebrew university of Jerusalem, the Genome Center at the Weizmann Institute of Science, and the Goethe Institute, Jerusalem.

8. Israeli obstetricians Vered H. Eisenberg and Joseph G. Schenker of the Hadassah University Medical Center: "Eugenic genetics is purely theoretical at present and is likely to remain so for a long time.... Still, society must be concerned about the possibility that gene therapy will be misused in the future. Gene therapy should only be used in ways that maintain human dignity. The best insurance against mi-

---

[1] Hashiloni-Dolev, 2007, 86.
[2] Waldman, 1997.
[3] Vol. 11, 3-4, 1998.

suse is a public well informed and not necessarily frightened. With proper safeguards imposed by society, gene therapy can be ethically used."[1]

## Jewish Wariness toward Eugenics

9.  Jewish-American bioethicist Paul Root Wolpe: "We are in the process of a fundamental change in the nature of the self. The problem is not the cloning of an army of Hitlers and the solution is not the Luddite reaction against technology. Rather, we must monitor the slow, fundamental change in our conceptions of ourselves and our place in the world. We must be wary of the temptation of the possible, and we must draw from the deep fount of accumulated human wisdom to temper and judge developments that can so profoundly alter the nature of our existence."[2]

## 1998

## Context

1.  IsraelWire: "...Based on a study of 306 Jewish men in Israel, Canada and England, the researchers discovered that the 106 Jews who had identified themselves as kohanim shared genetic markers in their Y chromosomes that members of the general Jewish population did not.... The study also found a predominance of certain chromosome features in kohanim of both Ashkenazi and Sephardi origin.... But Jonathan Marks, a biological anthropologist at the University of California at Berkeley, has difficulty accepting the study's results. 'I'm a skeptic,' he said. 'What they're doing is Mickey Mouse social science.' The problem, he said, is their interpretation of the facts.... Besides, he continued, 'there's no reason to think that there even was a priestly Aaron. It's an origin myth. To take at random something from the deep hoary past as if it's literally true and use that as your starting point, there's a problem with that. It's not science.'... Michael Hammer, a geneticist at the University of Arizona who worked on parts of the study, said Marks' criticisms 'were fairly irrelevant. It was a test case for genetics to see if the Y chromosome can be consistent with patrilineal descent.'"[3]

---

[1] Eisenberg/Schenker, 1997, 314.
[2] Wolpe, 1997, 227.
[3] "Jewish Priestly Line Maintains Legacy – and Genetic Marker," *IsraelWire*, September 23, 1998; Excerpted in http://khazaria.com/genetics/abstracts-cohen-levite.html, accessed Sept. 26.

2.  Geneticist and Nobel Prize winner James Watson states that the time has come "to put Hitler behind us": "Those of us who venture forth into the public arena to explain what Genetics can or cannot do for society seemingly inevitably come up against individuals who feel that we are somehow the modern equivalents of Hitler. Here we must not fall into the absurd trap of being against everything Hitler was for.... Common sense tells us that if scientists find ways to greatly improve human capabilities, there will be no stopping the public from happily seizing them."[1] He also chides Germany for not having purged its discredited geneticists, whom he accuses of bad science.[2] "Those disposed to see a cloud in every silver lining have managed to cast doubt on research about the causes of and possible cures for genetic diseases prevalent among Ashkenazi Jews.... Those Jewish spokesmen who challenge such research raise the specter that the research will give rise to theories about the genetic inferiority of Jews. But it is a dishonest use of history to fail to recognize the differences between racist Nazi eugenics and responsible scientific research to relieve human suffering. One can only hope that the Jewish community will prove more sober than some of its leaders, who imagine stigma and harm to Jewish self-image where none exist."[3]

3.  Boca Raton Orthodox Rabbi Kenneth Brander is asked for counsel regarding a child born by egg donation. Brander's response reflects the unresolved nature of the topic at the time: "I couldn't even articulate the question, let alone process an answer. I couldn't be the spiritual caregiver I wanted to be, because I didn't have any understanding of egg donations or reproductive physiology." The concern is over the child's Jewishness if the donor is not Jewish, and also that the practice might unwittingly lead to marriage between family members. After spending a year in Israel studying reproductive technologies, Brander reports that Orthodox authorities are split on the subject, but in practice most Orthodox rabbis perform a conversion on the infant, "just in case." Brander comments: "We can tell them: 'Don't worry. You can embrace the gift of science, which is a gift from God, without having to worry about the issue of your child

---

[1] Watson, 1997, 636.
[2] Koenig, 1997, 892.
[3] Waldman, 1998.

being Jewish. And celebrate that Judaism can embrace this with enthusiasm.'"[1]

4.  Dean and Founder of the Jewish Institute of Bioethics Rabbi David
    M Feldman: "The laws against incest and consanguinity in the Old
    Testament would seem to have a rationale in eugenics...."[2]

5.  Geneticists Neil Bradman and Mark Thomas: "Notwithstanding the
    identification of the CMH [Cohen modal haplotype], it is not possi-
    ble to say that those are the markers of a 'true' Cohen or whether,
    indeed, there was a 'first Cohen' – be it Aaron or somewhere else.
    In a similar way, there is no Jewish haplotype and genetics cannot
    'prove' whether someone is a Jew; that is a matter for religious au-
    thorities. Nor can genetics decide whether a particular community is
    or is not Jewish."[3]

6.  Evolutionary psychologist Kevin MacDonald: "The entire enter-
    prise [i.e., Boasian anthropology] may thus be characterized as a
    highly authoritarian political movement centered around a charis-
    matic leader. The results were extraordinarily successful. [As noted
    by anthropologist George W. Stocking,] 'The profession as a whole
    was united within a single national organization of academically
    oriented anthropologists. By and large, they shared a common un-
    derstanding of the fundamental significance of the historically con-
    ditioned variety of human cultures in the determination of human
    behavior.' Research on racial differences ceased, and the profession
    completely excluded eugenicists and racial theorists like Madison
    Grant (1865-1937) and Charles Davenport (1866-1944). By the
    mid-1930s the Boasian view of the cultural determination of human
    behavior had a strong influence on social scientists generally. The
    followers of Boas also eventually became some of the most influen-
    tial academic supporters of psychoanalysis. Marvin Harris notes that
    psychoanalysis was adopted by the Boasian school because of its
    utility as a critique of Euro-American culture, and, indeed, as we
    shall see in later chapters, psychoanalysis is an ideal vehicle of cul-
    tural critique. In the hands of the Boasian school, psychoanalysis

---

[1] Berck, 2006.

[2] Feldman, 1998.

[3] Bradman, Thomas; Thomas, Mark. 1998. "Genetics: The Pursuit of Jewish History
by Other Means," *Judaism Today* 10 (Autumn), 4 -6. Excerpted in
http://khazaria.com/genetics/abstracts-cohen-levite.html, accessed Sept. 26, 2008.

was completely stripped of its evolutionary associations and there was a much greater accommodation to the importance of cultural variables."[1]

7.   MacDonald again: "There is an eery sense in which National Socialist ideology was a mirror image of traditional Jewish ideology. As in the case of Judaism, there was a strong emphasis on racial purity and on the primacy of group ethnic interests rather than individual interests. Like the Jews, the National Socialists were greatly concerned with eugenics. Like the Jews, there was a powerful concern with socializing group members into accepting group goals and with the importance of within-group altruism and cooperation in attaining these goals. Both groups had very powerful internal social controls that punished individuals who violated group goals or attempted to exploit the group by freeloading. The National Socialists enacted a broad range of measures against Jews as a group, including laws against intermarriage and sexual contact, as well as laws preventing socialization between groups and restricting the economic and political opportunities of Jews. These laws were analogous to the elaborate social controls within the Jewish community to prevent social contact with gentiles and to produce high levels of economic and political cooperation. Corresponding to the religious obligation to reproduce and multiply enshrined in the *Tanakh*, the National Socialists placed a strong emphasis on fertility and enacted laws that restricted abortion and discouraged birth control. In a manner analogous to the traditional Jewish religious obligation to provide dowries for poor girls, the National Socialists enacted laws that enabled needy young couples to marry by providing them loans repayable by having children."[2]

8.   Legal scholar Sheila A.M. McLean: "In particular, the Human Genome Diversity Project has raised profound ethical concerns about what it tells us of the Western World's attitude to ethnic minorities."[3]

---

[1] MacDonald, 1998b, 28; the references are to Stocking's *Race, Culture, and Evolution: Essays in the History of Anthropology*, and Harris's *The Rise of Anthropological Theory: A History of Theories of Culture.*
[2] MacDonald, 1998a, 161.
[3] McLean, 1998, 687.

9.  Israeli geneticist Raphael Falk: "There is no doubt that Jewish communities live in various degrees of reproductive endogamy and isolation from their neighbors in the countries of their dispersal, and from other Jewish communities.... The history of Zionism and the biology of the Jews expose... the futility of any attempts to demarcate groups by 'racial' or 'genetic' or 'DNA' characteristics *within* the human species, even in a context intended for liberation from a history of discrimination and persecution."[1]

10. An angry exchange of views takes place between Canadian physician Eyal Cohen and Canadian historian Wilhelm Kreyes:

    - Cohen: "Although most German physicians did not participate in the heinous crimes attributed to the Nazi Doctors, *collectively as a profession* [emphasis added], they were not just victims of the oppressive rule of Hitler and his collaborators. Attracted to the biologically based tenets of Nazi doctrine and benefiting from its gracious treatment of medicine, physicians played an integral part in the orchestration of the Nazi state."[2]

    - Kreyes accuses Cohen of character assassination: "The victims here are the German colleagues who cannot defend themselves because they are now dead. These are the same colleagues, as I respectfully remember, who put their lives on the line, who lost their health and lives during the Second World War, and who suffered more casualties than any other military unit in their attempts to help the wounded – no matter which army they belonged to. It eludes me how they could have 'collaborated in the infamous Nazi programs,' yet at the same time display through their selfless and self-sacrificing actions such exemplary morals?"[3]

11. Molecular geneticist Michel Revel of the Weizmann Institute of Science: "Is Human Cloning Feasible?... We cannot entirely eliminate maternal influence on genetic programming during pregnancy. More scientific research is required before thinking about safely applying the procedure to humans. This is why research must be allowed to proceed."[4]

---

[1] Falk, 1998, 603, 605.
[2] Cohen, 1998, 339.
[3] Cohen, 1998.
[4] Revel, 1998.

12. Rambam Hospital in Haifa, Israel, supplies human embryos to the University of Wisconsin which are used to produce the first human cell lines.[1]

13. At the General Assembly of United Jewish Appeal Federations of North America Rabbi Adin Steinsaltz maintains that Jewish traits, "ranging from pushiness to intelligence," are inbred by natural selection. He is countered by geneticist Robert Pollack, who stresses environment over genetic determinism.[2]

14. Sociologist Barbara Katz Rothman: "I'm not made of stone. My eyes fill when the childless woman cries out her need, when the mother keens over the body of her child dead of some genetic disease, when the young man watches his father die of a disease he and his newly born son share, linking them across generations in a family tragedy in far more than three acts. My children will not be led to genetic technology in chains and shackles, or crowded into cattle cars. It will be offered to them. It will be sold to them. So why am I so afraid of a technology that offers to solve so many problems? Why am I so profoundly skeptical? How can I explain, justify, defend my sense of distrust?"[3]

15. An Israeli-American medical team grows human stem cells in culture.[4]

16. New York Rabbi Tzvi Flaum: "There is a broad consensus of the *poskim* [authorities on Jewish law] regarding the pre-embryo. At the insertion of the zygote [fertilized egg] into the uterus, issues of abortion begin. The concept of destruction of the fetus/embryo is based upon the fact that left untouched it will develop into a full human." An embryo that has not been implanted and will thus not develop fully and thus may be frozen, discarded or used for experimentation.[5]

---

[1] Traubmann, 2004a.
[2] Rabinovich, 1998.
[3] Rothman, 1998, 502.
[4] Wahrman, 2002, 55-56.
[5] Wahrman, 1998.

## Jewish Secular Advocacy of Eugenics

17. An Israeli team working with the Negev Arab Bedouin community: "a well-targeted approach in terms of the identification of families at risk for devastating genetic diseases, coupled with appropriate genetic education and counseling, can make good use of the available genetic technology."[1]

18. Noting that "Israel is engaged in cloning, mapping, sequencing, and evaluating several human genes, primarily of pathological implications and with Jewish orientation," Yossi Segal of the Israeli Academy of Sciences and Humanities is cautious in predicting future use of the four potential areas for the application of genetic engineering designed to insert a gene into a human: somatic cell therapy, germline gene therapy, enhancement genetic engineering, and eugenic genetic engineering.[2]

19. Israeli geneticist Raphael Falk: "Racial and eugenic notions have persisted, though in a thinly disguised mode, in post-World War II Israel…. Whether or not the human species should be more alert to the qualitative, and not only to the quantitative consequences of its reproductive policies, whether the impact of modern techniques of genetic manipulations should also be examined by their effects on the gene pool, are important issues towards the twenty-first century."[3]

20. Jewish-Russian geneticist Vladimir Pavlovich Efroimson: "I believe in transhumanism. If one day there will be enough people capable of striving toward this goal, the human species will find itself on the threshold of a new way of life – one that will be as different from ours as ours is from that of Sinanthropus [Peking man], and man will finally begin to fulfill his true destiny."[4]

## Jewish Religious Advocacy of Eugenics

21. Professor of medicine and specialist on Jewish medical ethics Fred Rosner of the Mount School of Medicine: "It is prohibited in Jewish law to marry a woman from a family of epileptics or lepers (Yeba-

---

[1] Carmi, *et al.*, 1998, 395.
[2] Segal, 1998, 24, 29.
[3] Falk, 1998, 604-605.
[4] Efroimson, 1998, 288.

mot 64b; Maimonides' Mishneh Torah, Issurei Biyah 21:30; Karo's Shulchan Aruch, Even Haezer 2:7) lest the illness be genetically transmitted to future generations. According to Rashi (Yebamot 64b), any hereditary disease is included in this category."[1]

22. Noam J. Zohar, a professor of philosophy at Bar-Ilan University in Israel, responds to rabbi Max Reichler's 1910 essay "Jewish Eugenics." He notes that Reichler's emphatically pro-eugenics views are "shared... by more than a few Judaic circles today": "A program of individualized eugenics... would seem to be consonant with an attitude that was, at the very least, tacitly endorsed by traditional Judaic teachings.... To work out a Judaic response to the sort of new eugenics now looming on our horizon it will be necessary to evaluate the various specific means that might serve a modern individualized eugenics. I hope that some of the groundwork for that has been laid in this examination of traditional Judaic voices."[2]

23. Rabbi Moshe Tendler of Yeshiva College complains that many Jewish newspapers have accepted advertisements seeking Ashkenazi families to study the biological basis for bipolar disorder and schizophrenia, fearing that such materials will reawaken the idea that "Jews carry genes that are polluting the world. That's the basis of eugenics. If you have a [disease] gene, don't you owe it to society not to propagate that gene?"[3]

24. Conservative rabbi and bioethicist Elliot N. Dorff (b. 1943) advocates repairing mutated genes: "Jews have the duty to try to prevent illness if at all possible and to cure it when they can, and that duty applies to diseases caused by genes...."[4]

### Jewish Rejection of Eugenics

25. The 'neoconservative' Jewish magazine *Commentary* publishes an anti-eugenics article by clergyman (originally Lutheran, later Roman Catholic) Richard John Neuhaus (1936-2009), who warns that eugenics is back "with a vengeance." Readers' comments are numerous, for the most part attacking eugenics, but two readers point out that Neuhaus has, as is commonly done, associated eugenics

---

[1] Rosner, 1998, 408.
[2] Zohar, 1998, 584-585.
[3] Wahrman, 1998.
[4] *Matters of Life and Death*,1998, 157; cited in Wahrman, 2002, 178.

with practices that have no relation to reproduction, such as pro-longing the lives of the irreversibly comatose.[1]

26. Anti-eugenics activist Jeremy Rifkin: "the new eugenics bears little resemblance to the shrill cries of racial purity that culminated in the Holocaust. The old eugenics was motivated by fear and hate; the new eugenics is spurred by market forces and consumer desire. Genetic engineering is coming to us not as a sinister plot, but rather as a social and economic boon...."[2]

27. When the Jewish-American anti-eugenics activist Barry Mehler accuses psychologist Raymond B. Cattell (1905-1998) of "fascism," "racism," and "eugenics," the American Psychology Association postpones granting Cattell a lifetime achievement award.[3] Cattell, who has authored or co-authored over 50 books, 500 articles, and over 30 standardized tests, and is rated in one ranking as the six-teenth most influential and eminent psychologist of the twentieth century,[4] dies a few months later.

## 1999

### Context

1. Professor Ze'ev Herzog, Department of Archaeology and Ancient Near Eastern Studies at Tel Aviv University, writing in *Ha'aretz*: "Following 70 years of intensive excavations in the Land of Israel, archaeologists have found out: The patriarchs' acts are legendary stories, we did not sojourn in Egypt or make an exodus, we did not conquer the land. Neither is there any mention of the empire of David and Solomon. Those who take an interest have known these facts for years, but Israel is a stubborn people and doesn't want to hear about it."[5]

2. The journal *Human Immunology* publishes a study of genetic distances between Israeli ethnic groups, concluding that Jews share a common ancestry.[6] Yale University geneticist Mazin Qumsiyeh

---

[1] Neuhaus, 1998.
[2] Rifkin, 1998.
[3] Hilts, 1997, A10.
[4] S. J. Haggbloom *et al.* (2002), "The 100 most eminent psychologists of the 20th century", *Review of General Psychology*, 6(2), 139-152.
[5] October 29, 1999.
[6] Amar *et al.*, 1999, 723.

protests in a letter to the Society of Histocompatibility and Immunology: "Valid scientific research must not be shunned by political pressure groups intent on preventing any rational discussion and stifling apparent conflict with the aims of Zionism. Similarly, scientists should not be allowed to publish statements and conclusions not supported by the data simply because they appear 'politically correct' at the moment or do not generate an outcry. A statement such as that by Amir et al. that 'We have shown that Jews share common features, a fact that points to a common ancestry' should not be allowed to stand. The correct statement from their own data is that some Jews (Sephardim) are more similar to Palestinians than either group is to other Jews (Ashkenazim or Ethiopian Jews)."[1]

3.   Tim Cole, former Paul Resnick Resident Scholar at the Center for Advanced Holocaust Studies at the United States Holocaust Memorial Museum: "At the end of the twentieth century the 'Holocaust' is being bought and sold. $168,000,000 was donated to pay for the building of the United States Holocaust Memorial Museum on a plot of Federal Land in Washington, D.C. Millions of dollars more have financed memorial projects throughout the United States, ranging from the installation of holocaust memorials to the establishing of University chairs in Holocaust Studies. Steven Spielberg's 1993 movie *Schindler's List* netted over $221 million at foreign box offices and seven Academy Awards. In short, 'Shoah business' is big business."[2]

4.   The Lemba, a largely Christian endogamous tribe (some Muslims) resident in Zimbabwe and South Africa, is found to observe certain Semitic customs – a fact that attracts great attention in the United States, even though they are physically indistinguishable from their Venda or Shona neighbors. Tudor Parfitt of the European Association for Jewish Studies and Yulia Egorova studying under him at the School of African Studies, University of London, comment that such assertions "may represent attempts by some Jewish groups to traverse barriers that can appear insurmountable when dealing with more closely situated black communities at home in New York, Washington, and elsewhere. For these groups, then, the genetic studies on the Lemba may be presumed to have great ideological value, as they may be used to prove that Zionism or Judaism are not, as

---

[1] http://eaazi.blogspot.com/2005_07_01_archive.html.
[2] Cole, 2000, 1.

claimed by their detractors, racist." Even though most Lemba vehemently reject the claim, one rabbi writes that others have abruptly declared themselves "determined to re-affirm their Jewishness and their allegiance to Judaism." Other African tribes follow suit, arguing that they "should be admitted as a matter of course and urgency into '*kelal Yisrael*,' the family of Israel."[1]

5.  Editorial in *The Gazette*, Montreal: "Christian Identity...presents a neat theological package in which minorities are not even human and Jews are literally the sons and daughters of Satan. Therefore, any assault on pre-human 'mud people' or demonic Jews is not only desirable but divine. Minorities, who are depicted as responsible for crime and other social ills, are seen as the tool of the Jew-devils in their war against the white Aryans. Nothing less than survival of the white genotype is at stake."[2]

6.  *Washington Times* columnist Suzanne Fields attacks Jewish bioethicist Peter Singer: "The crippled and the lame, on crutches and in wheelchairs, walking slowly with friends and family, will gather at Princeton University on Tuesday, Sept. 21 to bear witness for humanity. Students with muscular, athletic bodies as well as the bowed and infirm will register their outrage. Protestants, Catholics and Jews (in the season of the New Year when the Jewish people pray for their names to be written down in The Book of Life) will join hands to express their fury at the presence of a professor on their campus whose intellectual coldness and academic credentials have led him to a prestigious chair as tenured professor. This is a 'scholar,' whose ideas, if they had prevailed, would have denied many of them a life on the planet. Why should such a man be invited to teach ethics at Princeton?"[3]

## Jewish Advocacy of Eugenics

7.  Bioethicist Jonathan R. Cohen: "In time, we may well see a world in which many people will be cloned or genetically engineered, while others will be created through traditional means. Perhaps both will be pleasing in God's eyes."[4]

---

[1] Parfitt/Egorova, 2005.
[2] Greenebaum, 1999.
[3] Fields, 1999.
[4] Cohen, 1999, 11-12.

8. The Michael Reese Health Trust awards a three-year grant to establish the Chicago Center for Jewish Genetic Disorders.[1]

9. Ruth Schwartz Cowan, professor of the history and sociology of science at the University of Pennsylvania, travels to Cyprus to study thalassemia, which is prevalent on the island, and also the local genetic-screening program intended to prevent new cases by aborting active carriers. An article in the Chronicle for Higher Education describes the program as "eugenics" – a term that participants in the program, including Cowan, resolve to avoid.[2]

10. The Knesset imposes a five-year moratorium on human cloning. Rabbi Dr. Avraham Steinberg, head of the Medical Ethics Program at Hebrew University, calls it "a very bad law. So who are we, the small Knesset of small Israel to do such an overriding prohibition that forever any research involved should be outlawed?"[3]

11. Jewish-German-American Arno (Arnold) Motulsky in an article entitled "If I had a gene test, what would I have and whom would I tell?" provides a lengthy list of genetic illnesses.[4]

## The International Scientific Consensus

12. The Directors of the American Society of Human Genetics claim to condemn coercive measures but at the same time essentially restate the classical eugenics platform, noting that they decided to "de-emphasize" the word 'eugenics' in their "Statement": "Many governments support programs, in the interests of improving the odds that children will be healthy. Some are mandatory. In our view, none involve the misuse of genetic information. Examples include:
   - programs to encourage or discourage the number of births among the entire population;
   - laws that try to protect the fetus from environmental harm (e.g., warnings on cigarette packages about the risk of smoking during pregnancy);
   - laws that implement newborn genetic screening programs;

---

[1] "Chicago Center…" 2005, 61.
[2] Guterman, 2003.
[3] "Cloning – Jewish Medical Ethics," lecture delivered at Congregation Shomrei Tirah, Fair Lawn, NJ, Jan. 5; cited in Wahrman, 1998.
[4] Motulsky, 1999.

- laws or regulations that fund genetic services, including genetic counseling, genetic testing, prenatal diagnosis, and the provision of special dies for newborns with certain inborn errors of metabolism; laws forbidding marriage between first cousins and consanguineous unions."[1]

## 2000

### Context

1. A large proportion of contemporary Jewish Kohanim are found to share a set of Y chromosomal genetic markers, known as the Cohen Modal Haplotype. Although this haplotype is not limited to Jews, the finding attracts a considerable amount of attention in lay circles.[2]

2. The President of California State University at Long Beach receives a number of demands that evolutionary psychologist Kevin MacDonald be stripped of tenure over his sociobiological study of the Jews.[3] New York editor and cultural columnist for the online magazine *Slate* Judith Shulevitz attacks MacDonald's three-volume sociobiological study of the Jews: "it is the job of a scholarly association not just to foster discussion but also to police the boundaries of its discipline."[4]

3. Israeli geneticists discover that the Indian group Bene Israeli, as well as Ethiopian and Yemeni Jews, have a higher frequency of Haplogroup 9 than does the population of India, leading to popular speculation that they may be at least partly of Jewish descent. The Bene Israel exult: "We always knew we were Jewish, now we know we are Cohens."[5]

4. Article in Boston Globe: "Jews Fear Stigma of Genetic Studies."[6]

5. Fred Rosner, Director of Medicine at the Long-Island Jewish-Hillside Medical Center: "Genetic screening, gene therapy, and oth-

---

[1] Directors, 1999, 337.
[2] Hammer *et al.*, 2000.
[3] Schneider, 2000.
[4] Shulevitz, 2000.
[5] Parfitt/Egorova, 2005, 214-215.
[6] Wen, 2000.

er applications of genetic engineering for the treatment, cure, or prevention of disease fulfills the biblical mandate to heal"[1]

6.  Norman G. Finkelstein (*b.* 1953), American historian and son of Holocaust survivors, launches a frontal attack on the Holocaust Memorial Movement in his book *The Holocaust Industry*, calling much of it "worthless as scholarship. Indeed, the field of Holocaust studies is replete with nonsense, if not sheer fraud."[2]

7.  From an interview granted by Israeli geneticist Raphael Falk to Dalia Karpel of *Ha'aretz* newspaper: "There is no biological way to say who is a Jew – that is, who is included in the population.... The Jews were generally isolated in their own 'reproductive circles' (or gene pools, as geneticists would say) from the populations among which they resided, because of their religion, their customs and the prejudices regarding them. Obviously, the isolation of reproductive populations is always relative and there is always 'leakage' of genes to and from a population...."

8.  The Anti-Defamation League attacks the Federation for American Immigration Reform (FAIR), chaired by Dan Stein: "FAIR opened itself to such criticism with unretracted offensive statements by several FAIR leaders, and by its willing acceptance of financial support from the Pioneer Fund, a controversial foundation with a tainted history that was established to promote the discredited "science" of eugenics, and that continues to financially support questionable research into the comparative intelligence of ethnic minorities."[3]

### Jewish Advocacy of Eugenics

9.  Student Naomi Stone: "I am an Ashkenazi Jew, and I know that it is my obligation to be acutely aware of my heightened risk factor for the disease [Tay Sachs].... Dor Yeshorim is a program of eugenics, whatever its overriding aims and moral purposes may be. There is an unspoken irony: not simply that a conservative group clings to such an inherently bold and controversial genetic project, but in the inevitable linkage of eugenics to the persecutors of those who now embrace Dor Yeshorim."[4]

---

[1] Rosner, 2000.
[2] Op. cit. 41-42, 55.
[3] Anti-Defamation League, 2000
[4] Stone, 2000.

10. *B'Or Ha'Torah: Journal of Science, and Modern Life in the Light of the Torah* asks bioethicist Fred Rosner: "Does halakha [Jewish religious law] sanction eugenics? From Rosner's response: "Genetic screening, gene therapy, and other applications of genetic engineering for the treatment, cure, or prevention of disease fulfills the biblical mandate to heal. If Tay-Sachs Disease, diabetes, hemophilia, cystic fibrosis, Huntington's disease, or other genetic diseases can be cured or prevented by "gene surgery," it is permitted in Jewish law."[1]

11. Israeli bioethicists Rafi Cohen-Almagor and Merav Shmueli: "The state's interest in preserving life is a most meaningful interest, but it is not an absolute interest. Therefore, in certain cases it is possible to evaluate life and to determine that a certain characteristic could make it better or worse in comparison with other lives. The Kantian view that conceives of people as ends rather than means leads us to conclude that life is not sanctified when the continuation of life harms human dignity and contradicts the patient's best interests."[2]

12. Rabbi Byron L. Sherwin: "whether genetics serves as a source of blessing or as a curse for humankind ultimately depends upon whether we use wisdom and humility in utilizing the knowledge and the power acquired through decoding the secret language of life encoded into our every cell by the author of life. Or, as the Talmud might put it: from the bee, one can receive either its honey or its sting. Genetics may provide us either with hell or healing – but, the danger of the sting ought not scare us away from the sweetness of the honey."[3]

13. Philip J. Boyle, editor of *Bulletin of the Park Ridge Center* (Chicago's Spertus Institute for Jewish studies): "Genetics' progress in our new millennium, for all the hope it promises, needs a faith context. Faith traditions can sit on their theologies and let genetic progress define what is most valuable in health and healing, or they can embrace and transform the opportunities brought by nanomedicine. Happy New Year."[4]

---

[1] Rosner, 2000.
[2] Cohen-Almagor/Shmueli, 2000, 133.
[3] Sherwin, 2000.
[4] Boyle, 2000.

## Jewish Rejection of Eugenics

14. Rabbi Yitzchok Adlerstein, supporting the anti-eugenics views of Leon Kass, writes of "our mandate as Torah Jews to raise the banner of moral freedom aloft, to forever insist upon the specialness of a being who cannot be reduced to mere substance."[1]

15. Washington Times columnist Ben Wattenberg: "Eugenics theory helped slam shut the doors of immigration in the 1920s, but such pseudoscience is in the trash can now. In 1965, Americans reopened the immigration flow, this time allowing persons from around the world to share and shape our liberty."[2]

16. Journalist Danny Katz: "This Sir Francis Galton guy believed that certain kinds of people should be genetically eradicated from the world, like handicapped people and black people and Jewish people – and if you happened to be a handicapped, black Jew, you were definitely not going to be on his Christmas card list. I got the feeling that if Sir Francis Galton were alive today, he probably wouldn't be listening to a lot of Sammy Davis Junior records."[3]

## 2001

### Context

1. Psychologist and historian Richard Lynn (b. 1930) of the University of Ulster: "Hitler believed that the Jews and the Aryans were the two most talented races and that they were in competition to secure world supremacy. Thus, he wrote in *Mein Kampf* that the Jews are 'the mightiest counterpart to the Aryan.' He feared that the outcome of the struggle between these two people might easily be 'the final victory of this little nation.' This was the reason that Hitler was determined to destroy the Jews. He believed that if he could achieve this, the Aryans would remain as the unchallenged master race."[4]

2. Israeli sociologist Larissa I. Remennick and Amir Hetsroni: "Whereas American Jews form the mainstay of the prochoice camp, Jews

---

[1] Adlerstein, 2000?
[2] Wattenberg, 2000.
[3] Katz, 2000.
[4] Lynn, 2001, 239.

in Israel are clearly divided along the lines of religiosity, ethnicity, socioeconomic status, and political views." [1]

3. When the journal *Human Immunology* publishes a study establishing genetic links between the Palestinians and Sephardic Jews, but not between Palestinians and Ashkenazi Jews, the journal is forced to remove the article from its Web site, but the print version has already been distributed.[2] Yale University geneticist Mazin Qumsiyeh protests in a letter to the Society of Histocompatibility and Immunology: "The paper demonstrated with ample evidence the similarity of certain Jewish populations to Palestinians. After some pressures because the data appears inconsistent with Zionist ideology and mythology (including the preposterous claims that Palestinians are recent immigrants to the 'land of Israel' and Jews as a distinct race), the paper was pulled from web pages and the society took an unprecedented and in my humble opinion illegal action of penalizing an author (removing him from the editorial board) to satisfy a political constituency within the society."[3]

4. *Ha'aretz* reports that the new Israeli law "Families Blessed by Children" pays larger allowances from the fifth child in the family.[4]

5. Testifying before the National Bioethics Advisory Committee on Cloning, Rabbi Moshe Tendler, a medical ethicist at Yeshiva University, states that if he could get the DNA of the people killed in the Holocaust, he'd clone them tomorrow because their genetic lineage was unfairly cut off by the Nazis.[5]

6. Psychologist Jefferson A. Singer attacks psychologist Kevin MacDonald's sociobiological study of the Jews and Jewish eugenics *Separation and its Discontents* in *Shofar: An Interdisciplinary Journal of Jewish Studies*: "Ultimately, I believe that this book is written out of a deep and destructive hatred for Jews. In the preface to the book, the author attempts to diffuse any charges of antisemitism with regard to the polemical claims of the book by stating that such charges are simply further evidence of 'intellectual defenses' that

---

[1] Remennick/Hetsroni, 2001, 420.
[2] Arnaiz-Villena *et al.*, *2001.*
[3] http://eaazi.blogspot.com/2005_07_01_archive.html.
[4] Cited in Landau, 2003, 70.
[5] Mono, 2001.

have supported Jewish 'evolutionary strategies' throughout history."[1]

7.  Alan Mintz, professor of Hebrew literature at the Jewish Theological Seminary in New York City: "I found myself increasingly unea-
sy with the way in which works of Holocaust culture were talked
about as if they formed a world unto themselves, possessing their
own laws and poetics.... The exceptionalist model is based on a
conception of the Holocaust as an unprecedented event that transformed our understanding of the world and produced a literature
that can be understood only internally and by reference to itself."[2]

8.  Physician Fred Rosner: "Preimplantation of in vitro fertilized eggs
with the discarding of affected zygotes, if any, avoids the issue of
pregnancy termination since pregnancy in Judaism does not begin
until zygote implantation into the wall of the uterus."[3]

9.  Geneticist Michael Hammer: "The haplotypes of all but Ethiopian
Jews shared a similar pattern. This means we are not descended
from one person or 12 tribes but 13 founder males."[4]

10. Although Germany's *Embryonenschutzgesetz* ("Embryo Protection
Law") prohibits harvesting embryonic stem cell (ESC) in Germany,
it does not regulate importing such cells. When Wolfgang Clement,
Prime Minister of the German state of North Rhine-Westphalia, travels to Israel to explore the possibilities of a potential future collaboration of the University of Bonn and the University of Haifa in
ESC research with the goal of importing human ESC lines from
Israel to Germany, the endeavor results not only in harsh criticism
from the Church, but also from his own (Social Democratic) party,
the German mass media, and a large number of philosophers, bioethicists, and other public figures. The German weekly magazine *Die
Zeit* quotes geneticist R. Schnabel: "It is an almost macabre irony of
fate that Israeli scientists – Israel of all nations! – currently are
putting German bioethicists under pressure. Isn't it the traumatic
experience of the German past, the breeding fantasies of the Nazis,
and the killing of six million Jews, which poses a particular load on

---

[1] Singer, 2001, 166.
[2] Mintz, 2001, ix-xi, 84.
[3] "Genetic Screening, Genetic Therapy and Cloning in Judaism," *B'Or Ha'orah* 12E,
17-29; quoted in Wahrman, 2002, 103.
[4] Epstein, 2001.

Germany's shoulders in the (bio)ethical debate? . . . And now it is Jewish reproductive medicine practitioners who evidently have no scruples to deliver these controversial embryonic stem cells to the University of Bonn, thereby provoking the breach of a taboo...." [1]

11. The Israeli National Committee of Science writes regulations governing stem cell research, stipulating what kind of embryos can be used for research and how consent should be procured from families who are no longer using the embryos as part of in vitro fertilization treatments. Recognizing Israel's lead, the U.S. National Institute of Health (NIH) approves Technion as one of just 10 academic institutes and companies worldwide for federally funded research. [2]

## Jewish Advocacy of Eugenics

12. Israeli physician Ali Ben Abraham joins Italian gynecologist Severino Antinori and reproductive scientist Panayiotis Zavos in announcing their intent to clone human beings in an "unidentified Mediterranean country" rumored to be Israel. They propose transferring the nucleus of a cell from the man or woman into a woman's egg cell, which would then be stimulated to divide. The developing embryo would then be implanted in the woman's uterus. The technique is similar to that used to produce Dolly, the first cloned mammal, and has proven successful in the cloning of some primates. [3]

13. Gideon Bach, head of Genetics at the Hadassah-Hebrew University Medical Center in Jerusalem: "We now know that most, if not all, human disorders have a genetic background, and we're acquiring the tools to study, treat and eventually prevent or cure them.... Israel, with many inbred ethnic groups, has proven a rich human laboratory for genetic detectives. It's far easier to trace genetic anomalies in inbred groups with homogeneous pedigrees." [4]

14. Israeli scientist and businessman Avi Ben-Abraham invites Cypriot-American fertility specialist Panayiotis Zavos to visit Israel to discuss the establishment of a human reproductive cloning company.

---

[1] Prainsack, 2006.
[2] Steinberg, 2005, 67.
[3] Reaves. 2001. Also "Cloning: Cloning of Humans Planned," *Applied Genetics News*, March.
[4] Elliman, 2001.

During the visit, which is paid for via Ben-Abraham from unspecified sources, Zavos meets with prominent Mizrahi Haredi Rabbi Yitzchak Kaduri (late 19th century-2006), who encourages Zavos to continue his project "for the good of humanity." Zavos also has a lengthy meeting with Israel's President Moshe Katzav (b. 1945), who is equally encouraging. Kaduri's son, grandson, and great grandson discuss financial details of a possible arrangement. Zavos also has shorter meetings with 10-15 Knesset members. The meetings attract popular attention, and the contacts are broken off.[1]

15. Bioethicist Jonathan Glover: "...to renounce positive genetic engineering would be to renounce any hope of fundamental improvement in what we are like.... Preserving the human race as it is will seem an acceptable option to all those who can watch the news on television and feel satisfied with the world. It will appeal to those who can talk to their children about the history of the twentieth century without wishing they could leave some things out."[2]

### Jewish Rejection of Eugenics

16. Jewish-Swedish-American historian Elof Axel Carlson: "Judged by the Holocaust, all efforts at human betterment are tarnished by a eugenic brush."[3]

### Palestinians

17. From a study of the Jewish gene pool published by an Israeli-German-Indian group of researchers in the *American Journal of Human Genetics*: "We propose that the Y chromosomes in Palestinian Arabs and Bedouins represent, to a large extent, early lineages derived from more-recent population movements. The early lineages are part of the common chromosome pool shared with Jews (Nebel *et al*, 2000).[4]

### United States Immigration Policy

18. In an article unusual for its frankness, Former Director of National Affairs at the American Jewish Committee Stephen Steinlight ar-

---

[1] Zavos/Glad, 2009b.
[2] Glover, 2002.
[3] Carlson, 2001, 388.
[4] Nebel *et al*, 2001, 1105.

gues that the traditional Jewish support of unrestricted immigration may have "dire implications for Jews and America," and that many Jews are secretly "terrified" at the ongoing transformation into a non-white society: "Is the emerging new multicultural American nation good for the Jews? Will a country in which enormous demographic and cultural change, fueled by unceasing large-scale non-European immigration, remain one in which Jewish life will continue to flourish as nowhere else in the history of the Diaspora? In an America in which people of color form the plurality, as has already happened in California, most with little or no historical experience with or knowledge of Jews, will Jewish sensitivities continue to enjoy extraordinarily high levels of deference and will Jewish interests continue to receive special protection? Does it matter that the majority [of] non-European immigrants have no historical experience of the Holocaust or knowledge of the persecution of Jews over the ages and see Jews only as the most privileged and powerful of white Americans?... Does it matter that most Latino immigrants have encountered Jews in their formative years principally or only as Christ killers in the context of a religious education in which the changed teachings of Vatican II penetrated barely or not at all? Does it matter that the politics of ethnic succession – colorblind, I recognize – has already resulted in the loss of key Jewish legislators...and that once Jewish 'safe seats' in Congress now are held by Latino representatives?" Steinlight predicts that "disproportionate [Jewish] political power" will erode in a few decades, that nearly "80 percent of the entire foreign aid budget will [no longer] go to Israel," that the United States may cease to tolerate "dual loyalty" or even accept Jews seeking refuge in the future. Steinlight is particularly concerned with the immigration of Muslim immigrants who regard the founding of Israel as a "catastrophe."[1]

## 2002

### Context

1.  According to Russian government statistics, 76.5% of migrants to Israel from the Russian Federation are not Jewish. Israel claims that only 57% are not Jewish.[2]

---

[1] Steinlight, 2001.
[2] Tolts, 2003.

2.  Biologist Miryam Z. Wahrman: "Whether statuary or stature, fossil or finery, Jewish tradition does speak to us today, and the writings of the past reverberate with relevant messages. It is our challenge to unearth those messages and examine their links to the modern world, addressing how ancient traditions relate to new technologies. With a set of Jewish bioethical principles in hand, we proceed to analyze this brave new world."[1]

3.  At least 178 books are published on eugenics over the course of the year. And although some of them are still quite shrill, such as Bruce R. Dain's *A Hideous Monster of the Mind: American Race Theory in the Early Republic* (Harvard University Press), the general tone has become more calm and scholarly. The peak of 'eugenics bashing' is roughly the 1990s, plus 2004.

4.  Demographer Sergio DellaPergola of the Hebrew University: "It should be emphasized... that the elaboration of a worldwide set of estimates for the Jewish populations of the various countries is beset with difficulties and uncertainties. Users of Jewish population estimates should be aware of these difficulties and of the inherent limitations of their estimates.... In general, the amount and quality of documention on Jewish population is far from satisfactory."[2]

5.  United Jewish Communities and the Jewish Federation system release some of the doomsday findings of the *National Jewish Population Survey 2000-01.* The Survey fails to confirm 1960s optimistic hypotheses about supposedly 'converging' Jewish fertility patterns which would wipe out or at least diminish the negative correlation between educational level and the birth rate. Other findings include an ageing population marrying at later ages with fertility rates below replacement levels. Jews are found to donate more to non-Jewish charities than to Jewish, and 65% have never set foot in Israel. After a preliminary release of key conclusions, the Survey is so broadly criticized that its final release has to be delayed. After the definition of who is Jewish is broadened so that the intermarriage rate, estimated at 52% in the 1990 NJPS Survey, is lowered to 43%, the report is approved for publication. Despite the manipulation and censorship surrounding the Survey's findings and even its release, its authors optimistically, and perhaps naively, express hope

---

[1] Wahrman, 2002, 23.
[2] DellaPergola, 2002.

that its themes will "serve as the basis of important policy discussions in the American Jewish community."[1]

6.  Psychologist Steven Pinker: "When it comes to explaining human thought and behavior, the possibility that heredity plays any role at all still has the power to shock. To acknowledge human nature, many think, is to endorse racism, sexism, war, greed, genocide, nihilism, reactionary politics, and neglect of children and the disadvantaged. Any claim that the mind has an innate organization strikes people not as a hypothesis that might be incorrect but as a thought it is immoral to think."[2]

7.  Pablo V. Gejman and Ann Weilbaecher in the *Israeli Journal of Psychiatry and Related Sciences*: "There is a danger in viewing eugenics as a purely historical phenomenon with no relevance to current medical genetic practices. On the contrary, eugenic concepts are being employed today by medical genetics."[3]

8.  Columnist for *America on Line* Miryam Z. Wahrman maintains that cloning may actually be a preferred method of reproduction in some instances: "The problems with many reproductive technologies involve the procurement and use of sperm. There is concern by many rabbinic authorities that artificial insemination as well as in vitro fertilization would violate the prohibition of the 'wasting of seed.'[4] In addition, use of a donor sperm for a married woman carries the risk of technical adultery and might result in her offspring being of questionable lineage. Cloning would obviate those problems because semen is not used at all in the process."[5]

9.  Scripps Howard News Service: "for the 2,000-strong vegetarian and polygamous black Hebrew community, as they are widely known here, living in Israel is the fulfillment of a scriptural promise to create what they call the 'Kingdom of Yah,' or God on earth. Calling themselves the African Hebrew Israelites of Jerusalem, their origins are rooted in their charismatic leader, Ben Ammi Ben Israel. Ben Ammi, who was a foundry worker named Ben Carter in Chicago, had a vision in 1966 that his African ancestors were descended

---

[1] Updated 2004 version.
[2] Pinker, 2002, viii.
[3] Gejman/Weilbaecher, 2002, 229.
[4] Gejman/Weillbaecher, 2002, 229.
[5] Wahrman, 2002, 72-73.

from one of the 10 lost tribes of Israel.... Israeli authorities reject the claims that the black Hebrews are authentic Jews, and have insisted in vain that they convert to Judaism so they can be recognized as full citizens."[1]

10. Political scientist Michael L. Gross of the University of Haifa writes that Gaucher disease, a rare, chronic, ethnic-specific genetic disorder affecting Jews of Eastern European descent, "offers no grounds for abortion."[2]

11. Rabbi, Provost, and professor of philosophy at the University of Judaism in Los Angeles Elliot N. Dorff: "The potential of embryonic stem cell research for creating organs for transplant and cures for diseases is, at least in theory, both awesome and hopeful. In light of our Divine mandate to seek to maintain life and health, I would argue that, from a Jewish perspective, we have a duty to proceed with this research."

## Jewish Advocacy of Eugenics

12. Bioethicist Vardit Ravitsky of Bar Ilan University: "Just as parents should not be allowed to intentionally raise illiterate children who will not be able to become effectively integrated in other cultures, so they should not have the liberty to make genetic choices which will result in children with limited physical or cognitive abilities. Any choice that will obviously trap the future individual in her community of origin is thus impermissible."[3]

13. According to researchers at the University of Haifa and the Sheba Medical Center, donor insemination is "highly curtained" in Israel and "camouflaged" as a treatment for male infertility, but is in reality a "popular mode of conception" even among singles. Israel has 16 sperm banks, and the greatest number of in vitro fertilizations per capita of any country in the world. Israeli Jewish women tend to choose sperm from tall Ashkenazi men with light brown hair and light colored eyes. In the words of the researchers, "these ideals are couched in the hegemonic discourse of Israeli Ashkenazi groups."[4]

---

[1] Scripps Howard, 2002.
[2] Gross, 2002.
[3] Ravitsky, 2002.
[4] Birenbaum-Carmeli, 2002.

14. Jewish-Russian geneticist and eugenicist Vladimir Èfroimson's book *The Genetics of Genius* is published posthumously in Moscow.[1]

15. Israeli rabbi Michael Graetz in *Conservative Judaism*: "perhaps it is our moral duty to use the scientific knowledge that we have gained in order to 'perfect' the human species. This approach resonates in Jewish tradition. God, or Nature, endowed humans with wisdom. All that we find in the created world is 'raw material,' and human wisdom is bestowed upon us precisely in order for human beings to 'perfect' the raw material into something better."[2]

16. From *The Chosen Body: The Politics of the Body in Israeli Society*, by sociologist Meira Weiss of the Hebrew University of Jerusalem: "Today, Israeli society is still obsessed with fertility. The prestate zeal for community eugenics and the post-independence craving for quantity and quality have found their contemporary consummation in genetic screening, testing, and counseling.... The chosen body, as described in this book, is a masculine body; I therefore discussed the construction of Israeli manhood as bound to, and by, the bodily practices of soldiering, war, and the 'fatherland.'"[3]

### Jewish Rejection of Eugenics

17. Leon Kass, a firm opponent of eugenics, is appointed Chair of the President's Council on Bioethics.

18. In Jerusalem, Israel's Chief Rabbi Israel Meir Lau (b. 1937) states that in principle, Judaism favors technological developments and medical progress that can help save a life or solve infertility problems, but rejects the artificial creation of life: "The moment medical science tries to take upon itself duties and areas which are not its responsibility such as shortening life, cloning, or creating life in an unnatural way we must set down borders in order not to harm the basic belief that there is a creator of the universe in whose hands life and death are placed."[4]

---

[1] Èfroimson, 2002.
[2] Graetz, 2002, 4445; citing Genesis Rabbah 11 (Theodore-Albeck edition).
[3] Weiss, 2002, 2, 28, 140.
[4] Worldwide Religious News, 2002.

19. Rabbi Lawrence Troster (b. 1953) in *Conservative Judaism*: "When it comes to genetic enhancement, the greatest danger may be a modern version of cross-generational retribution. While originally only God was given this power, in this world of human freedom, it is we who may be both committing the sin and visiting the punishment on our descendants. In the arrogance of our assuming to be partners in Creation with godlike freedoms and powers, our sin will be to attempt to create a kind of genetic utopia for our children and descendants."[1]

20. Rabbi Paul Root Wolpe in *Conservative Judaism*: "Some Jewish leaders have already recommended that Jews not participate in genetic research, both for the stigma it can cause and for the possibility of insurance discrimination. In a Boston Globe article entitled 'Jews Fear Stigma of Genetic Studies,' one Jewish spokesman gives the geschrei that seems to lurk just beneath the surface of every Jew: 'Why are we being singled out?'"[2]

21. Wolpe again: "It is not state-sponsored eugenics that will most likely control genetic engineering in the future; it is the market, which is almost as scary a proposition."[3]

22. American historian Richard Weikart concedes that while "there was no direct path from Darwinism to the Nazi's mass extermination," he argues that what "*all* [emphasis mine, JG] German eugenicists did, was to embrace an outlook that led, after many twists and turns, to the slave-labor and death camps of Auschwitz."[4]

23. Political activist Jeremy Rifkin: "As a Jew and a progressive, I was shocked at the news that the Union of Orthodox Jewish Congregations of America had announced their support of so called 'therapeutic' cloning.... Who would have believed that just a half century after the Holocaust, our own religious leaders would give their blessing to a new kind of medical research with commercial eugenics implications that are likely to be every bit as evil in the long run as the social eugenics dogma of an earlier era?"[5]

---

[1] Troster. 2002, 39.
[2] Wolpe, 2002, 21-22.
[3] Wolpe, 2002, 23.
[4] Weikart, 2002.
[5] Kaiser, 2002.

24. Author and editor Dan Seligman (1924-2009): "A fair critique of Galton and other eugenics enthusiasts of his era is that they never had a political strategy for implementing their vision in a democratic society. The argument that their ideas somehow culminated in Hitler's atrocities is ludicrous...."[1]

## 2003

### Context

1. The Russian newspaper *Nezavisimaya gazeta* interviews Zinovii Lvovich Kogan, Chairman of the Congress of Jewish Religious Organization in Russia, who declares that Judaism is a religion that blesses life and that cloning is both inevitable and acceptable: "We will create a person in our own image."[2]

2. In late July the Israeli Knesset votes to enact a law denying citizenship to Palestinians from the occupied territories who marry Israeli citizens. The law is criticized as a racial-hygiene replication of Germany's 1935 Law for the Protection of German Blood and German Honor.[3]

3. Steven Kaplan in the *Journal of Modern Jewish Studies*: "Although events in the twentieth century did much to discredit the racial paradigm, the problem of 'Who is a Jew?' became, if anything, more complex. The establishment of the State of Israel transformed it from a theoretical or existential question, to a practical issue with broad political implications. Under such circumstances, it is not at all surprising that while rejecting the idea of race, Jews continue to believe and act as if there were clear physical, historical and, later, genetic clues to membership of their people."[4]

4. Gary Rosen, managing editor of neoconservative Jewish magazine *Commentary*, writes a favorable article on the highly conservative bioethicist Leon Kass, Chair of the President's Council on Bioethics. While acknowledging that Kass is viewed as being "on the fringe of medical consensus" and that Kass is "a not altogether convincing prophet," Rosen nevertheless lauds him as "an indispensa-

---

[1] Seligman, 2002.

[2] *Nezavisimaya gazeta*, 2003.

[3] "Nationality and Entry into Israel Law (Temporary Order) – 2003," *Adalah*, 2003; May, 2003.

[4] Kaplan, 2003, 90.

ble teacher and administrator." Rosen's article is noteworthy in that its conservative-liberal unanimity reflects the gap separating the media from the scholarly community.[1]

5. The Jewish Telegraph Agency comments that the Nazis condemned IQ tests because Jews averaged higher scores than did gentiles.[2]

6. Journalist and founder of the Human Biodiversity Institute Steve Sailer points out the high frequency of cousin marriage in the Middle East – among Arabs and Jews.[3]

7. Jewish-American journalist Ben Wattenberg praises "designer immigration" and the "de-europeanization" of America: "I think we are the first universal nation, that the melting pot is working, and that we are creating – through immigration and intermarriage – a new folk that will be the model for mankind."[4] When asked by John Glad about his comment, Wattenberg replies: "I don't remember."

8. Cultural and literary scholar Sander L. Gilman argues that in the current "post-Zionist" period, Jews no longer define themselves through their relationship to the center (Israel) but as victims of the Holocaust: "Who are the Jews? Those who understood themselves as Jewish at specific moments in time. Does this definition change? It is constantly shifting and constantly changing, which is why absolute boundaries *must* (emphasis in original) be constructed...."[5]

9. Biochemist David Glick of the Hebrew University of Jerusalem: "Unlike the classical Mendelian traits, many specific aspects of behavior are, in part, determined by several genes. The corresponding abnormalities of behavior or deficiencies are therefore polygenic. New genetic techniques are leading to the discovery of these genes, and the techniques and knowledge developed in the Human Genome Project make it possible to screen the genome of any individual for the presence of known polymorphisms. This raises great hopes for diagnosis and the individualization of therapy. However, the genetic prediction of unacceptable behavior can further lead to social and occupational discrimination and enforced therapy. This

---

[1] Rosen (Gary), 2003.

[2] Jewish Telegraph Agency, 2003.

[3] Sailer, 2003.

[4] http://www.vdare.com/pb/control_borders.htm, accessed July 19, 2008.

[5] Gilman, 2003, vi-25.

raises serious concerns about how this information will be collected and who will have access to it."[1]

10. An international group of geneticists studies the Y chromosomes of three Jewish castes (Cohen, Levi, and Israelite) in which membership is determined by paternal descent. They conclude that while 70% of the Cohanim (plural of Cohen), and also a relatively large percentage of the Israelites, each display their own similar chromosomal lineages, the lineages of Sephardi Levites and Ashkenazi Levites are genetically diverse. (It is estimated that Cohanim and Levites each comprise 4% of the Jewish people.) The team speculates that the non-Jewish "introgression" may have been accounted for by Turkic converts to Judaism from the Khazar Khaganate, located in the northern Caucasus, who fled westward from the Golden Horde and adopted Yiddish as a Slavic tongue relexified with a German vocabulary, possibly via Sorbian. The "introgressors," the team hypothesizes, may have numbered between 1 and 50 men. (It is commonly accepted among geneticists that the Ashkenazi Jewish community started from a very small base which as late as 1500 C.E. may have numbered only about 30,000 people.) One of the principal researchers, Karl Skorecki of Technion and Rambam Medical Center in Haifa, comments: "If someone has a non-Jewish haplotype, it doesn't mean that person is not Jewish."[2]

## Jewish Advocacy of Eugenics

11. The *Times* of London reports that, whereas Tay-Sachs disease was "rife" in the Jewish community prior to the establishment of Dor Yeshorim in New York and Israel in 1983, of ten babies born with the condition in North America this year, none is Jewish and only one child with Tay-Sachs is born in Israel.[3]

12. Laurie Zoloth, Director of the Program in Jewish Studies at San Francisco State University: "If the concept of prenuptial and prenatal screening is Halachically acceptable for Tay-Sachs, and the technology exists to uncover more and more diseases, then the process shifts perilously close to the eugenic imperative."[4]

---

[1] Glick, 2003.
[2] Behar *et al.*, 2003; Bloch, 2004.
[3] Henderson, 2010.
[4] Quoted by Rosen (Christine), 2003.

13. Bioethicists Michael L. Gross and Vardit Ravitsky on the guidelines for genetic research in Israel: "the United Nations International Convention against the Reproductive Cloning of Human Beings poses an interesting challenge for Israel. Although it wishes to contribute to international efforts to control what is now a medically and ethically unsafe method of reproduction, Israel, as a Jewish state, cannot support limitations on cloning for therapeutic research and will be reluctant to support a ban on *reproductive* cloning if the sole objection is built on the belief that cloning offends human dignity."[1]

14. Dr. Harvey Stern, head of the Genetics and IVF Institute's PGD program in Fairfax, Virginia: "I do see that in the future, every embryo [produced in the course of IVF cycles] will be tested."[2]

15. In contrast to Germany, where the estimated rate of legal abortions resulting from embryopathic indication is estimated at only 2-4%, the corresponding rate in Israel is 17%.[3]

16. Sociologist Ruth Landau of the Hebrew University: "Israel has the highest rate of in vitro fertilization clinics in the world per capita, and is the only country explicitly to legalize surrogacy. Although the relevant data on fertility treatments are inconsistently collected, the data gathered between 1993 and 1996 by the Israeli Ministry of Health reveal that 2% of all births in Israel are the result of in vitro fertilization (Ministry of Health, 1999). The equivalent for USA is 0.2% (McClure, 1996), and for the UK about 1% (Human Fertility and Embryology Authority, 1996)."[4]

17. Historian Christine Rosen: "If ever there was a clear case for using our knowledge of human genetics to end suffering, Tay-Sachs, a killer of children, is it. There is no cure for the disease.... The question is no longer whether we will practice eugenics. We already do. The question is: Which forms of eugenics will we tolerate and how much will we allow the practice of eugenics to expand?"[5]

---

[1] Gross/Ravitsky, 2003, 251.

[2] Healy, 2007?

[3] Hashiloni-Dolev, 2007; citing www.health.gov.il.

[4] Landau, 2003, 68.

[5] Rosen (Christine), 2003.

18. American Rabbi David Fuld donates $700,000 to Sha'are Zedek Hospital in Israel for the prevention of births of children with congenital defects in Jewish families.[1]

19. From the *Palestine Solidarity Review*:

    • Veronica Ouma in her review of Susan Kahn's *Reproducing Jews: A Cultural Account of Assisted Conception in Israel* (Duke University Press, 2000) argues that "state and religious support for reproductive technologies is part of a broader eugenics movement in Israel."[2]

    • Kahn protests: "Jewish identity is not being conceptualized in genetic or racial terms in the discourse surrounding the appropriate uses of new reproductive technology – a fact that Ms. Ouma totally ignores and which is in complete opposition to her 'eugenics' argument."[3]

    • Ouma responds: "Ms. Kahn, I have not misread your book, in fact, I have read it very carefully. The Zionist state would never actively encourage Palestinian access to reproductive technologies for their fertility is seen as a direct threat to the state."[4]

20. Rabbi James M. Gordon (*An Overview of Halachic Issues Pertaining to Jewish Genetic Disorders*): "Using Advanced Technology to Sustain Life is Not an Interference with G-d's Will. We learn from the *Ramban's (Nachmanides)* interpretation of *"v'kivshua"* ("and subdue it") (*Genesis 1:28*), that humankind is given the license to make positive use of all of G-d's creations. Using advanced technology, such as genetic screening devices, is not an interference with G-d's Will. Rather, it is a positive usage of G-d's creations."[5]

21. Carron Sher, Orly Romano-Zelekha, Manfred S. Green, and Tamy Shohat of the Institute of Medical Genetics, Asaf Harofeh Medical Center, Zerifin, Israel: "94.4% of the [Jewish Israeli] secular women, 36.4% of the religious, and none of the ultrareligious women older than 35 years performed amniocentesis."[6]

---

[1] Traubmann/Reznick, 2005.

[2] Ouma, 2003a.

[3] Kahn, 2003.

[4] Ouma, 2003b.

[5] http://jewishgeneticscenter.org/rabbis/overview/, accessed May 15, 2008.

[6] Carron *et al.*, 2003.

## 2004

### Context

1. French law (as of August 6): ARTICLE 214-1, "The implementing of any eugenic practice aimed at organizing the selection of persons is punished by thirty years' criminal imprisonment and a fine of €7,500,000." ARTICLE 214-2, "Carrying out any procedure designed to cause the birth of a child genetically identical to another person whether living or deceased is punished by thirty years' criminal imprisonment and a fine of €7,500,000." ARTICLE 214-3, "The offences provided for by articles 214-1 and 214-2 are punished by criminal imprisonment for life and a fine of €7,500,000 if they are committed by an organized gang."[1] (Read literally, these laws apply not only to cloning but also to fertility techniques aimed at producing identical twins, as well as to eugenics in general.)

2. Physician Avraham Steinberg and physiologist John Locke of the Dr. Falk Schlesinger Institute for Medical-Halachic Research: "[I]t seems wise to adopt a moratorium on germ line genetic manipulation in humans but not an a priori prohibition."[2]

3. Temple University creates the Center for Afro-Jewish Studies. The center does not focus exclusively on Afro-Jewish populations; Chinese, Latin American, and other Jewish populations are included, in addition to Ashkenazim and Sephardim. The Jewish daily *Forward* reports: "[Philosophy professor Lewis] Gordon, whose mother is a Jamaican Jew and whose maternal grandfather's family left Jerusalem in the 19th century... noticed that when he presented himself as a 'black Jew,' he was often met with shock and disbelief. According to Gordon, people's ideas about Jews are skewed and there are things about Jewish communities that even fellow Jews don't understand. 'You imagine Jews are people who came from Europe, and that is absurd,' he said. 'It's just not correct.' Gordon said that the term 'black Jew' is something 'externally imposed upon Jewish communities."[3]

---

[1] "Crimes in Relation to Eugenics and Reproductive Cloning," legifrance, http://195.83.177.9/code/liste.phtml?lang=uk&c=33&r=3683.

[2] Steinberg/Loike, 2004, 189.

[3] McKigney, 2007.

4.  Lifesitenews.com: "The Israeli Health Ministry has given the go-ahead to cloning of human beings for experimentation. The Health Ministry committee decided to approve the experiments and is being criticized for going forward without consulting other bodies and without public debate on the controversial issue.... Professor Yosef Itzkovitz, director of the women's ward in Rambam Hospital, made a request to the committee asking for approval for his experiments in cloning.... Criticism of the committee's decision has also come from the Ombudsman for Future Generations in the Knesset. Retired judge Shlomo Shoham, called it 'scandalous,' saying that a decision that affects 'all of society' must be made with public scrutiny. Shoham vowed 'to demand the relevant documents and protocols of the committee's debates, and an explanation why they did not tell the public.'"[1] The Knesset extends the moratorium on cloning until March 2009.[2]

5.  Emory University professor of law and religion Rabbi Michael J. Broyde: "Jewish law insists that new technologies – and particularly new reproductive technologies – are neither categorically prohibited nor categorically permissible. Rather, they are subject to a case-by-case, method-by-method analysis of the consequences of the new technology as well as the methodology employed, and both need to be permissible for new technology to be proper in the eyes of Jewish law."[3]

### Jewish Advocacy of Eugenics

6.  Medical historians Nadav Davidovich and Shifra Shvarts in *Israel Studies*: "The eugenic outlook became intertwined with colonialist practices that presented the white European body as the 'right' model. The Zionist movement, with its European foundations and aspirations to forge a 'New Jew,' fit in well with this approach. Public health policy toward immigrants in Israel was founded on a similar belief in public health of the individual, but also on the level of the nation as a whole."[4]

---

[1] Lifesitenews.com. 2004
[2] Prainsack/Firestine, 2006, 36.
[3] "Pre-Implantation Genetic Diagnosis, Stem Cells and Jewish Law," *Tradition* 38(1), 56; quoted in "Complex, Controversial Decisions Call for Rabbinical Consultation, *Jewish Genetic Disorders*, Sept., 65.
[4] Davidovich/Shvarts, 2004, 154.

7.   Science reporter for *Ha'aretz* Tamara Traubman: "[Dr. Joseph Its-
     kovitz-Eldor of Rambam Hospital in Haifa, Israel] is at the forefront
     of a scientific pressure group lobbying the Israeli government for
     permission to clone human embryos.... Doctors from Sheba Medi-
     cal Center, near Tel Aviv, headed by Dr. Jacob Levron were but a
     step away from cloning embryos... All this research has been car-
     ried out without any public debate....Reproductive cloning is not
     perceived as a taboo, and is often condoned.... The 'Helsinki
     Committee,' which is appointed by the Israel Ministry of Health,
     neither has the authority nor the ability to monitor or supervise the
     experiments it approves."[1]

8.   Dr. Frida Simonstein of the Ben-Gurion University in an article in
     the *Israel Medical Association Journal*: "Using GLE [germ-line en-
     gineering] in order to 'self-evolve' (when it becomes safe) is not on-
     ly inevitable but also morally justified.... Trying to free future gen-
     erations from late-onset diseases (such as Alzheimer's for instance)
     may be considered as 'eugenics' but, if pursued freely and justly, is
     a noble goal."[2]

9.   *Ha'aretz* quotes Sachlav Stoler-Liss of Ben-Gurion University,
     "Eugenic thinking is alive and well [in Israel] today."[3]

10.  The New-York weekly *Jewish Press* publishes an Op-Ed by John
     Glad exposing myths about eugenics and pointing out that the at-
     tacks on eugenics by certain Jewish writers is a suicidal rush to
     doom.[4] The following week a letter to the editor by Gilbert Sapers-
     tein is published: "The *Jewish Press* deserves an enormous amount
     of credit for publishing an article on eugenics that did not take the
     intellectually dishonest approach, so prevalent in academia and the
     media, of tarring an entire scientific discipline with the brush of
     Nazism. Dr. Glad masterfully laid out the historical facts and, while
     making no effort to airbrush the blemishes, presented a picture of
     eugenics and eugenicists that was both fair and enlightening. Given
     the unthinking allegiance to political correctness and limp liberal-
     ism manifested on a weekly basis by other Jewish publications, I
     doubt we'll ever see the day when your competitors carry this type

[1] Traubmann, 2004a.
[2] Simonstein, 2004.
[3] Traubmann, 2004b.
[4] Glad, 2004.

of honest appraisal of an issue as controversial – and as encrusted and overlaid with myth – as eugenics."[1]

11. In honor of a Jewish-German feminist and eugenicist, the Henriette Fürth Prize is established by universities in Hessen, Germany, for the best Bachelor's or Master's thesis on gender studies.[2]

### Jewish Rejection of Eugenics

12. Sachlav Stoler-Liss, who is researching the topic "eugenicist Zionists" at Ben-Gurion University, comes across a card file with notes written by the editors of a collection of Joseph Meir's writings, published in Israel in the mid-1950s, where the editors call the article "problematic and dangerous" and comment that "Now, after Nazi eugenics, it is dangerous to publish this article."[3]

13. From an internet blog: "It would be hard to imagine a more offensive charge to hurl at Jews, whatever their political persuasion, whose collective (and often family) history includes victimization by the eugenics policies that accompanied the Nazi genocide."[4]

14. At least 131 books are published on eugenics, and the tone again becomes shrill:

- Susan D. Bachrach and Dieter Kuntz, *Deadly Medicine: Creating the Master Race*, United States Holocaust Memorial Museum and University of North Carolina Press.

- *Daylanne K. English, Unnatural Selections: Eugenics in American Modernism and the Harlem Renaissance, University of North Carolina Press.*

- *Martin Brookes, Extreme Measures: The Dark Visions and Bright Ideas of Francis Galton,* Bloomsbury.

- John P. Jackson and Nadine M. Weidman, *Race, Racism, and Science: Social Impact and Interaction*, ABC-CLIO.

- Gretchen Engle Schafft, *From Racism to Genocide: Anthropology in the Third Reich*, University of Illinois Press.

---

[1] Saperstein, 2004.
[2] FFZ.
[3] Traubmann, 2004a.
[4] http://aulula.blogspot.com/2004/08/my-letter-to-presbywebcom.html.

- John V. Van Cleve, *Genetics, Disability, and Deafness*, Gallaudet University Press.

- Paul Weindling, *Nazi Medicine and the Nuremberg Trials: From Medical War Crimes to Informed Consent*, Palgrave Macmillan.

- Adam Jones, *Gendercide and Genocide*, Vanderbilt University Press.

- Edwin Black, *War against the Weak: Eugenics and America's Campaign to Create a Master Race*, Four Walls Eight Windows.

- Wolfgang Freidl and Werner Sauer, *NS-Wissenschaft als Vernichtungsinstrument*: Rassenhygiene, *Zwangs-sterilisation, Menschenversuche und NS-Euthanasie in der Steiermark* [*Nazi Science as an Instrument of Destruction: Racial Hygiene, Compulsory Sterilization, Human Experimentation, and National-Socialist Eutha-nasia in the Steiermark*], Facultas.

- Gerald Randolph Revelle, *My Enemy's Child*, Smultron Publications.

15. Historian Todd M. Endelman concludes an essay on Jewish racialist scientists by implying that questions of biology and science be resolved by political considerations: "[Redcliffe] Salaman, however, unlike other Anglo-Jewish scientists, was too invested in genetics to forswear completely the language of racial descent even after Nazism demonstrated its potential for deadly abuse. While he modifed his views and no longer trumpeted them in community forums, he never abandoned his belief in a biological component to Jewishness."[1]

16. *Dimensions*, an online "Journal of Holocaust Studies," produced by the Anti-Defamation League, quotes Holocaust historian Yehuda Bauer: "The Nazi movement was based on the pseudoscience of raceology or eugenics. For the first time in history, Bauer stresses, the Nazis set out to create a society based on a racial hierarchy – this was revolutionary. In fact, Bauer continues, National Socialism

---

[1] Endelman, 2004, 84.

was the only real revolutionary movement of the twentieth century."[1]

17. Jewish-American geneticist and eugenics opponent Jerry Hirsch (1922-2008) attacking Murray and Herrnstein's Bell Curve: "Herrnstein had neither dark skin nor kinky hair. He was a white, Jewish, Harvard professor. Otherwise, such scholarly incompetence, as has here been revealed, might, if one were to apply his professed high standards, have had to be interpreted as an unmistakable sign of his own genetic inferiority."[2]

## 2005

## Context

1. The size of world Jewry is assessed at 13,034,000 by the *Encyclopedia Judaica*, Jews constituting 2.04 per 1,000 of the world's total population of 6,396 million (one in about 490 people in the world).[3]

2. Tudor Parfitt and Yulia Egorova of the University of London examine the impact of genetic research on the religious identity of the Bene Israel Indian Jewish community and the Lemba Judaising group of southern Africa.[4]

3. Sephardi Chief Rabbi of Israel Shlomo Amar (b. 1948) announces Israel's recognition of the Bnei Menashe from India as part of the lost tribe of "Menashe," legalizing their immigration to Israel under the Law of Return, but only after a complete Jewish conversion.[5] This is not the same group as the Bene Israel, mentioned above.

4. Abe Foxman, National Director of the Anti-Defamation League, on an article by Gregory Cochran, Jason Hardy, and Henry Harpending, claiming Jews possess superior intelligence: "If it's a genetic condition, it's not for us to embrace or reject. It is what it is, and that's the way the genetic cookie crumbles." By contrast, historian Sander Gilman in an interview with the New York magazine, calls the paper "insulting" and "bullshit."[6]

---

[1] Anti-Defamation League, 2004.
[2] Hirsch, 2004, 313.
[3] Schmelz/DellaPergola, 2007.
[4] Parfitt/Egorova, 2005.
[5] http://en.wikipedia.org/wiki/Shlomo_Amar, accessed january 25, 2009.
[6] Senior, 2005.

5.  Germany stiffens regulations on Russian-Jewish immigration, requiring that applicants have an invitation, and the Jewish Central Committee (Zentralrat der Juden) is accused of inaction.[1]

## Jewish Advocacy of Eugenics

6.  Shifra Shvarts, Nadav Davidovitch, Rhona Seidelman, and Avishay Goldman of Ben-Gurion University discuss the debate over medical selection of immigrants by the Israeli government: "It is our claim that the debate was shaped to a large extent by the combining of Zionist ideology and eugenic influences – two intellectual forces that had interacted with each other well before the creation of the Israeli State in the first half of the 20th century."[2]

7.  Robert Desnick of the Department of Human Genetics at New York's Mount Sinai Hospital writes that Tay-Sachs disease appears to have disappeared almost completely from among the Jewish nation. The Blog Pytheas Online comments: "Tay-Sachs is mostly confined to a specific ethnic group – in this case, Ashkenazi Jews. Tay-Sachs may also be the first genetic disorder almost totally eliminated by voluntary eugenics.... Despite the intellectuals' disaffection with genetics, mounting scientific evidence eventually turned the tide in favor of the genetic basis of human behavior. Eugenics has been slower to recover and is still viewed by many as a dangerous, or at least publicly unmentionable, idea. And yet, eugenics – selective breeding – remains a simple idea whose effectiveness has been borne out by five thousand years of agriculture and animal husbandry. The fact that the Nazi misused a particular scientific idea doesn't vitiate its validity."[3]

8.  Behavioral scientist Aviad E. Raz of Ben-Gurion University of the Negev, Be'er Sheva on the positive attitude in Israel toward genetic testing and eugenics: "Israeli interviewees did not share critical attitudes similar attitudes held by some disability rights in the United States and the United Kingdom. Indeed, Israeli respondents generally viewed prenatal diagnosis favorably. Moreover, most of the interviewees were not just protesting but also in favor of terminating a pregnancy after the detection of fetal abnormality. In other words, *the majority of respondents subscribed to the eugenic ideals* criti-

---

[1] *Spiegel*, "Lockrufe aus Tel Aviv," No. 40, 66.
[2] Shvarts *et al.*, 2005, 6.
[3] Pytheas, 2005. Source of statements by Desnick and Zlotogora not indicated.

cized by disability rights advocates in the United States and the United Kingdom." [Emphasis added] [1]

## Jewish Rejection of Biology

9. Objecting, in advance of publication, to "Natural History of Ashkenazi Intelligence," by Cochran, Hardy, and Harpending, medical ethicist Michael Grodin accuses the researchers of raising "the specter of eugenics": "To suggest IQ test scores are a function of genetics is just not true. Intelligence is not even in a small way contributed to by genetics. It is a function of socio-economic status, of the environment in which one grows up."[2] Cochran comments to *Forward*: "I predict that if we turn out to be right, we'd change the minds of maybe 100 people worldwide."[3]

## Conflation of Eugenics with Political Agendas

10. Political conservative Jeff Ballabon in *Forward*: "As a child of a Holocaust survivor, I am terrified by what the left is doing here and grateful beyond words for the Republican response to the Terri Schiavo tragedy. It may look like a Christian agenda in Missouri, but in my neighborhood it looks profoundly Jewish."[4]

11. Internet author Daniel L. Abrahamson attacks Jewish-American linguist and political thinker Noam Chomsky (b. 1928): "Chomsky and other gatekeepers claim to maintain the moral high ground, but then advocate the eugenics agenda pushed by elite roundtable groups like the Bilderberg and Club of Rome. How can Chomsky claim to be an advocate for the Third World while simultaneously pushing for the managed murder agenda favored by Henry Kissinger?... The New Freedom Initiative, bolder than Stalin's wildest dreams, is just a small part of the eugenics agenda.... Key aspects of the scientific eugenics movement such as population control, abortion legalization, poisonous vaccinations, and stem cell research find their most vocal advocates on the far Left of the managed political debate. Leftist college majors like sociology advocate programs like parental licensing by the state, state-controlled child care, and Chinese style childbirth laws with forced sterilization as a penal-

---

[1] Raz, 2005, 185.
[2] Siefer, 2005.
[3] *Forward*, 2005.
[4] Ballabon, 2000.

ty.... All of this is designed to create a tight spectrum, a masterfully crafted false paradigm to enslave the mind and give the people false choices. Thus at either extreme of the spectrum and all points in between, from Chomsky to Horowitz, one finds they are endorsing total enslavement and global government. This is the genius of the New World Order, their complete castration of free political will through carefully managed propaganda agents. The rest of the media jackals serve as willing accomplices, mere useful idiots and power hungry sycophants with massive egos and more concern for their career than the truth."[1]

## 2006

### Context

1.  In an interview in *Ha'aretz*, contemporary Israeli geneticist Raphael Falk of the Hebrew University of Jerusalem comments on the four-volume study *Ideas on the Philosophy of Human History* (*Ideen zur Philosophie der Geschichte der Menschheit*) by Herder (1774-1803): "The German philosopher Johann Gottfried Herder spoke of the idea of the Volk, the 'folk-nation' that viewed a people as an organic unit. And not just in the cultural sense. In time, this also came to include race. From this movement of the concept of the Volk, you get Zionism developing on the one hand, and German nationalism, which later evolved into Nazism, on the other hand. This is an uncomfortable fact, but a fact nonetheless."[2]

2.  Historian Rena Selya: "Paradoxically, the increased acceptability of teaching evolution to Jewish teenagers over the past twenty years coincides with an increasing turn to the right among Modern Orthodox Jews."[3]

3.  Sociologist Steven M. Cohen and historian Jack Wertheimer decry "the new 'globalist' consciousness much touted by Western and especially European intellectuals. In the name of eliminating 'boundaries' between and among people, whether national, ethnic, or religious, this quintessentially postmodern movement celebrates the trans-national, trans-cultural individual. It urges us to sample civilizational offerings wherever they may be found, and from these to

---

[1] Abrahamson, 2005.
[2] Karpel, 2006.
[3] Cantor/Swetlitz, 2006, 207.

assemble our own private identities. Rejecting 'essentialist' claims of all kinds, it upholds the virtues of 'hybridity,' stressing that even the most homogeneous-seeming cultures are but manufactured admixtures of numerous influences. Absorbed into the mindset of educated Jews, this cluster of ideas works powerfully to undermine the concept of a distinctive Jewish people with its own culture, its own separate interests, and its own unique obligations." Tahl Raz, editor of internet magazine *Jewcy*, responds: "If the revitalization of peoplehood implies dismantling the modern project of securing a universal human rights and returning to a primitive state of tribalism, as it apparently does for Steven Cohen and Jack Wertheimer, than I – and I suspect a large chunk of my generation here and in Israel – will want no part in it."[1]

4.  Rabbi and Yeshiva College biologist Carl Feit: "...I maintain that there is no Jewish 'problem' with the science of evolution.... Abraham Isaac Kook and Joseph B. Soloveitchik, the two most important and influential Orthodox Jewish thinkers of the twentieth century, who based their analyses on fairly traditional readings of classic Jewish texts, not only dismissed the notion of any conflict between modern science and Torah, but actually found contemporary scientific notions of evolution and cosmology to be harmonious with classic rabbinic thought."[2]

5.  Rabbi Lawrence Troster: "While the scientific challenge of evolution forces us to confront our ideas of divine action and of providence, we can nonetheless endorse [Roman Catholic theologian John F.] Haught's sentiment that this challenge is Darwin's gift to theology: 'Evolutionary biology not only allows theology to enlarge its sense of God's creativity by extending it over measureless eons of time; it also gives comparable magnitude to our sense of the divine participation in life's long and often tormented journey.'"[3]

6.  British psychologist Richard Lynn (b. 1930) reports a mean IQ of 107-115 for British and American Jews, 103 for Israeli Ashkenazic Jews, and 91 for "Oriental Jews."[4]

---

[1] Cohen/Wertheimer, 2006; Raz/Rosner, 2007.
[2] Cantor/Swetlitz, 2006, 224.
[3] Cantor/Swetlitz, 246.
[4] Lynn, 2006, 93-95.

7. Advertisement in the *Harvard Crimson*: "Jewish ovum donor needed for infertile couple willing to pay a fee of $3,500 plus expenses." The *New York Times* comments: "Jewish donors are hard to find and because of the tradition of matrilineal descent, some Jewish women believe that the donor has to be Jewish for the child to be considered Jewish."[1]

8. B'nai B'rith International President Joel S. Kaplan hosts a presentation in Washington, D.C. by Bernard Siegel, Co-Director of Floridians for Stem Cell Research and Cures (FSCRC) and President of the Genetics Policy Institute.[2]

9. Professor of Philosophy at Hofstra University John Teehan: "Beginning with the stories of the patriarchs, we can see these as embodiments of the logic of kin selection. Jews are all children of Abraham. 'Israel' is not merely the ancestral home of the Jewish people (bequeathed by God) but was the father of the twelve tribes. All Jews through their tribal lineage are members of one, extended family. This extended family is the basis of Judaism and Jewish morality. It is the basis but of course not the whole. As we saw, kin selection can do only so much in binding a complex society. As we turn to the Mosaic Law, we can see, at least in part, that it functions to extend the force of this basic tribal ethic."[3]

10. The Longitudinal Israeli Study of Twins (LIST) traces children's prosocial development from phenotypic, genetic, and environmental perspectives, focuses on measuring prosociality with a multi-trait multi-method approach, and relates it to children's general cognitive and sociocognitive abilities, and to parenting in the family. Other variables of interest are children's temperament and parental values.[4]

11. Physician and population geneticist Doron Behar of the Technion-Institute of Technology: "Yes, all Jews are related. But in the same way that the male population in the world is related to a common ancestor.... This is remote ancestry."[5]

---

[1] October issue of HC; cited by Conklin, 1996.
[2] Siegel, 2006.
[3] Teehan, 2006, 747.
[4] Knafo, 2006.
[5] Pash, 2006.

12. The Baltimore Jewish Times receives a flyer announcing the "historic reunion" in Jerusalem in May 2007 of the Jewish descendants of King David. The event also marks the opening of the worldwide Davidic Dynasty Genealogic Center and Museum there, and is being hosted by the Esthet Chayil Foundation, in Union, New Jersey, of which Susan Roth, "a descendant of the Biblical king," the flyer states, is the Chief Executive Officer. Bennett Greenspan, founder of the largest genealogy testing company in the world *Family Tree DNA* comments: "I have a number of male clients who claim to be descended from King David. But they don't match other." [1]

13. *Wall Street Journal*: "It all began with a 'serendipitous feeling' that hit him while he stumbled through Auschwitz in 2000. Like *most* visitors John Haedrich was deeply moved by what he saw. But this was something different. A kind of epiphany. Though raised a Christian, for reasons Mr. Haedrich cannot quite articulate, he began to suspect that he might be Jewish. Gradually, this hunch became too vital to ignore. He decided to investigate his origins by taking a DNA test, the results of which confirmed that he had, according to the test conclusion, 'a rather populous pedigree of Ashkenazi Polish Jews.'" Hedrich petitions the Israeli government for citizenship under the Law of Return, but his appeal is denied on grounds that DNA does not prove Jewish identity. [2]

14. Jewish-American geneticists Jon Beckwith and Joseph S. Alper in solidarity with the egalitarian views of Jewish-American paleontologist Steven Jay Gould maintain that "a biological or genetic conception of race is meaningless" and "bankrupt," and that "research into the genetics of group differences will provide fodder for racist views." [3]

15. Immigration of Russian Jews to Germany shrinks from 60,000 in 2000 to 20,000. [4]

### Jewish Advocacy of Eugenics

16. Political scientist Barbara Prainsack of the University of Vienna and Israeli attorney Ofer Firestine: "Biotechnology regulation in Israel

---

[1] Pash, 2006.
[2] Goldstein, 2006.
[3] Beckwith/Alper, 2006.
[4] *Spiegel*, "Lockrufe aus Tel Aviv, No. 40, 66.

is characterized by a relatively permissive approach and a low regulatory density. It is a field of non-controversy; moral and ethical objections to research that is highly controversial elsewhere is virtually absent in the public debates in Israel.... With regard to human cloning, for example, virtually all Israeli bioethicists agree that even reproductive cloning should not be opposed in principle. Anecdotal evidence suggests this is also representative of the general public."[1]

17. Sociologist Larissa Remennick of Bar-Ilan University: "It seems that the Israeli public is ready to accept eugenic applications of new genetics as long as it remains uneducated and uninvolved in its public policy aspects."[2]

18. Lutfi Jaber and Gabrielle J. Halpern of the Bridge to Peace Community Pediatric Center, Taibe, Israel: "Future goals are to expand the educational programs aimed at the Israeli Arab community and to promote the uptake of genetic counseling and prenatal testing where available in order to reduce the health problems even further. Ongoing research in order to identify specific genes will enable more conditions to be detectable early in pregnancy. We expect that the willingness of families to agree to termination of affected pregnancies will reduce the number of babies born with these conditions."[3]

19. The Chicago Center for Jewish Genetic Disorders publishes *A Young Couple's Guide to Jewish Genetic Disorders and Screening*: "Given the array of new technologies, early detection of carrier status allows couples like you to consider a wide range of reproductive options." The organization's Web site contains a glossary of several hundred technical terms referring to genetic counseling; 'eugenics' is not one of them.[4]

### Jewish Rejection of Eugenics

20. Historians Geoffrey Cantor and Marc Swetlitz: "The concept of 'race' became discredited among biologists and social scientists – Jews playing leading roles in that effort – and social and cultural

---

[1] Prainsack/Firestine, 2006, 33-34, 37.
[2] Remennick, 2006.
[3] Jaber/Halpern, 2006.
[4] http://jewishgeneticscenter.org/, accessed May 15, 2008.

explanations became prominent in the social sciences, where Jews continued to work in large numbers."[1]

21. The United States Holocaust Memorial Museum sends a major exhibit entitled *Deadly Medicine: Creating the Master Race* to Dresden, claiming to "follow the German science establishment down the slippery slope from heredity research to eugenics and ultimately to the genocidal Final Solution."[2] One of the exhibits is "a teddy bear from a Nazi-SS home for Aryan children." The exhibit at the home institution of the Holocaust Museum in Washington, D.C. has received 720,000 visitors. The exhibit is so expensive that the Museum was at first reluctant to stage it.

22. Physician Leon Kass: "Genetic knowledge, we are told, is merely providing information and technique to enable people to make better decisions about their health or reproductive choices. But our existing practices of genetic screening and pre-natal diagnosis show that this claim is at best self-deceptive, at worst disingenuous. The choice to develop and practice genetic screening and the choices of which genes to target for testing have been made not by the public but by scientists – and not on liberty-enhancing but on eugenic grounds."[3]

## The International Scientific Consensus

23. Without mentioning the word 'eugenics,' anthropologists Henry Harpending and Gregory Cochran restate a central thesis of the eugenics movement: "We can think of genetic burden as the net contribution of genetic diversity to disease in either sense. The burden may be apparent, meaning that it is responsible for medical disease, or it may be unapparent meaning that it does not lead [to] a diminished quality of human life.... From this perspective a major goal of prenatal diagnosis and selective abortion is to convert apparent burden to unapparent burden. The ethical problems surrounding this field are complex, of course, but from the viewpoint of allocating

---

[1] Cantor/Swetlitz, 2006, 15.
[2] Curry, 2006.
[3] Kass, 2006.

burden, human intervention in some ways mirrors evolutionary processes."[1]

## 2007

### Context

1. A Jewish haplotype map (HapMap) project is launched under the umbrella of the human genetics program at NYU's medical school and in collaboration with the Albert Einstein College of Medicine. The project will examine DNA samples, including the Y chromosome and mtDNA, from Jews across the three major Jewish groupings: Ashkenazic (descendant from Central and Eastern Europe Jews), Sephardic (descendant from the Jews of Spain), and Mizrachic (descendant from the ancient Jewish communities of the Middle East). The aim – with potentially far-reaching implications – is to understand Jewish migration and community formation.[2]

2. Charles Murray, co-author of *The Bell Curve*: "it is currently accepted that the mean [Jewish IQ] is somewhere in the range of 107 to 115, with 110 being a plausible compromise."[3] Michael Bagraim, Chairman of the South African Jewish Board of Deputies, calls Charles Murray's theory of 'Jewish genius' racist: "There were clever Catholics, clever Zulus...."[4]

3. Hanna David of Tel Aviv University and Richard Lynn of the University of Ulster publish a paper in the *Journal of Bioscience* proposing that Oriental Jews in Israel have an average IQ 14 points lower than that of European (largely Ashkenazi) Jews.[5]

4. *Forward* reader David L. Nilsson: "It is not necessary to exaggerate the pace of exogamy, as some alarmist American rabbis have done, to see that something very special fades away after a rich Jew marries a blonde and begets offspring. Biodiversity isn't just for weeds and bugs."[6]

---

[1] Harpending/Cochran, 2006, 161.
[2] Bloch, 2007.
[3] Murray, 2007, 30.
[4] Huisman, 2007.
[5] David/Lynn, 2007.
[6] Bloch, 2007.

5.  Claude Maurice Marcel Vorilhon, founder of the religion known as Raëlism that maintains that humans were created according to instructions from alien beings and advocates cloning and human engineering, adopts the Star of David intertwined with a swastika as the official symbol of the Raëlian Movement:

6.  More gloom and doom: according to the five-year National Jewish Population Survey (NJPS) the number of Jews in the United States fell by 5 percent over the last decade, from 5.5 million to 5.2 million. Children now account for 19 percent of the Jewish population, down from 21 percent 10 years earlier. The median age of American Jews rose from 37 in 1990 to 41 in 2000. Fifty-two percent of Jewish women ages 30 to 34 have not had any children, compared with 27 percent of all American women. Jewish women nearing the end of their reproductive years (ages 40 to 44) have had approximately 1.8 children, below the replacement level of around 2.1.[1]

7.  Temple University's Center for Afro-Jewish Studies conducts a seminar on "Jews and Race." Melanie Kaye / Kantorowicz of Queens College speaks about "contemporary notions of whiteness (including Jewish whiteness) in the U.S."[2]

8.  As a result of high fertility, the Indian group Bene Israel numbers 50,000 in Israel.[3]

9.  Joël Zlotogora, Sjozef van Baal, and George P. Patrinos of the Department of Community Genetics, Ministry of Health, Ramat Gan, Israel, report the construction of the Israeli National Genetic database (Available at: www.goldenhelix.org/israeli to document the sheer genetic heterogeneity found in the Jewish and non-Jewish populations in Israel.[4]

---

[1] www.ujc.org.
[2] Jews and Race, 2008.
[3] Bar-Giora (note by Shalva Weil, 2007, 339.
[4] Zlotogora *et al.*, 2007.

10. Professor of Jewish Philosophy and Mysticism, Spertus Institute of Jewish Studies, Chicago, Byron L. Sherwin: "the dominant Halachic view is that as long as a golem remains a golem it does not have the status of a human being. Consequently, destroying it is not murder. This view has direct implications not only for abortion but also for embryonic stem cell research. If a human embryo or pre-embryo is a golem, it is not a human person. This approach rejects the claim of various religious and secular bioethicists, including some members of the U.S. President's Commission on Bioethics, who have declared that embryos, even at very early stages, are human persons 'like us.'"[1]

11. Journalist Gus Tylor in the Jewish daily Forward: "In America, at the present time, a Jewish-black intermarriage is not uncommon. One of the oldest such marriages in recorded history was that of Moses to an Ethiopian woman named Zipporah."[2]

12. Israeli Ambassador to Turkey Gabby Levy: "In the last 10-15 years we have brought within our country a new element: the Afro-Jews, people from Ethiopia. We need more time to create a final breed of society which will have definite characteristic." [sic][3]

13. The newspaper The Forward publishes its annual guide to some of the more serious Jewish genetic diseases, which it attributes to "centuries of inbreeding": Bloom's syndrome, Canavan disease, congenital hyperinsulinism, dystonia, familial dysautonomia, Fanconi anemia, Gaucher disease, Mucolipidosis type 4, Niemann-Pick, infantile Tay-Sachs, and late-onset Tay-Sachs: "There are about 20 'Ashkenazic diseases,' not counting the higher rates of at least four cancer-related genes."[4]

### Jewish Advocacy of Eugenics

14. The *Cleveland Jewish News* reports a speech given by Case Western University School of Law professor Maxwell J. Mehlman in which he warns that association of eugenics with Nazi extremism may have caused society to shy away from identifying policies as

---

[1] Sherwin, 2007, 139.
[2] Tylor, 2007.
[3] Levy, 2008.
[4] Annual Guide..., 2007.

eugenics and from engaging in dialogue about the (positive) role of eugenics in the modern world.[1]

15. American historian Mitchell B. Hart: "Jewish genetic identity over thousands of years; Jewish diseases passed on hereditarily; a Jewish advantage intellectually, passed on genetically, that explains the statistical overrepresentation of Jews in the arts and sciences – these are all images and ideas that are circulating again."[2]

16. Eugenics researcher Yael Hashiloni-Dolev of the Academic College of Tel Aviv-Yaffo, Israel: "In Israel there is no wide and continuous discussion regarding the ethical issues involved in medical genetics.... The Israeli legal system does not generally protect fetuses.... Cost-benefit analysis related to genetic abnormalities is not ethically rejected in Israel.... Regarding history, I showed that the start of applied human genetics in Israel after the establishment of the state in 1948 was not controversial.... [A]t the prenatal stage, nontolerance towards the genetically deviant is the norm among Israeli [genetic] counselors.... [Abortion] is not a controversial issue in Israeli politics.... [Israeli law] permits abortion on the grounds of extramarital sex, which is based on the Halachic fear of giving birth to what religious law calls a 'mamzer'.... [T]he Israeli-Jewish fetus is not considered to be a 'life' right from conception, and it has no rights.... [E]gg and embryo donation... embryo wastage...surrogate motherhood, preimplantation diagnosis, sex selection...fertilization with sperm retried post mortem... are not prohibited in Israel.... *Wrongful life*: such claims have been rejected by the overwhelming majority of courts around the world. Thus, Israel is a world exception."[3]

17. Anonymous Israeli pediatrician-counselor: "You cannot imagine how sad it is to see a child plead for his death or have his parents ask you not to do everything to keep him alive. Such situations have made me sure that selective abortions are the right choice."[4]

---

[1] Lutz, 2007.
[2] Hart, 2007, 199.
[3] Hashiloni-Dolev, 2007, 49, 51, 53, 60, 73, 97, 100-102, 123.
[4] Cited in Hashiloni-Dolev, 2007, 77-78.

## 2008

## Context

1.  A *New York Times* article discusses the increasingly rigid attitude of Israel's rabbinical courts in issuing marriage licenses. The Ultra-Orthodox mistrust even Orthodox rabbis, and the number of American Jews who are not recognized in Israel as Jews exceeds 100,000.[1]

2.  Rabbi Baruch (Barry) Leff: "There were converts who chose to join us. How else would one explain the fact that Polish Jews look like people from Poland, and Iranian Jews look like other Iranians? If we had been strictly endogamous for the last 2,000 years, we would all look like Arabs. But we don't. But just because there were some converts and some intermingling of the blood lines, does not mean that there is no real connection to the land. I believe the Jews of today mostly have some genetic component from the Jews who stood at Sinai, and they have a lot of outside genetic material as well. But does it matter that the Jews are not all 100% pure descendants of the Jews who stood at Sinai?"[2]

3.  *Ha'aretz* prints a book review of *Gizanut beyisrael* (Racism in Israel), edited by sociologist Yehouda Shenhav (b. 1951) and Ben Gurion University lecturer Yossi Yonah (b. 1953?): "Two Israeli women want to have a child. They apply to a sperm bank. Due to the shortage of sperm currently available at public stocks, they choose a private bank operating out of Rishon Letzion. For a fee of several hundred shekels, they are allowed to browse a catalog of donors. The first detail they learn about an anonymous donor is his parents' ethnic origin. Subsequent details in the entry include height, weight, hair color, skin tone and eye color. The catalog the couple received by e-mail lists the details of 22 donors. Among their 44 parents, 38 are of Ashkenazi origin (mainly Eastern Europe and the former Soviet Union) or are sabra (native-born) Israelis. There is no donor on the list both of whose parents are of Mizrahi origin (Jews of Middle Eastern origins). That is, the donors are all Ashkenazi, sabras or mixed. There also is not a single donor of

---

[1] Gorenberg, 2008.
[2] Leff, 2008.

Ethiopian origin on the list." The reviewer, Orna Coussin, wonders if "Israel is not submerged up to its neck in eugenics."[1]

4. Journalist Masha Gessen discusses in *Blood Matters* how she discovered that she had she inherited a genetic mutation that greatly increases her risk of breast cancer – a risk especially high among Ashkenazis. She also cites evidence that some such recessive genes common among Ashkenazis may actually heighten IQ when only one is inherited.[2]

5. Director of the Center for Bioethics at the University of Pennsylvania Arthur Caplan on Ben Stein's documentary film *Expelled*: "To lay blame for the Holocaust upon Charles Darwin is to engage in a form of Holocaust denial that should forever make Ben Stein the subject of scorn not because of his nudnik concern that evolution somehow undermines morality but because in this contemptible movie he is willing to subvert the key reason why the Holocaust took place – racism – to serve his own ideological end. Expelled indeed."[3] Historian Richard Weikart responds to Caplan's criticism of Stein: "Historians may differ over how to draw the intellectual connections between Darwinism and Nazism, but denying such connections is absurd."[4]

6. Ashkenazi women test disproportionately for the BRCA mutation, which is associated with an 85% probability of breast cancer and 50% probability of ovarian cancer. Upon learning that she is such a "previvor," journalist Masha Gessen writes: "My body had turned against me. All I could do now was declare war on it myself."[5]

7. The online version of John Glad's *Future Human Evolution: Eugenics in the Twenty-First Century* (preface by Seymour Itzkoff) has been downloaded over one million times, making it by far the most widely read book on eugenics ever written.

8. Results of a Google search for "eugenics" plus the following words[6]:

---

[1] Coussin, 2008.
[2] Gessen, 2008.
[3] Caplan, 2008.
[4] Weikart, 2008.
[5] Gessen, 2008; quoted in Senior, 2008.
[6] Feb. 24.

| Nazi | 70,900 |
| Racism | 359,000 |
| Anti-Semitism | 125,000 |
| Criminal | 337,000 |
| Holocaust | 231,000 |
| Human Ecology | 5,860 |
| Praiseworthy | 4,820 |

9. Gary Tobin, President of the Institute for Jewish & Community Research in San Francisco: "Do we want to enter the competition armed with our wonderful 3,000-year-old history, or kvetch about assimilation, intermarriage and our dwindling numbers? Those who choose to join the Jewish people will enrich us with their ideas, energy and passion. And born Jews who choose to embrace their Judaism in an open marketplace also will enrich Jewish life. It is time to embrace the America in which we live. We must abandon the paradigm that our children and grandchildren are potential gentiles and promote the new belief that America is filled with potential Jews."[1]

10. From an exchange of letters to the editor in the *Chronicle of Higher Education*:

- Brian Baute of Burlington, North Carolina: "[Professor of History and Sociology of Science] Ruth Schwartz Cowan rejects classical eugenics, with its lofty goal of 'improvement of the race,' and instead advocates a new micro-eugenics with a smaller, more selfish chief goal: protecting the comfort and independence of the parents of children prenatally diagnosed with genetic abnormalities by ending the life of the child. Professor Cowan eagerly advocates 21st-century genetic testing, prenatal diagnosis, and pregnancy termination, yet she clings to an outdated and discriminatory early-20th-century view of those with disabilities, calling people with Down syndrome 'chronically dependent' and 'suffering.' We should all reject Professor Cowan's outdated and discriminatory view that death is the appropriate outcome for those with prenatally diagnosed genetic abnormalities."

---

[1] Tobin, 2008.

- Ruth Schwarz Cowan: "I applaud those parents who, like Mr. Baute, have chosen to take on the extra burden not only of raising but also of advocating for a disabled child – but I also applaud those parents who have made the equally caring decision that they do not have the financial or emotional resources to do the same thing. Medical genetics is not eugenics, I argue in my new book, *Heredity and Hope: The Case for Genetic Screening*, precisely because both sets of choices are available to parents today."[1]

11. Leon Mellul (b. ca. 1949) of the 'Raëlian' movement (that maintains that humans were created according to instructions from alien beings and advocates cloning and human engineering): "I'm the National Guide for Israel of the Raëlian Movement, which is the equivalent of 'Bishop' or 'Grand Rabbi' of the Raëlian Movement – of the Raëlian branch of Judaism in Israel.... We consider ourselves as the real Jews." The *Jerusalem Post*, in a news brief entitled "Limp Finish for Planned Tel Aviv Orgy," announces death threats have forced the group to cancel plans for an international celebration of World Orgasm Day on December 21 by staging a mass orgy to help achieve a simultaneous global orgasm for world peace.[2]

12. *Jerusalem Post* editorial: "Obama and his advisers do not care that Jewish fertility rates are the fastest rising in the world. They do not care that by arguing for a complete halt in 'natural' growth, they are effectively adopting a eugenics argument the likes of which no US policy-maker has dared to advance since before the Holocaust."[3]

13. Cornell University geneticists create what is believed to be the first genetically engineered human embryo, which critics immediately brand as a step toward "designer babies." A gene for a fluorescent protein is inserted into a single-celled human embryo, giving the embryo three sets of chromosomes instead of two. After the embryo divides for three days, all the cells in the embryo glow. "This particular piece of work was done on an embryo that was never going to be viable," says Dr. Zev Rosenwaks, director of the Center for Reproductive Medicine and Infertility at New York-

---

[1] *Chronicle of Higher Education*, 2008.
[2] Hoffman, 2009.
[3] Glick, 2009.

Presbyterian/Weill Cornell hospital, whose ethics board approved the privately financed project.[1]

14. Geneticist David B. Goldstein: "I cannot claim the evidence proves a Khazari connection. But it does raise the possibility, and I confess that, although I can not prove it yet, the idea does now seem to me plausible, if not likely."[2]

## Jewish Advocacy of Eugenics

15. American-Jewish historian and eugenicist Seymour Itzkoff (b. 1928) advances the thesis that humanity is currently living off the temporary benefits of resources put in the ground long before people walked the earth, and once those resources are depleted – which will happen relatively soon – a vastly overpopulated world will be forced to rely on its "human capital." While the intelligent minority may (or may not) survive, the scenario is far less optimistic for the majority. Even though intelligent individuals are to be found throughout humanity, the descendants of the Cro-Magnons, Asians, and especially the Jews have a disproportionate percentage of this talent, and the less fortunate, especially sub-Saharan Africans are in for bad times. The only solution is a) population reduction and b) selective management of human breeding resources, regardless of 'ethnicity.' In Itzkoff's view modern egalitarianism is preparing the way for unheard-of disaster, and *de facto* censorship threatens our common survival. Only a "paradigm shift" can save us.[3]

16. Laurie Zoloth, Director of the Program in Jewish Studies at San Francisco State University, responds to the question "What is Judaism's take on genetic enhancement?" in an interview published in *Forward*: "We think it's permissible to provide 20/20 vision and straight teeth; would it be impermissible to do this genetically if we could? If it's correct to avoid Tay-Sachs in a child, would it be incorrect to do that genetically, for your grandchild? It's hard to find why that's wrong in principle."[4]

---

[1] Zimmer, 2008, A14.
[2] Goldstein, 2008, 74.
[3] Itzkoff, 2008.
[4] Zoloth, 2008.

17. David Berman in the *Galton Institute* (former Eugenics Society) *Newsletter*: "Evidence that Galton's *Inquiries into the Human Faculty* (1883) is a scientific classic can be found in the most respected historians of psychology.... And while the *Inquiries* is not elegantly written, I would say it is written well enough, so that once gotten into, it grows increasingly more compelling. Like most classics, it repays re-reading and studying. Add to this, that Galton's book is more than just a canonic work in psychology; it also makes a contribution to anthropology and is the pioneering work in eugenics and what might be called evolutionary theology."[1]

18. Israeli-American historian Rakefet Zalashik: "Israel is a superpower in terms of pre-pregnancy tests and abortions. Abortions are performed here on the slightest pretext, including [correctable] aesthetic flaws such as a cleft palate. The notion that there are some babies that shouldn't be born is part of the eugenic philosophy."[2]

19. A nineteen year old Lithuanian-Jewish Orthodox woman confirms in a sociological survey that a genetic disease stigmatizes the family in the eyes of the community: "This is well known. You do Dor Yeshorim so that the genetics will be good. So that you don't have children who suffer. I have in my class a friend with three brothers who are sick with cystic fibrosis and she tells everyone: 'Do the test.'" [Question] "Is your friend married?" [Answer] "No, and it is difficult for her to find a match. Since both of her parents have the defective gene, it is very difficult for her. People do not understand this. She will have to compromise." [Question] "How about you, would you consider a match with such a family?" [Answer] "No, I wouldn't. I know it's a stigma. But – I wouldn't go for it because it is frightening."[3]

## Jewish Rejection of Eugenics

20. Letter to editor of *Jerusalem Post*: "Sir, - Free will, as taught by Judaism over thousands of years, is still under attack. This time not by simpletons such as reductionists and philosophers, but by mod-

---

[1] Berman, 2007/2008, 6.

[2] Feldman, 2009; referring to Zalashik's 2008 book *Ad Nefesh: Refugees, Immigrants, Newcomers and the Israeli Psychiatric Establishment*, (*Hakibbutz Hameuchad* in Hebrew).

[3] Raz/Vizner, 2008, 1366.

ern-day eugenics (Gene variant may determine if you'll be a Scrooge December 6)."[1]

21. Los Angeles blogger Larry Fafarman: "The evidence of a connection between Darwinism and Nazi eugenics is overwhelming. The easiest way to minimize the connection between Darwinism and the holocaust is to minimize the holocaust itself! So, Darwinists, you should be grateful for the help you are getting from holocaust deniers and revisionists, and it would be in your best interests to become holocaust deniers and revisionists yourselves."[2]

## 2009

### Context

1. Upon the recommendation of Chair of the Knesset Science Committee MK Seer Sheetrit and MK Rachel Adato the cloning moratorium is extended by 7 years. The special Jewish view of the benefits of science precludes an outright ban.[3]

2. *Beshvil Hazikaron*, the periodical of the Yad Vashem Holocaust commemoration authority's school of Holocaust studies is reported to be about to publish an article by Sergio DellaPergola, Director of the Division of Jewish Demography and Statistics at the Hebrew University in Jerusalem, estimating that, if not for the Holocaust, there would be as many as 32 million Jews worldwide, instead of the current 13 million.[4]

3. Tel Aviv University historian Shlomo Sand (b. 1946) argues that most Jews actually descend from converts from disparate groups and that the existence of a Jewish people is an invention inspired by European nationalism. Israeli journalist and Historian Tom Segev (b. 1945) praises Sand's book *The Invention of the Jewish People*, which was on Israel's bestseller list for nineteen weeks, and maintains that "there never was a Jewish people, only a Jewish religion, and the exile also never happened – hence there was no return." Jewish-British historian Tony Judt (b. 1948) likewise supports Sand's analysis, maintaining that Sand exposed "the implausible

---

[1] Moshe M. Van Zuiden, Jerusalem, Dec. 9, pg. 14.
[2] Fafarman, 2008.
[3] Prainsack, 2009; Raz, 2009c.
[4] Ilani, 2009.

myth of a unique nation with a special destiny – expelled, isolated, wandering and finally restored to its rightful home.... The self-serving and mostly imaginary Jewish past that has done so much to provoke conflict in the present is revealed, like the past of so many other nations, to be largely an invention."[1]

4. Behavioral scientist Aviad Raz of Ben Gurion University: "It is arguably the time to ask whether our perception of eugenics is undergoing a tremendous shift – from its previous dystopian labeling to a utopian view, from a tragic nightmare to an inspiring dream.... Indeed, if we ask 'can we afford not to evolve?' then eugenics (like it or not) appears to be a necessary tool. This, in a nutshell, is what propagandists of 'eugenics in the 21st century' (Glad, 2006) are arguing: not only that biotechnological enhancement is going to happen, but we have a moral obligation to make it happen."[2]

5. A Gallup Poll indicates that only 39% of Americans "believe in the theory of evolution," the rest either rejecting it or having no opinion.[3]

### Jewish Advocacy of Eugenics

6. Israeli anthropologist Dafna Hirsch: "Today ideas about the benefits of mixing between Ashkenazi and Misrahi [oriental] Jews persist in Israel in a widely held assumption that a constantly growing rate of intermarriage between them will eventually lead to the creation of a new Israeli Jewish type, free of any ethnic branding, and so the solution of the 'ethnic problem' – i.e. the state of inequality between Ashkenazim and Mizrahim."[4]

### The Continuing Assault on Eugenics

7. Some book titles:

- Bender, Daniel. *American Abyss: Savagery and Civilization in the Age of Industry.* Ithaca, New York: Cornell University Press.

---

[1] Sand, 2009, back cover.
[2] Raz, 2009a, 603, 608.
[3] Corley, 2009.
[4] Hirsch, 2009, 605.

- Black, Edwin. *Nazi Nexus: America's Corporate Connections to Hitler's Holocaust.* Washington, D.C.: Dialog Press.

- Clarke, Julie. *The Paradox of the Posthuman: Science Fiction/Techno-Horror Films and Visual Media.* Saarbrücken: VDM Verlag Dr. Müller.

- Dorr, Gregory Michael. *Segregation's Science: Eugenics and Society in Virginia.* Charlottesville, University of Virginia Press.

- Kampmeyer-Käding, Margret; Kugelmann, Cilly. *Tödliche Medizin: Rassenwahn im Nationalsozialismus (Deadly Medicine: Race Madness in National Socialism.* Göttingen: Wallstein; Berlin: Jüdisches Museum.

- Lingard, Ann. *The Embalmer's Book of Recipes.* Brighton: Indepenpress.

- Mass, Ad. *Scientific Research in World War II: What Scientists Did in the War.* London/New York: Routledge.

- McWhorter, Ladelle. *Racism and Sexual Oppression in Anglo-America.* Bloomington, Indiana: Indiana University Press.

- Pichot, André. *The Pure Society: From Darwin to Hitler.* London/New York: Verso.

- Spring, Claudia. *Zwischen Krieg und Euthanasie: Zwangssterilisationen in Wien 1940-1945 (Between War and Euthanasia: Forced Sterilizations in Vienna 1940-1945).* Vienna: Böhlau.

- Stackelberg, Roderick. *Hitler's Germany: Origins,* Interpretations, Legacies. London/New York: Routledge.

- Weikart, Richard. *Hitler's Ethic: The Nazi Pursuit of Evolutionary Progress.* New York: Palgrave Macmillan.

8.   In a review of *Darwin's Racists: Yesterday, Today and* Tomorrow, by Sharon Sebastian and Raymond G. Bohlin, the influential Christian website Movieguide.org denounces the "eugenics nightmare" and maintains that the book "exposes the real Charles Darwin: a racist, a bigot and 1800's naturalist whose legacy is mass murder. This well written book shows that Adolf Hitler, along with other genocidal mass murderers, was influenced by Darwin's half-baked

Theory of Evolution. This book exposes Darwin's Theory of Evolution for what it is: an elitist and racist dogma that has infiltrated our every area of culture thereby undermining sense and sensibility." Bohlin is indicated as holding a doctorate in cell and molecular biology.[1]

### Jewish Exogamy

9. *New York Times*: "Rabbi Capers Funnye celebrated Martin Luther King Day this year in New York City at the Stephen Wise Free Synagogue, a mainstream Reform congregation, in the company of about 700 fellow Jews – many of them black. The organizers of the event had reached out to four of New York's Black Jewish synagogues in the hope of promoting Jewish diversity, and they weren't disappointed. African-American Jews, largely from Brooklyn, the Bronx and Queens, many of whom had never been in a predominantly white synagogue, made up about a quarter of the audience. Most of the visiting women wore traditional African garb; the men stood out because, though it was a secular occasion, most kept their heads covered. But even with your eyes closed you could tell who was who: the black Jews and the white Jews clapped to the music on different beats."[2]

10. Venezuelan-born Israeli historian Ariel Segal on Peruvian descendants of Ashkenazi men and Indian women in the Amazon Basin who now wish to emigrate to Israel: "The notion of a Jew who looks like an Indian and lives in a poor house in a small city in the middle of the jungle is, at best, an exotic footnote to the official history of Peru's Jewry as Lima sees it."[3]

11. In an event of historic importance, Alyssa Stanton, a 45 year old woman of mixed African-European heritage is ordained as a rabbi at Hebrew Union College, a mainstream Jewish seminary, and assumes leadership of the overwhelmingly white Congregation Bayt Shalom in Greenville, North Carolina. "I feel awe and a healthy dose of fear about being first," she comments, "I try to keep it simple. I am a Jew, and I will die a Jew." The *New York Times* article reporting the event notes that Rabbi Capers Funnye is a first-cousin, once removed, of First Lady Michelle Obama and contains an esti-

---

[1] Baehr, 2009.
[2] Chavets, 2009.
[3] Romero, 2009.

mate that "probably no more than 2 percent of the American Jewish community is made up of black Jews."[1]

12. The Moscow newspaper *Kommersant* reports that many non-Jews in Ukraine, Russia, and Belorussia are claiming to be Jewish, "attempting to capitalize on mythical family relations."

13. An article entitled "Asians: The New Shiksas" on the Web Site JewishJournal.com: "Anecdotal evidence abounds. Take a look around your temple, family bar mitzvahs, even Hollywood parties: That nice, successful Jewish boy has a willowy Asian woman on his arm."[2]

14. Conservative Rio de Janeiro Rabbi Nilton Bonder: "There are no accurate numbers. One core group is those affiliated with synagogues. They send their kids to Jewish day schools, pay dues to Jewish institutions. Their rate of intermarriage is probably lower than that in the United States. For the others, the unaffiliated, it's maybe 80 to 90 percent. The average of the two groups is probably 40 to 50 percent – comparable to the United States."[3]

15. The Web Site *convertingtojudaism.com*: "Throughout history there have been men and women who joined the Jewish people through conversion.... In the Talmudic period several well-known rabbis and scholars were themselves converts, or descended from converts, and there have also been interesting examples of group conversion to Judaism, like the entire kingdom of the Khazars in the eighth century. Talmudic Rabbi Eliezar ben Pedat said that the exile of Jews from Israel, the most terrible event in ancient Jewish history, had but one positive outcome, that 'the holy one, praised be he, exiled Israel, among the nations for the purpose of gaining converts' (Pesachim 87B).... Perhaps if it had been possible to be more open to Jewish conversion, some of the great tragedies of the Jewish people such as the Inquisition and pogroms, and even the Holocaust may have never happened, because Jews would not be such a vulnerable minority. Only now, in modern times have people begun to choose Jewish conversion, and replenish the ranks of the Jewish people that were lost in such tragedies. Among hundreds of

[1] Maag, 2009, A14.
[2] Oh, 2009.
[3] Weingart, 2009.

thousands more each year are such celebrity converts as Tom Arnold, Connie Chung, Isla Fisher, Mary Hart, Marilyn Monroe, Elizabeth Taylor, and Sammy Davis, Jr."[1]

16. Geneticists Eitan Friedman of Sheba Medical Center and Harry Ostrer of New York University Medical School launch a Jewish "HapMap" project, intended to trace haplotypes, a group of closely linked genetic markers located on one chromosome and inherited together. Friedman claims that about 50% to 60% of Israelis are eligible to participate based on their background, but concedes that in another generation, that figure could decline to 20%.[2]

17. Israeli Prime Minister Binyamin Netanyahu recommends the construction of two massive fences along the long and porous southern border with Egypt. He says he wants to stem a growing flood of African asylum seekers and to prevent Islamist militants from entering the country. He argues that it is essential to preserve Israel's Jewish majority. Government ministers approve the plan on Jan. 10, 2010.[3]

### Jewish Opposition to Eugenics

18. Rabbi Moshe Botschko, head of the Heichal Eliyahu hesder yeshiva: "In my opinion a creature born through genetic duplication is not considered human. It is clear beyond all doubt that the life form created in some scientific institution will be an animal that walks on two feet, no more. Anyone who kills a creature of this type will not be indicted, because he has not killed a man." According to Botschko, the reason for this is that the Creator only gives man a soul at the moment when sperm meets ovum. But the cloning process involves no sperm.[4]

19. An article in *Ha'aretz* asks if a human child born from the womb of a surrogate cow is considered to be a human being or a cow. If a cow, can the child be slaughtered according to *Kashrut* (Jewish dietary laws)? Rabbi and professor of Jewish law and ethics J. David Bleich responds: "A creature resembles a man and is born from a

---

[1] *convertingtojudaism.co*, 2009.
[2] Itzkovich, 2009.
[3] "Fences ordered on Egyptian Border," *Associated Press*, cited in *New York Times*, Jan. 12, 2010, A6.
[4] Ilan, 2009.

beast. If it is born to a kosher beast, the Gemara asks if it is permissible to slaughter the offspring. This question indicates that the Gemara does not consider such a creature to be human."[1]

## Jewish Advocacy of Eugenics

20. Sociologist Aviad Raz of the Ben Gurion University: "Can the Israeli model of carrier screening also provide a model for other countries? Israel may have reached a place that other western countries are more slowly opting for. After all, enthusiasm concerning the 'liberal eugenic' prospects of reprogenetics is currently building up."[2]

## 2010

### Context

1. In a report to the Haifa-based feminist organization *Woman to Woman* on the use of the contraceptive Depo Provera (administered by injection every three months), 'eco-feminist' Hedva Eyal reports that 57 per cent of Depo Provera users in Israel are Ethiopian, even though the community accounts for less than two per cent of the total population: "The unspoken policy is that only children who are white and Ashkenazi are wanted in Israel."[3]

2. A letter to the editor of the *Gazette* (Montreal): "The current movement to decriminalize euthanasia and assisted dying is based on the person's own subjective judgment of his quality of life and prospects. The eugenics movement purported to express society's interest. The current euthanasia movement represents the patient's own interest."[4]

3. A carrier of Huntington's disease in Vancouver, Canada who opted for genetic testing: "I felt very strongly that I didn't want to pass on this…. [Huntington's] is done killing people in my family when I am gone."[5]

---

[1] Ilan, 2009.
[2] Raz, 2009a, 160.
[3] Cook, 2010
[4] Zollmann, 2010, A16.
[5] Marchionne, Marilynn. 2010, A12.

4. Medical anthropologist Sydney Ross Singer cautions against the theory of: 'invasion biology,' which teaches that non-native species should not be introduced into alien environments: "On the surface, even eugenics sounded somewhat reasonable given the problems caused by overpopulation and the need to keep humanity evolving in a 'healthy' way.... To the Palestinians displaced by Jews, they are the natives and the Jews the invaders.... *Surely, there is a time and place for weeding, selecting and controlling species and people* [italics added: JG]. But we must reject the very notion that some species should be eradicated simply because they are not 'native.'"[1]

5. Republican candidate for U.S. Senator from Arkansas Curtis Coleman: "Embryonic stem cell research is taking the concept of taking a life and using it to conduct experiments so we can temporarily extend somebody else's life. Let me tell you what I just described. I just described what the Nazis did to the Jews in the death camps of World War II."[2]

6. Israeli President Shimon Peres (b. 1923) reasserts the figure of six million Jewish Holocaust victims.[3] At the annual meeting of the American Israeli Political Affairs Committee (AIPAC) Israeli Prime Minister Binyamin Netanyahu repeats the same figure.[4]

7. An exchange of views:

- Jewish-American specialist on Middle East Martin Kramer (b. 1954) argues against sending humanitarian aid to the Palestinians: "Aging populations reject radical agendas, and the Middle East is no different. Now eventually, this will happen among the Palestinians too, but it will happen faster if the West stops providing pro-natal subsidies for Palestinians with refugee status."[5]

- The website The *Electronic Intifada* accuses Kramer of calling for genocide, pointing out that "the 1948 UN Convention on the Prevention and Punishment of the Crime of Genocide, created in the wake of the Nazi holocaust, defines genocide to include

---

[1] Singer, 2010, 16.
[2] DeMillo, 2010.
[3] Peres, 2010.
[4] May 22-24.
[5] Kramer, 2010.

measures 'intended to prevent births within' a specific 'national, ethnic, racial or religious group.'"[1]

- According to *The Electronic* Intifada, Beth Simmons, Director of the Weatherhead Center for International Affairs at Harvard University, where Kramer is a visiting fellow, at first characterizes Kramer's words as "appalling," but shortly thereafter issues a statement cosigned by acting directors Jeffry Frieden and James Robinson, retracting her former position: "Accusations have been made that Martin Kramer's statements are genocidal. These accusations are baseless."[2]

8. British psychologist Richard Lynn (b. 1930) in a study of Jewish intelligence estimates that there is a difference in IQ between the Ashkenazim (110) and the Ethiopian Jews (66) of 44 points.[3]

9. Two readers' letters to *NewYork Times*:
    - "I'm sure my half-Japanese half-Jewish ex-boyfriend will be glad to note that his mixed ancestry makes him "extra-beautiful."[4]
    - "As a mixed-race child of an intermarried Jewish stepfather, I can only applaud my own generation's neglect of atavistic obsessions with eugenics and racial preservation and embrace of tolerance. May we all keep interbreeding."[5]

10. A genetic study is published in the *American Journal of Human Genetics* of seven Jewish groups (Iraqi, Syrian, Italian, Turkish, Greek, and Ashkenazi) in comparison with non-Jewish groups. Its authors conclude that "the issue of how to characterize Jewish people as mere coreligionists or as genetic isolates that may be closely or loosely related remains unresolved.... Over the past 3000

---

[1] Report, 2010a.

[2] Report, 2010b.

[3] Lynn, 2010, 173.

[4] Responding to Ross Douthat's "The Crisis of Liberal Zionism: Or is it the crisis of liberal Judaism?" *New York Times*, June 3, 2010,
http://community.nytimes.com/comments/opinionator.blogs.nytimes.com/2010/05/11/kissing-cousins/?permid=93&scp=2&sq=eugenics%20jews&st=cse.

[5] Ibid.
http://community.nytimes.com/comments/douthat.blogs.nytimes.com/2010/05/18/the-crisis-of-liberal-zionism/?scp=1&sq=eugenics%20jews&st=cse.

years, both the flow of genes and the flow of religious and cultural ideas have contributed to Jewishness."[1]

## United States Immigration Policy

11. Former Director of National Affairs at the American Jewish Committee Stephen Steinlight argues that "an historic shift in American-Jewish opinion on immigration" has taken place and that the "American-Jewish establishment" has engaged in transparently fraudulent polling methods on immigration to misrepresent Jewish popular opinion before the Congress as continuing to support unrestricted immigration.[2]

12. During a two-hour presentation entitled "Coercive Medical Research and Practice during the Holocaust," given at the United States Holocaust Memorial Museum in Washington, D.C., the eight panelists only twice use the word 'eugenics,' and on both occasions fleetingly and in passing.[3] After four decades of feverish public assault on the eugenics movement, is this new, extraordinary reticence merely a lull in the storm, or is the campaign finally being tacitly downplayed, possibly even abandoned altogether by serious researchers and maintained strictly as a propaganda tool?

---

[1] Atzmon *et al.* 2010, 850, 858.

[2] Steinlight, 2010a, 2010b.

[3] August 20. Margit Berner, Curator, Department of Anthropology, Museum of Natural History, Vienna, Austria; Gabriele Czarnowski, Researcher, Institute for Social Medicine and Epidemiology, Medical University of Graz, Austria; Herwig Czech, Researcher, Center for Documentation of the Austrian Resistance, Vienna; Michael Grodin, Professor of Health Law, Bioethics, and Human Rights, Boston University School of Public Health, Massachusetts; Sabine Hildebrandt, Lecturer of Anatomy, University of Michigan Medical School, Ann Arbor; Sari Siegel, Independent Scholar affiliated with Columbia University School of Continuing Education, New York; Kamila Uzarczyk, Department of Humanistic Sciences in Medicine, Medical University of Wroclaw, Poland; Paul Weindling, Professor in the history of Medicine, Oxford University.

## Summing Up

The material presented here allows us to draw the following conclusions:

a.  Jews have pursued practical eugenics since Biblical times, were active and largely welcome participants in the eugenics movement right up until the late 1960s, when the movement was driven underground, and remain in the vanguard of eugenic practices today.

b.  Modern Jews and ancient Jews share no particular genetic relationship. Genetic commonalities linking modern Jews together are explained almost entirely by endogamy (inbreeding); furthermore exogamy has even surpassed endogamy.

c.  While initially accepting Darwinism as applicable to human beings, Jewish thinkers attempted to cushion the blow by advocating Lamarckian eugenics, then egalitarian utopianism, and finally reverted to a qualified Darwinism, i.e. not 'nature or nurture' but 'nature plus nurture.'

d.  Despite the enthusiastic support of many Jewish thinkers of universalist eugenics, Zionism has its origins in tribalist eugenics.

e.  No biological species can exist without selection. To the degree that it is successful, the Jewish assault on eugenics poses a mortal threat to Jewry, both in a qualitative and a quantitative sense.

f.  If humanity does not adapt the eugenic practices already practiced by much of the Jewish community, it is likewise threatened.

g.  Although a few Jewish scholars now cautiously discuss the historic and current eugenic thrust of Jewish culture, the one-sided presentation of eugenics was made possible by the virtual exclusion of non-Jewish scholars from this particular area of scholarly discourse.

h.  The contradictory success of the assault on eugenics combined with support for the theory of evolution was made possible by the existence of a quasi-monopoly of the mass media. But now the former "megaphone journalism" is being undermined by the infinite multiplicity of the Internet, and it is unclear how much the influence of this group will be diminished, or even how long exaggerated claims of human particularism can continued to be viewed as viable

i.      Dispassionate scholarly discourse is poorly compatible with the politically motivated manipulation of internse emotions, however understandable, justified, and even inevitable the latter may be. While the history of the eugenics movement is hardly free of blemish – What historical movement is? – the evidence contained in this book demonstrates overwhelmingly and unambiguously that the image presented to the public is, to a dismaying degree, based on myth, disproven science, and historical distortion, compromising the very legitimacy of Holocaust/genocide studies, so that a fundamental reexamination of the entire discipline is urgently required.

# Bibliography

"Abraham, Felix." German *Wikipedia*, http://de.wikipedia.org/wiki/Felix_Abraham, accessed Feb. 2, 2008; http://www.transhistory.net/history/TH_Abraham.html, accessed Feb. 2, 2008.

Abrahamson, Daniel L. 2005. "Noam Chomsky: Controlled Asset of the New World Order," *Educate Yourself: The Freedom of Knowledge: The Power of Thought*, http://educate-yourself.org/cn/noamchomskygatekepper26sep05.shtml, accessed September 22, 2008.

*Adalah*. 2003. "Knesset passes racist law barring family unification of Palestinians married to Israeli citizens," July 31, http://electronicintifada.net/v2/article1784.shtml.

Adam, Yehuda G. 2007. "Justice in Nuremberg: The Doctors Trial – 60 Years Later. A Reminder," *Israel Medical Association Journal*, vol. 9, 194-195.

Adams, Frederick Franklin. 1926. *Eugenics: (Well Born)*, a Mother's Day sermon delivered to three denominations at The United Church, Hinesburg, Vermont, May. 9.

Adams, Mark B. (ed.). 1990. *The Wellborn Science: Eugenics in Germany, France, Brazil, and Russia*. New York/Oxford: Oxford University Press.

Adams, Susan M.; Bosch, Elean; Balaresque, Patricia L.; Ballereau, Stéphane J.; Lee, Andrew D.; Arroyo, Eduardo; López-Parra, Ana M.; Mercedes, Aler; Gisbert-Grifo, Marina S.; Brion, Maria; Carracedo, Angel; Lavinha, João; Martínez-Jarreta, Begoña; Quintana-Murce, Lluis; Picornell, Antònia; Ramon, Miserercordia; Skorecki, Karl; Behar, Doron B.; Calafell, Francesc; Jobling, Mark A. 2008. "The Genetic Legacy of Religious Diversity and Intolerance: Paternal Lineages of Christians, Jews, and Muslims in the Iberian Peninsula," *The American Journal of Human Genetics*, Dec.12, 725-736.

Adlerstein, Yitzchok. 2000? "Genetic Engineering and Its Malcontents; Commentary: Examining Halacha, Jewish Issues and Secular Law," *Jewish Law,* http://www.jlaw.com/Commentary/genetic.html.

A-J World News. 2001. "Israel Denies Attempt at Cloning," March 12, http://www.avalanchejournal.com/stories/031201/wor_031201078.shtml, accessed February 2, 2009.

Aleksandrov, V. Ia. 1993. "Trudnye gody sovetskoi biologii: Zapiski sovremennika," http://vivovoco.rsl.ru/VV/BOOKS/ALEXANDROV/CHAPTER_1.HTM, accessed Jan. 14, 2008.

*Allgemeine Zeitung des Judenthums*. 1877. "Der Racenschwindel," March 27, 197-198; April 3, 213-215.

*Allgemeine Zeitung des Judenthums*. 1885. "Über die Abstammungsverhältnisse der Juden," Feb. 17, 121-123; Feb. 24, 135-139.

Almog, Shmuel. 1983. "Alfred Nossig: A Reappraisal," *Studies in Zionism*, vol. 7, 1-29.

Alper, Joseph; Beckwith, Jon; Bruce, Bertram; Crompton, Robin; Dusek, Val; Egelman, Edward; Gould, Stephen Jay; Hubbard, Ruth; Inouye, Hiroshi; Lange, Robert; Leibowitz, Lila; Lewontin, Richard; Salzman, Freda. 1979. "The

Politics of Sociobiology," *The New York Review of Books*, 26(9), May 31, http://www.nybooks.com/articles/7782, accessed July 22, 2008.

Alsberg. Moritz. 1891. *Rassenmischung in Judenthum* Hamburg: Verlangsanstalt und Druckerei A.-G.

Amar, A.; Kwon, O.J.; Motro, U.; Witt, C.S.; Bonne-Tamir, B.; Gabison, R.; Brautbar, C.

1999. "Molecular Analysis of HLA Class II Polymorphisms among Different Ethnic Groups in Israel," *Human Immuunology*, 60(8), 723-730.

American Eugenics Party. 1964. *American Eugenics Party Platform*. American Philosophical Society, Image 762, http://www.eugenicsarchive.org/html/eugenics/index2.html?tag=762.

American Eugenics Society membership lists. 1916-1973. American Philosophical Records, http://www.amphilsoc.org/library/mole/a/aes.htm#contact.

American Eugenics Society. 1939. Invitation to a Conference on the Relation of Eugenics and the Church, New York City, May 8; speakers include Rabbi Louis I. Newman, "The Modern Jew Looks at Programs of Human Betterment."

American Hebrew Publishing Company. 1884. "New Jews and Judaism," *The American Hebrew,* March 14, 66.

Amlinsky, Andrei. 2009. "Pyatyi punkt kak brènd," *Kommersant*, magazine section *Sekret firmy*, Feb. 2, No. 1-2. 283, http://www.kommersant.ru/doc.aspx?DocsID=1105027&print=true.

Andersen, Rebecca. 2005. "The Singular Moral Compass of Otto Krayer," *Molecular Interventions*, 5:324-329, http://molinterv.aspetjournals.org/cgi/content/full/5/6/324, accessed Feb. 25, 2008.

Anderson, Lloyd L.; Birch, Guy Irving; Evans, S. Wayne; Hansell, Marion; MacArthur, K.C. 1931. *Eugenical Panorama*, http://www.eugenicsarchive.org/html/eugenics/static/images/1614.html, accessed July 23, 2008.

Anderson, Walter Truett. 2005. "Why Eugenics Is Here to Stay," *JINN*, July 27, http://www.pacificnews.org/jinn/stories/columns/heresies/950727-eugenics.html, accessed Dec. 25, 2007.

Andrée, Richard. 1881. *Zur Volkskund der Juden*, Bielefeld und Leipzig, Velhagen & Klasing.

"Annual Guide to Jewish Genetic Diseases." 2007. The *Forward*, Aug. 22, http://www.forward.com/articles/11430/.

Anonymous. 1905. "The Jewish population of the United Kingdom," *Jewish Chronicle*, April 14, 22.

Anonymous. 1907. "From Far and Near," *Jewish Chronicle*, June 28, 2.

Anonymous. 1911a. "A Study of the Jewish Race," *New York Times*, Feb. 12, BR75. 2007 (Web site accessed April 23, 2007). "The Jewish Self-Conception of Intellectual, Moral, and Spiritual Superiority," *When Victims Rule,* http://www.jewishtribalreview.org/16supe/htm.

Anonymous. 1911b. "Sir James Barr on Jewish Eugenics: An Extraordinary Deliverance," *Jewish Chronicle*, Dec. 1, 16.

Anonymous. 1911c. "Sir James Barr's Opinions: Jews as Soldiers," *Jewish Chronicle*, Dec. 8, 8.

Anonymous. 1917. "Jewish Eugenics: Perpetuation of the Race Explained by Application of Sound Biological Principles – Marriage Held in High Esteem and Its Success Measured by Eugenic Standard," *Journal of Heredity*, vol. 8, 72-74.

Anti-Defamation League. 2000. "Is FAIR Unfair? The Federation for American Immigration Reform (FAIR)," http://www.adl.org/civil_rights/is_fair_unfair.pdf, accessed May 28, 2008.

Anti-Defamation League. 2004. "Yehuda Bauer: Historian of the Holocaust," 18(1), fall, http://www.adl.org/education/dimensions_18_1/portrait.asp, accessed June 3, 2009.

Arad, Yitzhak; Gutman, Israel; Margaliot, Abraham. 1999. *Documents on the Holocaust: Selected Sources on the Destruction of the Jews of Germany and Austria, Poland, and the Soviet Union*, translations by Lea Ben Dor. Lincoln/London/Jerusalem:Yad Vashem, University of Nebraska Press.

Arendt, Hannah. 1963. *Eichmann in Jerusalem: A Report on the Banality of Evil*. New York: Viking.

Arnaiz-Villena, Antonio; Elaiwa, Nagah; Silvera, Carlos; Rostom, Ahmed; Moscosco, Juan; Gómez-Casado, Eduardo; Allende, Luis; Varela, Pilar; Martínez-Laso, Jorge. 2001. "The Origin of Palestinians and Their Genetic Relatedness with Other Mediterranean Populations," *Human Immunology*, vol. 62. 889-900.

Arroyo, Raymond. 2007. *Mother Angelica's Little Book of Life Lessons and Everyday Spirituality.* Doubleday, March 6, http://www.booksonboard.com/index.php?BODY=viewbook&BOOK=579 24&v=excerpt, accessed July 18, 2008.

Asher, G.M. 1891. "Der arische Ursprung der Juden und ihre welthistorische Bestimmung," *Allgemeine Zeitung des Judenthums*, Mar. 7, 319; Mar. 14, 387-388; Sept. 10, 486-487.

Associated Press. 2003. "Cloned Baby Is in Israel, Witness Says," Jan. 30, http://www.newsday.com/topic/sns-cloning-baby,0,6669617.story, accessed July 6, 2009.

Atallah, Lilian. 1976. "Report from a test-tube baby," *New York Times*, April 18, 155.

Atzmon, Gil; Hao, Li; Pe'er, Itsik; Velez, Christopher; Pearlman, Alexander; Palamara, Francesco; Morrow, Bernice; Friedman, Eitan; Oddoux, Carole; Burns, Edward; Ostrer, Harry. 2010. "Abraham's Children in the Genome Era: Major Jewish Diaspora Populations Comprise Distinct Genetic Clusters with Shared Middle Eastern Ancestry," *American Journal of Human Genetics*, No. 86, June 11, 850-859.

Auel, Jean. 2004. Discussion Enclave, Science Fiction and Fantasy Talk and Real Science, posted Tuesday, July 13, www.geocities.com/Athens/6293/Jean Auel.htm?20085.

Babkov, Vasily Vasilievich. 1998. "Kak kovalas' pobeda nad genetikoi," *Chelovek*, no. 6, Dec., 82-90.

Baehr, Ted. 2009. "Book Review: *Darwin's Racists – Yesterday, Today and Tomorrow*," Sept. 9, http://www.movieguide.org/articles/1/463-book-review-darwins-racists-yesterday-today-and-tomorrow.

Bakh, A. N.; Keller, B. A.; Koshtoiants, Kh. S.; Komarov, V. L. 1939. "Lzheuchenym net mesta v Akademii Nauk," *Pravda*, January 11.

Ballabon, Jeff. 2000. "In the name of Values, not Politics" *Forward*, April 8, http://www.forward.com/articles/3259/.

Bar-Giora, Naftali. 2007. "Bene Israel: In Israel," *Encyclopedia Judaica*, vol. 3, 2nd edition, 338-339.

Baranovsky, Andrei. 2005. "Dva Sologuba, i odin pishet Leninu..." *Poslednie novosti ot ACADEMIA-1937*, http://www.ozon.ru/context/detail/id/2431259/, accessed March 10, 2008.

Barkan, Elazar. 1992. *The retreat of scientific racism: Changing concepts of race in Britain and the United States between the world wars*. Cambridge/New York/Melbourne: Cambridge University Press.

Barnett, Larry D. 1965. "Religious Differentials in Fertility Planning and Fertility in the United States," *The Family Life Coordinator*, 14(4), Oct. 161-170.

Barnoy, Sivia; Ehrenfeld, Malka, Sharon, Rina; Tabak, Nili. 2006. "Knowledge and Attitudes toward Human Cloning in Israel," *New Genetics and Society*, 25(1), April, 31.

Bashi, Joseph. 1977. "Effects of inbreeding on cognitive performance" (letter), *Nature*, March 31, vol. 266, 440-442.

Baylis, Françoise; Robert, Jason Scott. 2004. "The Inevitability of Genetic Enhancement Technologies," *Bioethics*, 18(1), 1-26.

Beckwith, Jon. 1993. "A Historical View of Social Responsibility in Genetics," *Bioscience*, 43(5).

Beckwith, Jon; Alper, Joseph S. 2006. "'Race', IQ and Genes," *Encyclopedia of the Life Sciences*, John Wiley & Sons, www.els.net.

Beddoe, John. 1861. "On the Characteristics of the Jew," *Transactions of the Ethnological Society of London*, 222.

Behar, Doron M; Thomas, Mark G.; Skorecki, Karl; Hammer, Michael F.; Bulygina, Ekaterina; Rosengarten, Dror; Jones, Abigail L.; Held, Karen; Moses, Vivian; Goldstein, David; Bradman, Neil; Weale, Michael E. 2003. "Multiple Origins of Ashkenazi Levites: Y Chromosome Evidence for Both Near Eastern and European Ancestries," *American Journal of Human Genetics*, October; 73(4): 768–779.

Benedict, George. 1926. *Eugenics*, Mother's Day sermon delivered June 4 at the Temple, Emanu-El in Roanoke, Virginia.

Berck, Judith. 2006. "When Science Aids Reproduction, Some Parents Wonder What It Takes to Be Jewish," *The New York Times*, July 1, A14.

Berg, Raissa L. 1990. "The Grim Heritage of Lysenkoism: Four Personal Accounts. In Defense of Timoféeff-Ressovsky," *Quarterly Review of Biology*, 65(4), 457-479.

Bergman, Jerry. 1999. "Darwinism and the Nazi Race Holocaust." First published in *Creation Ex Nihilo Technical Journal*, 13(2), 101-111.

Bergson, Henri. 1907. *L'évolution créatrice*, Félix Alcan, Paris. http://www.trueorigin.org/holocaust.asp. (last modified March 9, 2006).

Berman, David. 2008. "Enthusiasm in Galton's Inquiries into the Human Faculty," *Galton Institute Newsletter*, Dec. 2007/March 2008, Numbers 65& 66.

Bernstein, Victor. 1945. "Man Who Created Nazi Science of Eugenics: Meet Professor Rudin," *Palestine Post*, Oct. 19, 5.

Besser, Max. 1911. *Die Juden in der modernen Rassentheorie*. Köln/Leipzig: Jüdischer Verlag.

Bialik, Chaim Nachman. 1934. "The Present Hour," *Young Zionist,* vol. 4, 6-7.

Binding, K.; Hoche. 1920, 1922. A. *Die Freigabe der Vernichtung lebensunwerten Lebens. Ihr Maß und ihre Form.* Leipzig: Felix Meiner Verlag.

Birenbaum-Carmeli, Daphna; Carmeli, Yoram S.; and Cohen, Rita. 2000. "Our First 'IVF Baby': Israel and Canada's Press Coverage of Procreative Technology," *International Journal of Sociology and Social Policy,* 20(7), 1-37.

Birenbaum-Carmeli, Daphna; Carmeli, Yoram S.; Weissenberg, Yigal; Weissenberg, Ruth. 2002. "Hegemony and Homogeneity: Donor Choices of Israeli Recipients of Donor Insemination," *Journal of Material Culture,* vol. 7(1), 73-95.

Birnbaum. 1886. "Nationalität und Sprache," *Selbst-Emanzipation,* 16, February, 4.

Black, Edwin. 2003. *War Against the Weak: Eugenics and America's Campaign to Create a Master Race.* New York/London: Four Walls Eight Windows.

Black, Edwin (Black Home Page). 2007. http://www.edwinblack.com/travel.php, accessed Dec. 10, 2007.

Black, Edwin. 2008. Web site. http://www.edwinblack.com, accessed September 13.

Blair, Fred. 1946. *The Ashes of Six Million Jews.* Milwaukee: The People's Bookshop.

Blau, Joel. 1916. *The Defective in Jewish Law and Literature: A Paper Read before the New York Board of Jewish Ministers.* New York: Bloch Publishing Company.

Bleich, J. David. 1977. *Contemporary Halakhic Problems,* the Library of Jewish law and ethics, v. 4, 10, 16, 20. New York: Ktav.

Bloch, Talia. 2004. "A Skeleton in the Jewish Family Closet?" *Forward,* Aug. 20.

Bloch, Talia. 2006. "Building a Memorial from Strands of DNA," *The Forward,* Aug. 25, http://www.forward.com/articles/1473/.

Bloch, Talia. 2007. "One Big, Happy Family: Litvaks and Galitzianers, Lay Down Your Arms: Science Finds Unity in the Jewish Gene Pool," *The Forward,* Aug. 22, http://www.forward.com/articles/11444/.

Bloom, Etan. 2007. "The Administrative Knight: Arthur Ruppin and the Rise of Zionist Statistics," *Tel Aviv University Yearbook for German History,* XXXV, 183-203.

Bock, Gisela. 1984. "Racism and Sexism in Nazi Germany: Motherhood, Compulsory Sterilization, and the State," in *When Biology Became Destiny: Women in Weimar and Nazi Germany,* edited by Renate Bridenthal, Atina Grossmann, and Marion Kaplan, Monthly Review Press, New York, 271-296.

Bock, Gisela. 1986. *Zwangssterilisation im Nationalsozialismus: Studien zur Rassenpolitik und Frauenpolitik.* Opladen: Westdeutscher Verlag.

Bokser, Ben Zion. 1975. "Problems in Bio-Medical Ethics: A Jewish Perspective," *Judaism,* Spring, 24(2), 134-143.

Bonne-Tamir, B.; Frydman, M.; Agger, M.S.; Bekeer, R.; Bowcock, A.M.; Hebert, J.M.; Cavalli-Sforza, L.L.; Farrer, L.A. 1990. "Wilson's disease in Israel: A genetic and epidemiological study," *Annals of Human Genetics,* 54(May), Part 2, 155.

Borkenau, Peter; Riemann, Rainer; Agleittner, Alois; Spinath, Frank M. 2001. "Genetic and Environmental Influences on Observed Personality: Evidence from the German Observational Study of Adult Twins," *Journal of Personality and Social Psychology*, Vol. 80, No. 4, 655-668.

Boyle, Philip J. 2000. "Genetic Challenges to Faith," *Bulletin of the Park Ridge Center*, Spertus Institute of Jewish Studies, Chicago.

Bozeman, John M. 2004. "Eugenics and the Clergy in the Early Twentieth-Century United States," *Journal of American Culture,* 27(4), 422-430.

Braund, James; Sutton, Douglas G. 2008. "The case of Heinrich Wilhelm Poll (1877-1939): A German-Jewish Geneticist, Twin Researcher, and Victim of the Nazis," *Journal of the History of Biology*, No. 41, 1-35.

Brenner, Lenni (ed.). 2002. *51 Documents: Zionist Collaboration with the Nazis* Barricade: Fort Lee, New Jersey.

Bresler, Jack B.; Dunn, Frances E.; Urquhart, Helen L.; Smith, Elmer R. 1961. "The Human Biology of Academic Potential: A Proposed Investigation," *Eugenics Quarterly.*

Brieger, Lothat. 1922. *E. M. Lilien*. Berlin/Vienna: Benjamin Harz.

Brin, Louis B. Brinn. 1965. "Jews, Genetics and Disease," *Harofé Haifiri*, 261-275.

Broyde, Michael. 2004. "Cloning People and Jewish Law: A Preliminary Analysis," *Jewish Law: Articles Examining Halacha, Jewish Issues and Secular Law*, http://www.jlaw.com/Articles/cloning.html#Foot1, accessed July 8, 2004.

Burger, John S. 1974-1975. ""Rabbi Sternheim – Progressive," *American Jewish Historical Quarterly*, 64, 1-4, ABIS journal, 358.

Burstein, N. S. 1912. "Eugenics: Jews and Morality," *Jewish Chronicle*, Aug. 16, 6.

Campbell, Arthur; Whelpton, Pascal K.; Patterson, John E. 1960. *Growth of American Families*.
http://www.sscnet.ucla.edu/issr/da/Mobility/abs/USA60gaf.html.

Cantor, Geoffrey; Swetlitz, Marc (editors). 2006. *Jewish Tradition and the Challenge of Darwinism*. University of Chicago Press.

Capaldi, Nicholas. 1999. "What is Bioethics without Christianity?" *Christian Bioethics*, 5(3), 246-262.

Caplan, Arthur. 1989. "The Meaning of the Holocaust for Bioethics," *Hastings Center Report*, 19(4), July/Aug., 2-3.

Caplan, Arthur. 2008. "Intelligent design film far worse than stupid: Ben Stein's so-called documentary 'Expelled' isn't just bad, it's immoral," MSNBC, http://www.msnbc.msn.com/id/24239755/, accessed May 11, 2008.

Caplan, Arthur L.; McGee, Glenn; Magnus, David. 1999. "What is immoral about eugenics?" *BMJ*, vol. 319, November 13.

Caplan, Gerald. 1960. "Emotional Implications of Pregnancy and Influences on Family Relationships," *The Healthy Child*, Harvard University Press, Boston.

Carlson, A.; Riederer, P.; Stern, G. 1997. "Walther Birkmayer – The Man behind the Name," *Journal of Neural Transmission*, Feb. 104(2-3), V-VIII.

Carlson, Elof Axel. 1986. Review of *In the Name of Eugenics* by Daniel Kevles, *Science*, New Series, vol. 232, No. 4794, April, 531-532.

Carlson, Elof Axel. 2001. *The Unfit: History of a Bad Idea*. Cold Spring Harbor, New York: CSHL Press.

Carmen, Ira H. 1986. Review of *In the Name of Eugenics: Genetics and the Uses of Human Heredity* by Daniel J. Kevles, *The American Political Science Review*, 80(3), Sept., 1007-1008.

Carmi, Rivka; Elbedour, Khalil; Wietzman, Dahlia; Sheffield, Val; Shoham-Vardi, Ilana. 1998. "Lowering the Burden of Hereditary Diseases in a Traditional, Inbred Community: Ethical Aspects of Genetic Research and Its Application," *Science in Context* 11(3-4), 391-395.

Carron, Sher; Romano-Zelekha, Orly; Green, Manfred S.; Shohat, Tamy. 2003. "Factors Affecting Performance of Prenatal Genetic Testing by Israeli Jewish Women," *American Journal of Medical Genetics*, July 30, 120A, (3), 418-422.

Carvutto, Susan Bulba. 2004. "Choose Life: The Abortion Dilemma from a Jewish Perspective," Aug. 31; cited by Faith Streams Network, June 29, 2008, http://www.faithstreams.com/ME2/Sites/dirmod.asp?sid=5F4E345683D84 92B9B56CBC49802F459&nm=Get+the+News&type=news&mod=News& mid=9A02E3B96F2A415ABC72CB5F516B4C10&SiteID=123E5391AB7 24C0487CE5AA61CA7ED75&tier=3&nid=0ACDC28757F24AB3BF5D8 5694AC6E538, accessed June 29, 2008.

Cavanaugh-O'Keefe, John. 1995. *Introduction to Eugenics,* Jan., http://reducetheburden.org/?p=504, accessed June 2, 2009.

Cavanaugh-O'Keefe, John. 2000. "Chapter Nine: World War II and the Nazi Holocaust, *The Roots of Racism and Abortion: An Exploration of Eugenics,* http://www.eugenics-watch.com/roots/chap09.html.

Cavanaugh-O'Keefe, John. 2001? "Whitewashing Eugenics," a review of *From Chance to Choice: Genetics & Justice,* by Allen Buchanan, Dan W. Brock, Norman Daniels, & Daniel Winkler (Cambridge: Cambridge University Press, 2000), http://www.eugenics-watch.com/wwash.html http://www.eugenics-watch.com/wwash.html, accessed March 16, 2008.

Chajes, Benno. 1921. *Kompendium der sozialen Hygiene.* Berlin: Fischer's Buchhandlung.

Charpa, Ulrich; Deichmann, Ute. 2007. *Jews and Sciences in German Contexts: Case Studies from the 19th and 20th Centuries.* Tübingen: Mohr Siebek.

Chafets, Zev. 2009. "Obama's Rabbi." *New York Times*, April 2, Magazine Section, http://www.nytimes.com/2009/04/05/magazine/05rabbi-t.html?pagewanted=1&ref=us.

Chavets, Zev. 2009. "Obama's Rabbi," *New York Times*, April 2, http://bechollashon.org/about/News/4-2-2009.php.

Cherry, Shai. 2003. "Three Twentieth-Century Jewish Responses to Evolutionary Theory," *Aleph,* vol. 3, 247-290.

"Chicago Center for Jewish Genetic Disorders Serves All Segments of the Community," 2005. *Jewish Genetic Disorders*, Sept. 61.

Childs, Donald J. 2001. *Modernism and Eugenics: Woolf, Eliot, Yeats, and the Culture of Degeneration.* London/New York: Cambridge University Press.

*Chronicle of Higher Education.* 2008. 54(40), "Eugenics in the Past and Eugenics Today," B25-B26, 2p, June 13, http://web.ebscohost.com.proxy-um.researchport.umd.edu/ehost/detail?vid=1&hid=104&sid=d0e344d5-c9b8-42df-bd6f-ba1cd533ccef%40sessionmgr109&bdata=JmxvZ2lucGFnZT1Mb2dpbi5hc3Amc2l0ZT1laG9zdC1saXZl#db=tfh&AN=32993503.

CNN. 2008. "McCain Rejects Minister's Endorsement," May 22, http://www.cnn.com/2008/POLITICS/05/22/mccain.hagee/index.html.

Cochran, Gregory; Hardy, Jason; Harpending, Henry. 2005. "Natural History of Ash-kenazi Intelligence," *Journal of Biosocial Science,* 1-35. Discussion: "Study's Claim on Intelligence of Ashkenazim Spurs a Debate," *Forward,* June 10, http://www.forward.com/articles/3297.

Cohen, Eyal. 1998. "The Nazification of German Physicians, 1918-1937," *Annals of Royal College of Surgeons and Physicians of Canada*, 31(7), 336-340. Subsequent correspondence between Cohen and Wilhelm Kreyes in 32(1), 1999, February, 40-41.

Cohen, Jonathan R. 1999. "In God's Garden: Creation and Cloning in Jewish Thought," *Hastings Center Report* 29, no. 4, July-August, 7-12.

Cohen, Naomi W. 1984. "The Challenges of Darwinism and Biblical Criticism to American Judaism," *Modern Judaism*, vol. 4, no. 2, May, 1984, 121-157.

Cohen, Noam. 2003. "Word for Word: The Raëlian Agenda; And You Figured Clon-ing a Human Would Be Their Biggest Challenge," Jan. 5, http://query.nytimes.com/gst/fullpage.html?res=9403E1D91F3FF936A357 52C0A9659C8B63&sec=&spon=&pagewanted=2, accessed May 14, 2008.

Cohen, Steven M.; Wertheimer, Jack. 2006. "Whatever Happened to the Jewish people," *Commentary*, June, http://www.commentarymagazine.com/viewarticle.cfm/whatever-happened-to-the-jewish-people-10079.

Cohen, T.; Levine, C.; Yodfat, Y.; Fidel, J.; Friedlander, Y.; Steinberg, A. G.; Brautbar, C. 1980. "Genetic Studies on Cochin Jews in Israel," *American Journal of Medical Genetics*, 6(1), 61-73.

Cohen-Almagor, Raphael; Shmueli, Merav. 2000. "Can Life Be Evaluated? The Jew-ish Halachic Approach vs. the Quality of Life Approach in Medical Ethic: A Critical View," *Theoretical Medicine and Bioethics*, 21(1), 117-137.

Cole, Tim. 2000. *Selling the Holocaust: From Auschwitz to Schindler, How History is Bought, Packaged, and Sold.* New York: Routledge.

Colloquy Live. "Ethical Eugenics?" *Chronicle of Higher Education*, April 30, http://chronicle.com/colloquylive/2003/04/eugenics/, accessed Jan. 2, 2007.

Committee on Energy and Commerce (Subcommittee on Oversight and Investiga-tions). 2001. "Issues Raised by Human Cloning Research," U.S. House of Representatives, March 28, http://archives.energycommerce.house.gov/reparchives/107/action/107-5.pdf.

Committee on Patrilineal Descent. Undated. "Reform Movement's Resolution on Patrilineal Descent: The Status of Children of Mixed Marriages," *Jewish Virtual Library,* March 15, http://www.jewishvirtuallibrary.org/jsource/Judaism/patrilineal1.html, ac-cessed April 18, 2008.

Conklin, J. 1996. "Giving the greatest gift of all: Infertile couples find hope in UW's egg donor program." *The Online Daily,* University of Washington Student Newspaper.

Cook, Jonathan. 2010. "Israel's Treatment of Ethiopians Racist," *Ethiopian Review*, Jan. 13. http://www.ethiopianreview.com/articles/31493, accessed Jan. 29, 2010. Connor, Steve. 2009. "Fertility Expert: 'I can clone a human being,'" *The Independent*, April 22.

convertingtojudaism.com, 2009, http://www.convertingtojudaism.com/Historical-
Context.htm.

Corley, Matt. 2009. "On Darwin's 200th birthday, only 39 percent of Americans believe in evolution," *Think Progress,*
http://thinkprogress.org/2009/02/11/darwin-200.

Coussin, Orna. 2008. "Racial Test," *Ha'aretz*, Jan 12,
http://www.haaretz.com/hasen/spages/1045035.html.

Cowan, Ruth Schwartz. 2008. *Heredity and Hope: The Case for Genetic Screening.*
Cambridge, Massachusetts: Harvard University Press.

Crosland, H. R. 1940. Review of Morris Siegel's *Population, Race, and Eugenics,*
*American Journal of Psychology*, 53(1), Jan., 156-157.

Crozier, Ivan. 2001. "Becoming a Sexologist: Norman Haire, the 1929 London
World League for Sexual Reform Congress, and Organizing Medical
Knowledge about Sex in Inter-War England, *History of Science*, xxxix,
299-329.

Culliton, Barbara J. 1975. "Harvard Researcher under Fire Stops Newborn Screening," *Science,* 188(4195), 1284-1285.

Curran, Doug. 2004. "Nazi Eugenics at the Holocaust Museum," *The Georgetown
Independent*, no. 10,
http://www.thegeorgetownindependent.com/home/index.cfm?event=displa
yArticle&ustory_id=27071aa2-01c7-407c-86e8-ee9aed1b236b&page=2,
accessed Dec. 30, 2007.

Curry, Andrew. 2006. "US Holocaust Museum Travels to Germany," *Spiegel On
Line*, October 13,
http://www.spiegel.de/international/0,1518,442516,00.html, Feb. 3, 2008.

Daniels, Cynthia R.; Golden, Janet. 2004. "Procreative compounds: Popular eugenics, artificial insemination and the rise of the American sperm banking industry," *Journal of Social History*, Sept. 22, 38(1), 5-27.

Danzis, Max. 1930. "The Jew in Medicine from Biblical to Modern Times," *Journal
of the Medical Society of New Jersey,* 27(10), 763-776.

David, Hanna; Lynn, Richard. 2007. "Intelligence Differences between European and
Oriental Jews in Israel," *Journal of Biosocial Science,* vol. 39, 465-473.

Davidov, B.; Goldman, B.; Atstein, E.; Barkai, G.; Legum, C.; Dar, H.; Ronem, Y.;
Amiel, A.; Cohen, H.; Bach, G. 1994. "Prenatal testing for Down syndrome
in the Jewish and non-Jewish populations in Israel," *Israel Journal of Medical Science*, Aug. 30(8), 629-633.

Davidovich, Nadav, and Schvarts, Shifra. 2004. "Health and Hegemony: Preventive
Medicine, Immigrants and the Israeli Melting Pot," *Israel Studies,* 9(2),
150-178.

Davids, Leo. 1983. "What's Happening in the Israeli Family? Recent Demographic
Trends," *Israel Social Science Research: A Multidisciplinary Journal,* 1(1),
34-38, Hubert H. Humphrey Center for Social Ecology, Ben-Gurion University of the Negev, Beer-Sheva, Israel.

Davis, Dena S. 1994. "Method in Jewish Bioethics: An Overview," *Journal of Contemporary Law*, 20(2), 325-52, http://www.lexisnexis.com.proxy-
um.researchport.umd.edu/us/lnacademic/search/homesubmitForm.do, accessed April 22, 2008.

Dawidowicz, Lucy S. 1977. "The Failure of Himmler's Positive Eugenics," *The Hastings Center Report*, 7(5), Oct., 43-44.

Deichmann, Ute. 1996. *Biologists under Hitler*, translated by Thomas Dunlop. Cambridge, Massachusetts/London: Harvard University Press.

DellaPergola, Sergio. 2002. "World Jewish Population 2002," *American Jewish Year Book,* http://www.jafi.org.il/education/100/concepts/demography/demjpop.html, accessed August 13, 2008.

DellaPergola, Sergio. 2003. *Jewish Out-Marriage: A Global Perspective*, International Intermarriage Roundtable Conference, Hadassah-Brandeis Institute December 17-18, http://www.brandeis.edu/hbi/pubs/conferencepapers.html.

DellaPergola, Sergio. 2005. "Was It the Demography? A Reassessment of U.S. Jewish Population Estimates, 1945-2001." *Contemporary Jewry*. 85-131.

DellaPergola, Sergio; Dror, Yehezkel; Wald, Shalom S. *The Jewish People 2005: Facing a Rapidly Changing World,* Executive Report, Annual Assessment. Jerusalem: The Jewish People Policy Planning Institute.

Demick, Barbara. 2001. "Scientist Look for a Nation to Host Human Cloning Project," *Knight Ridder Tribune Business News,*" March 17, http://global.factiva.com.proxy-um.researchport.umd.edu/ha/default.aspx.

DeMillo, Andrew. 2010. "Coleman: No Apology for Stem Cell-Nazis Comparison," Associated Press, http://www.biosciencetechnology.com/News/FeedsAP/2010/02/coleman-no-apology-for-stem-cell-nazis-comparison.

Dershowitz, Alan M. 1997. *The Vanishing American Jew*. New York: Simon and Schuster.

Directors, Board of, American Society for Human Genetics (Statement). 1999. "Eugenics and the Misuse of Genetic Information to Restrict Reproductive Freedom," *American Journal of Human Genetics*, NO. 64, 335-338.

Dolan DNA Learning Center. 2006? (accessed December 22). "Image Archive on the American Eugenics Movement," http://www.eugenicsarchive.org/eugenics/list3.pl/ Cold Spring Harbor Laboratory, Cold Spring Harbor, New York.

Donor program. 1996. *The Online Daily of the University of Washington*, January 16.

"Dor Yeshorim." 2007 (latest update). Chicago Center for Jewish Genetic Disorders, http://www.jewishgeneticscenter.org/genetic/doryeshorim/, accessed Dec. 30, 2007, accessed April 4, 2005.

Dorff, Elliot N. 2002. "Embryonic Cell Research: The Jewish Perspective," *United Synagogue of Conservative Judaism Review*, Spring.

Doron, Joachim. 1980. "Rassenbewusstsein und naturwissenschaftliches Denken im deutschen Zionismus während der Wilhelminischen Ara," Ja*hrbuch der Institut für deutsche Geschichte, Universität Tel Aviv*, vol. 9, 389-427.

Doron, Joachim. 1983. "Classic Zionism and Modern Anti-Semitism," *Studies in Zionism*, Autumn, no. 8, 169-204.

Drouard, Alain. 1999. *L'eugénisme en questions: L'exemple de l'eugénisme 'français.'* Paris: Ellipses.

Dubin, Lois C. 1995. "Pe'er Ha'adam of Vittorio Hayim Castiglioni: An Italian Chapter in the History of Jewish Response to Darwin," *The Interaction of*

*Scientific and Jewish Cultures in Modern Times*, Edwin Mellen Press, Lewiston, UK, 87-101.

Dubinin, Nikolai. 1968. "As the Geneticist Sees It," *Eugenics Quarterly*, June, 15(2), 144-145.

Èfroimson, Vladimir. 1998. *Genial'nost' i genetika.* Moscow: Russkii mir.

Èfroimson, Vladimir 2002. *Genetika genial'nosti: Biosotsial'nye faktory I mekhanizmy naivysshei intellektual'noi aktivnosti.* Taideks Ko., Moscow.

Efron, John M. 1994. *Defenders of the Race: Jewish Doctors and Race Science in Fin-de-Siècle Europe.* New Haven/London: Yale University Press.

Efron, John M. 2001. *Medicine and the German Jews: A History*, New Haven/London: Yale University Press.

Ehrenburg, Ilya. 1944. "Pomnit'!" *Pravda*, Dec, 17, 3.

Eisenberg, Vered H.; Schenker, Joseph G. 1997. "Reproductive Health Care Policies Around the World: Genetic Engineering: Moral Aspects and Control of Practice." *Journal of Assisted Reproduction and Genetics,* vol. 14(6), 297-316.

Elliman, Wendy. 2001. "Statistical probabilities and probable cures," *Jerusalem Post*, February 27.

Ellul, Jacques. 1967. *Histoire de la propagande.* Paris: Presses Universitaires de France.

Elmer-DeWitt, Philip; Bjerklie, David. 1993. *Time*, Nov. 8, 142(19), 64-68.

Endelman, Todd M. 2004. "Anglo-Jewish Scientists and the Science of Race," *Jewish Social Studies,* 11(1), 52-92.

Entine, John. 2007. *Abraham's Children: Race, Identity, and the DNA of the Chosen People.* New York/Boston: Grand Central Publishing.

Epstein, Nadine. 2001, 82(5), *Hadassah.*

Erlenmeyer-Kimling, L.; Falek, Arthur; Jarvik, Lissy F.; Rainer, John D.; Sank, Diane. 1965. "Franz J. Kallmann, 1897-1965," *Eugenics Quarterly*, Sept., 12(3), 123.

Essner, Cornelia. 1995. "Qui sera 'Juif'?" *Genèses,* vol. 21, Dec., 4-28.

"Eugenic Marriages Urged for Jersey." 1915. *New York Times*, Nov. 20, 1.

"Eugenics Timeline," Museum of disABILITY, http://www.museumofdisability.org/html/exhibits/society/Timeline_eugenics.htm, accessed Jan. 18, 2008.

Eugenics Watch Web site, http://www.eugenics-watch.com.

"Ezra." 2007. Wikipedia, http://en.wikipedia.org/wiki/Ezra.

Fafarman, Larry. 2008. "Holocaust denial/revisionism helps Darwinism, *I'm from Missouri*, http://im-from-missouri.blogspot.com/2008/04/holocaust-denialrevisionism-helps.html, accessed May 10, 2008.

Falek, Arthur. 1965. Book review of *Man and His Future, Eugenics Review*, March, 12(1), 50-51.

Falk, Candace; Cole, Stephen; Thomas, Sally. Undated. "Emma Goldman: A Guide to Her Life and Documentary Sources: Chronology (1901-1919)," *The Emma Goldman Papers*, http://sunsite.berkeley.edu/Goldman/Guide/chronology0119.html, accessed April 17, 2008.

Falk, Raphael. 1998. "Zionism and the Biology of the Jews," *Science in Context,* 11(3-4), 587-607.

Falk, Raphael. 2003-2004. "Nervous Diseases and Eugenics of the Jews: A View from 1918," *KOROT*, vol. 17, 23-46.

Falk, Raphael. 2002. "The Eugenic Dimension of the Settlement of Palestine," *Alpayim*, vol. 23, 179-198 (in Hebrew).

Falk, Raphael. 2007. "Three Zionist Men of Science: Between Nature and Nurture," 129-154, in *Jews and Scientists in German Contexts: Case Studies from the 19th and 20th Centuries*, ed.: Ulrich Charpa and Ute Deichmann, Mohr Siebeck, Tübingen.

Falk, Raphael; Pfaff, William. 1998. "Eugenics Denied," *New York Review of Books*, vol. 45, no. 1, January 15, http://www.nybooks.com/articles/956, accessed Dec. 10, 2007.

Fasten, Nathan. 1930. "What Do Eugenicists Aim For?" *Eugenics: A Journal of Race Betterment*, vol. 3, 225.

Feldman, David M. 1995. "Eugenics and Religious Law," *Encyclopedia of Bioethics*, Revised Edition. Macmillan, New York, 777-779.

Feldman, David M. 1998. "Eugenics and Religious Law: Judaism," *Bioethics: Sex, Genetics and Human Reproduction*. New York: Macmillan Library Reference.

Feldman, William Moses. 1917. *The Jewish Child: Its History, Folklore, Biology, & Sociology*. London: Baillière, Tindall and Cox.

Feldman, William Moses. 1921. "Eugenics: Rabbinic and Contemporary," *Jewish Chronicle Supplement*, Jan. 28, ii-iii.

Feldman, William Moses. 1939. "Ancient Jewish Eugenics," *Medical Leaves*, vol. 2, 28-37.

Feldman, Yotam. 2009. "Eugenics in Israel: Did Jews Try to Improve the Human Race too?" *Ha'aretz*, May 30, http://www.haaretz.com/hasen/spages/1085596.html.

Fenske, Hans; Mertens, Dieter; Reinhard, Wolfgang; Rosen, Klaus. 1987. *Geschichte der politischen Ideen: Von Homer bis zur Gegenwart*. Fischer Taschenbuch Verlag, Frankfurt am Main.

Fenton, Elizabeth. 2006. "Liberal Eugenics and Human Nature: Against Habermas," *Hastings Center Report,* 36(2), 35-42.

Fertility Alternatives, Inc. "Jewish Fertility." Murrieta, California. http://www.fertilityalternatives.com/jewish.html.

FFZ (Gemeinsames Frauenforschungszentrum der Hessischen Fachhochschulen), http://www.gffz.de/1_8.html, accessed April 19, 2008.

Fields, Suzanne. 1999. "Princeton's "Professor Death: Singer's Nazi Intellectual Roots," *Washington Times*, Sept. 20, A21.

Filc, Dani. 2010. *The Political Right in Israel*, London / New York: Routledge.

Finkelstein, Daniel. 2008. "Choosing a deaf baby is criminal: It is amazing how good people can have bad ideas through muddled thinking," *The Times*, March 12, http://www.timesonline.co.uk/tol/comment/columnists/daniel_finkelstein/article3533147.ece, accessed April 21, 2008.

Finkelstein, Norman G. 2003. *The Holocaust Industry: Reflections on the Exploitation of Jewish Suffering*. London/New York: VERSO.

Fishberg, Maurice. 1911. *The Jews*. New York/Melbourne: Walter Scott Publishing Co. Quoted according to reprint of 1975 by Arno Press, "A New York Times Company."

Fishberg, Maurice. 1917. "Rassenzuchtung der Juden," *Statistik der Juden: Eine Sammelschrift*, Bureau für Statistik der Juden, 70-86.

Florinsky, Vasily Markovich. 1866. *Usovershenstvovanie i vyrozhdenie chelovecheskogo roda*, St. Petersburg.

*Forward*. 2005. "Study's Claim on Intelligence of Ashkenazim Spurs a Debate," June 10, http://www.forward.com/articles/3297.

Frank, Gelya. 1997. "Jews, Multiculturalism, and Boasian Anthropology," *American Anthropologist*, New Series, vol. 99(4), 731-745.

Frank, Hans. 1953. *Im Angesicht des Galgens: Deutung Hitlers und seiner Zeit auf Grund eigener Erlebnisse und Erkenntnisse; Geschrieben in Nürenberger Justizgefängnis*, Friedrich Alfred Beck Verlag, München/Gräfeling.

Gitleman, Zvi. 1989. "Soviet Immigrant Resettlement in Israel and the United States," *163-185;* in Freedman, Robert Owen, *Soviet Jewry in the 1980s: The Politics of Anti-Semitism and Emigration*. Durham, North Carolina: Duke University Press.

Frenkel, David A. 2001. "Legal Regulation of Surrogate Motherhood in Israel," *Medicine and Law*, 20(4), 605-612.

Freud, Sigmund. 1931. *Modern Sexuality, Morality and Modern Nervousness*. New York: Eugenics Publishing Company.

Friedman, Dave. 2005. "Ashkenazi Jews and Eugenics," *Dave Friedman's Soul of Wit,* October 17. http://dfriedman.typepad.com/dave_friedmans_blog/2005/10/ashkenazi_jews_.html

"Furor about Sterilization" *Time Magazine,* Nov. 22, http://www.time.com/time/magazine/article/0,9171,806954,00.html, accessed March 4, 2008.

Fürth, Henriette. 1929. *Die Regelung der Nachkommenschaft als eugenisches Problem*. Stuttgart: Julius Püttmann.

Galili, Lily. 2002. "A Jewish Demographic State," *Ha'aretz*, June 28, http://www.bintjbeil.com/articles/en/020628_galili.html, accessed February 1, 2009.

Gallup Organization. 2001. "Public Favorable to Creationism," February 14.

Galton, Francis. 1910. "Eugenics and the Jew," *Jewish Chronicle*, July 29, available online at http://whatwemaybe.org/txt/txt0001/Galton.Francis.1910.Eugenics-and-the-Jew.reprint.pdf.

Gejman, Pablo V.; Weilbaecher, Ann. 2002. "History of the Eugenic Movement," *Israel Journal of Psychiatry and Related Sciences,* 39(4), 217-231.

Gelber, Mark. H. 2000. *Melancholy Pride: Nation, Race, and Gender in the German Literature of Cultural Zionism (Conditio Judaica)*. Tübingen: M. Niemeyer.

German, J.; Bloom, D.; Passarge, E.; Fried, K.; Goodman, R.M.; Katzenellenboigen, A.; Laron, Z.; Legum, C.; Levin, S.; Wahrman. 1977. "Bloom's syndrome. VI. The disorder in Israel and an estimation of the gene frequency in the Ashkenazim," *American Journal of Human Genetics*, Nov., 29(6), 553-562.

Gessen, Masha. 2008. *Blood Matters: From Inherited Illness to Designer Babies, How the World and I Found Ourselves in the Future of the Gene.* Harcourt, New York.

Gilbert, Martin. 2007. *Churchill and the Jews: A Lifelong Friendship.* Henry Holt and Company: New York.

Gilbert, Martin. 2009. "Churchill and Eugenics," The Churchill Centre, http://www.winstonchurchill.org/support/the-churchill-centre/publications/finest-hour-online/594-churchill-and-eugenics.

Gilman, Sander L. 1986. *Jewish Self-Hatred: Anti-Semitism and the Hidden Language of Jews.* New York/London: The Johns Hopkins University Press.

Gilman, Sander L. 1989. "Why and How I Study the German," *German Quarterly,* 62(2), 192-204.

Gilman, Sander. 1995a. Review of "*A People That Shall Dwell Alone: Judaism as a Group Evolutionary Strategy* by Kevin MacDonald," *Jewish Quarterly Review,* New Series, 86(1/2), 198-201.

Gilman, Sander. 1995b. "Looking at Eugenics before the Bell Curve: How 10th-century Scientists Argued That Jews Were a Master Race," *Forward,* LXXXXVIII (31,033), pg. 8. Review of John Efron, *Defenders of the Race: Jewish Doctors and Race Science in Fin-de-Siecle Europe.*

Gilman, Sander L. 1996. *Smart Jews: The Construction of the Image of Jewish Superior Intelligence.* Lincoln, Nebraska: University of Nebraska Press.

Gilman, Sander L. 2003. *Jewish Frontiers: Essays on Bodies, Histories, and Identities.* New York: Palgrave Macmillan.

Ginzburg, Benjamin; Stern, Curt. 1965. Letters to editor of *Science.* New Series, vol. 149, No. 3689, Sept., 10, 1171-1173.

Glad, John. 1998. "A Hypothetical Model of IQ Decline Resulting from Political Murder and Selective Emigration in the Former USSR," *Mankind Quarterly,* vol. XXXVIII, No. 3, Spring, 279-298.

Glad, John. 2004. "History, Eugenics, and the Jews," *Jewish Press,* May 13, http://www.jewishpress.com/page.do/16329/History%2C_Eugenics%2C_And_The_Jews.html.

Glad, John. 2006. *Future Human Evolution: Eugenics in the Twenty-First Century.* Schuylkill Haven, Pennsylvania: Hermitage.

Glad, John; Weissbort, Daniel. 1992. *Twentieth-Century Russian Poetry.* Iowa City, Iowa: University of Iowa Press, (updated version of *Russian Poetry: The Modern Period*).

Glass, Bentley. 1981. "A Hidden Chapter of German Eugenics between the Two World Wars," *Proceedings of the American Philosophical Society,* 125(5), 357-367.

Glass, Bentley. 1986. "Geneticists Embattled: Their Stand against Rampant Eugenics and Racism in America During the 1920s and 1930s," *Proceedings of the American Philosophical Society,* 130(1), 130-153.

Glick, Caroline B. 2009. "Avoiding an American Ambush," *Jerusalem Post,* July 7, pg. 15, http://www.lexisnexis.com.proxy-um.researchport.umd.edu/us/lnacademic/results/docview/docview.do?docLinkInd=true&risb=21_T7031351081&format=GNBFI&sort=BOOLEAN&startDoc-

No=1&resultsUrlKey=29_T7031351084&cisb=22_T7031351083&treeMa
x=true&treeWidth=0&csi=10911&docNo=1.

Glick, David. 2003. "Ethics, Public Policy and Behavorial Genetics," *Israel Medical Association Journal*, Feb., 5(2), 83-86.

Glick, Paul C. 1960. "Intermarriage and Fertility Patterns among Persons in Major Religious Groups, " *Eugenics Quarterly*, March, 7(1), 31-38.

Glover, Jonathan. 2002. "Questions About Some Uses of Genetic Engineering," *Biomedical Ethics Readings*, Feb 3, 12-25, http://mind.ucsd.edu/syllabi/02-03/01w/readings/biomed-readings.pdf, accessed June 3, 2008.

Goddard, Henry H. 1917. "Mental Tests and the Immigrant," *The Journal of Delinquency*, vol.11, no. 5, Sept., 243-277.

Goldberg, Jacob. 1954. "Heredity Counseling," *Eugenics Quarterly*, June, 1(2), 39-47.

Goldscheider, Calvin. 1965. "Socio-Economic Status and Jewish Fertility," *The Jewish Journal of Sociology*, vol. VII, Dec., 221-237.

Goldscheider, Calvin. 1967. "Fertility of the Jews," *Demography*, 4(1), 196-209.

Goldstein, Eric L. 1997. "'Different Blood Flows in Our Veins': Race and Jewish Self-Definition in Late Nineteenth Century America," *American Jewish History*, 85(1), 29-55.

Goldstein, Evan R. 2006. "Blood Brothers: Is There Such a Thing as Jewish DNA?", *Wall Street Journal*, Dec. 8.

Goldstein, David B. 2008. *Jacob's Legacy: A Genetic View of Jewish History*. New Haven & London: Yale University Press.

Goldstein, Sidney; Goldscheider, Calvin. 1966. "Social and Demographic Aspects of Jewish Intermarriages," *Social Problems*, 13(4), 386-399.

Golinkin, David. 1994. "Does Jewish Law Permit Genetic Engineering on Humans?" *Moment*, August, 28-29, 67.

Gordon, Albert J. 1956. "Anti-Zionist Jews Termed 'Stooges,'" *New York Times*, Nov. 17, 8.

Gorenberg, Gershom. 2008. "How Do You Prove You're a Jew," *New York Times*, March 2, 46-51.

Gottesman, Irving J. 2005. "Slater, Eliot Trevor Oakeshott," *Encyclopedia of Life Sciences*, John Wiley & Sons, Ltd., www.els.net.

Gottfredson, Linda, *et al.* 1994. "Mainstream Science on Intelligence," *Wall Street Journal*, A18.

Gould, Stephen Jay. 1977. "Review," *Isis*, 68(4), Dec., 626-627.

Gould, Stephen Jay. 1993 (reissued 1981 edition). *The Mismeasure of Man*. London/New York: W. W. Norton.

Graetz, Michael. 2002. "The Human Genome and Ethical Issues," *Conservative Judaism*, 54(3), 44-49.

Graham, Loren R. 1977. "Science and Values," *American Historical Review*, vol. 82, no. 5, 1133-1164.

Grant, Madison. 1916. *The Passing of the Great Race or the Racial Basis of European History*. New York Charles Scribner's Sons; http://www.churchoftrueisrael.com/pgr/pgr-toc.html, accessed March 5, 2008.

Green, Joseph. 1991. "Artificial Insemination in Israel – A Legal View: The Position of the Israeli Judicial System Regarding Artificial Insemination of a Married Woman," *Jewish Medical Ethics,* Dr. Falk Schlesinger Institute for Medical-Halachic Research, vol. 2(1), 21-28.

Green, Richard E.; Krause, Johannes; Briggs, Adrian W.; Maricic, Tomislav; Stenzel, Udo; Kircher, Martin; Patterson, Nick; Li, Heng; Zhai, Weiwei; Fritz, Marcus His-Yang; Hansen, Nancy F.; Durand, Eric Y.; Malaspinas, Anna-Sapfo; Jensen, Jeffrey D.; Marques-Bonet, Tomas; Alkan, Can; Prüfer, Kay; Meyer, Matthias; Burbano, Hernán A.; Good, Jeffrey M.; Schultz, Rigo; Aximu-Petri, Ayinuer; Butthof, Anne; Höber, Barbara; Höffner; Siegemund, Madlen; Weihman, Antje; Nusbaum, Chad; Lander, Eric S.; Russ, Carsten; Novod, Nathaniel; Affourtit, Jason; Egholm, Michael; Verna, Christine; Rudan, Pavao; Brajkovic, Dejana; Željko, Kucan; Gušic, Ivan; Doronichev, Vladimir B.; Golovanova, Liubov V.; Lalueza-Fox, Carles; de la Rasilla, Marco; Fortea, Javier; Rosas, Antonio; Schmitz, Ralf W.; Johnson, Phillip A. F.; Eichler, Evan E.; Fasush, Daniel; Birney, Ewan; Mullikin, James C.; Slatkin, Montgomery; Nielsen, Rasmus, Kelso, Janet; Mechmann, Michael; Reich, David, Pääbo. 2010. "A Draft Sequence of the Neandertal Genome," *Science*, 328/5979, 710-722.

Greenebaum, Gary. 1999. "The Making of a Racist," op/editorial in *The Gazette*, Montreal, Quebec, Aug. 13, B3.

Groen, J.J. 1964. "Gaucher's Disease: Hereditary Transmission and Racial Distribution." *Archives of Internal Medicine*, vol. 113, April, 543-549.

Gross, Michael L. 2002. "Ethics, policy, and rare genetic disorders: the case of Gaucher disease in Israel," *Theoretical Medicine and Bioethics*, 23(2), 151-170.

Gross, Michael; Ravitsky, Vardit. 2003. "Israel: Bioethics in a Jewish-Democratic State," *Cambridge Quarterly of Healthcare Ethics*, 12, 247-255.

Grossman, Vasilii; Erenburg, Ilya. 1981. *The Black Book: The Ruthless Murder of Jews by German-Fascist Invaders Throughout the Temporarily Occupied Regions of the Soviet Union and in the Death Camps of Poland During the War of 1941-1945*, translated from the Russian by John Glad and James Levine. New York: Holocaust Publications.

Grossman, W. 1929. "Eugenics in the Talmud and its Effects," *Eugenical News,* vol. 14, 104-106.

Grunwald, Max. 1930. "Biblische und talmudische Quellen jüdischer Eugenik," *Hygiene und Judentum: Eine Sammelschrift*. Verlag Jacob Sternlicht, Dresden, 57-61.

Günther, Hans F. K. 1924. *Rassenkunde des deutschen Volkes*, J. F. Lehman, Munich.

Gumplowicz, Ludwig. 1883. *Der Rassenkampf: Soziologische Untersuchungen.* Innsbruck: Wagner'sche Univ-Buchhandlung.

Guterman, Lila. 2003. "Choosing Eugenics: How Far Will Nations Go to Eliminate a Genetic Disease," *Chronicle of Higher Education*, vol. 49, Issue 34, A22, http://chronicle.com/free/v49/i34/34a02201.htm, accessed Jan. 2, 2007.

Hall, G. Stanley. 1915. "Yankee and Jew: An After-Dinner Address," *Menorah Journal,* Vol.1, 87-90.

Haller, Mark H. 1963, 1984. *Eugenics: Hereditarian Attitudes in American Thought.* New Brunswick, New Jersey: Rutgers University Press.

Halperin, Mordechai. 1991. "Post-Mortem Sperm Retrieval," *Journal of Medical Ethics*, 21-18.

Halperin, Mordechai. 1996. "In Vitro Fertilization (IVF), Insemination and Egg-Donation," *Jewish Medical Ethics*, Dr. Falk Schlesinger Institute for Medical-Halachic Research, Book 1, 1, 25-30, accessed June 10, 2005.

Hammer, M.F.; Redd, A.J.; Wood, E.T.; Bonner, M.R.; Jarjanazi, H.; Karafet, T.; Santachiara-Benerecetti, S.; Oppenheim, A.; Jobling, M.A.; Jenkins, T.; Ostrer, H.; Bonné-Tamir, B. 2000. "Jewish and Middle Eastern non-Jewish populations share a common pool of Y-chromosome biallelic haplotypes," *Proceedings of the National Academy of Sciences,* 97(2), 6769-6774.

Hanna, David; Lynn, Richard. 2007. "Intelligence Differences between European and Oriental Jews in Israel," *Journal of Bioscience*, 39, 465-473.

Harris Poll. 2006. "While Most U.S. Adults Believe in God, Only 58 Percent are 'Absolutely Certain'," October 31, #80, http://www.harrisinteractive.com/harris_poll/index.asp?PID=707, accessed July 25, 2008.

Harpending, Henry; Cochran, Gregory. 2006. "Genetic diversity and genetic burden in humans," *American Journal of Human Genetics*, No. 6, 154-162.

Hart, Mitchell. 1995. "Picturing Jews: Iconography and Racial Science," *Studies in Contemporary Jewry*, Issue 11, 159-175.

Hart, Mitchell B. 1999. "Racial Science, Social Science, and the Politics of Jewish Assimilation," *Isis*, vol. 90(2), 268-297.

Hart, Mitchell B. 2007. *The Healthy Jew: The Symbiosis of Judaism and Modern Medicine*. Cambridge/New York: Cambridge University Press.

Hashiloni-Dolev, Yael. 2006. "Between Mothers, Fetuses and Society: Reproductive Genetics in the Israeli-Jewish Context," *NASHIM: A Journal of Jewish Women's Studies and Gender Issues,* October, no. 12, 129-150.

Hashiloni-Dolev, Yael. 2007. *A Life (Un)Worthy of Living: Reproductive Genetics in Israel and Germany*. Dordrecht, The Netherlands: Springer.

Healy, Melissa. 2007? "Fertility's New Frontier," *Los Angeles Times*, July 21, http://geneticsandsociety.org/article.php?id=98, accessed Dec. 15, 2007.

Henderson, Mark. 2010, How a Community Stamped Out Their Bitter Inheritance," *The Times*, Feb. 8, National Edition, http://www.lexisnexis.com.proxy-um.researchport.umd.edu/us/lnacademic/results/docview/docview.do?docLin-kInd=true&risb=21_T8803992674&format=GNBFI&sort=RELEVANCE&startDocNo=1&resultsUrlKey=29_T8803992678&cisb=22_T8803992677&treeMax=true&treeWidth=0&csi=10939&docNo=4, accessed Mar. 13, 2010.

Hendricks, Melissa. 2006. "Raymond Pearl's 'Mingled Mess'," *Johns Hopkins Magazine*, vol. 58(2), April, http://www.jhu.edu/~jhumag/0406web/pearl.html.

Herbert, Solomon. 1910-1911. "The Making of a Nation: A Jewish Problem," *The Jewish Review,* vol. 1, 446-455.

"Hoover, Herbert." 1931. "President Herbert Hoover, Senate testimony on birth control, use of eye color inheritance in courts," *Eugenical Panorama,* New York Academy of Medicine, http://www.eugenicsarchive.org/eugenics/image_header.pl?id=

Herrnstein, Richard J.; Murray, Charles. 1994. *The Bell Curve: Intelligence and Class Structure in American Life*. New York: Free Press Paperbooks (Simon & Schuster).

Herskovits, Melville J.. 1925. "Brains and the Immigrant," *The Nation*, Feb. 11, 139-141.

Herskovitz, Melville Jean. 1955. *Cultural Anthropology*. New York: Alfred A. Knopf.

Herzl, Theodor. 1895. *Wenn ihr wollt, es ist kein Märchen*. Leipzig/Vienna: M. Breitenstein.

Herzl, Theodor. 1896. *Der Judenstaat: Versuch einer modernen Lösung der Judenfrage*. Leipzig/Vienna: M. Breitenstein.

Herzl, Theodor (writing under pseudonym 'Benjamin Sessel'). 1897. "Mauschel," *Die Welt*, October 15, No. 20, 1-2.

Hevesi, Dennis. 2008. "Rabbi David Lieber, Scholar and University President, Dies at 83," *New York Times*, Dec. 21, 34.

Heyd, David. 1995. "Prenatal Diagnosis: Whose Right?" *Journal of Medical Ethics*, vol. 21, 292-297.

Heyd, David. 2003. "Human Nature: An Oxymoron?" *Journal of Medicine and Philosophy*, 28(2), 151-169.

Hilchey, Tim. 2008. "Walther Birkmayer, 86; Treated Parkinson's," *New York Times*, Dec. 22.

Hilts, Philip J. 1997. "Group Delays Achievement Award to Psychologist Accused of Fascist and Racist Views," *New York Times*, Aug. 15, A10.

Hirsch, Dafna. 2009. "Zionist Eugenics, Mixed Marriage, and the Creation of a 'New Jewish Type,'" *Journal of the Royal Anthropological Society*, 15(3), 592-609.

Hirsch, Jerry. 2004. "Uniqueness, Diversity, Similarity, Repeatability, and Heritability," *International Journal of Comparative Psychology*, 17, 304-314.

Hirsch, Nathaniel David Mitron. 1926. *A Study of Natio-Racial Mental Differences*. Worcestor, Massachusetts: Genetic Psychological Monographs, Clark University.

Hitler, Adolf. 2003. *Hitler's Second Book: The Unpublished Sequel to Mein Kampf*, edited by Gerhard L. Weinberg. Enigma Books, New York.

"Hodann, Max." German *Wikipedia*, http://de.wikipedia.org/wiki/Max_Hodann, accessed Jan. 31, 2008.

Hoffman, Carl. "Out of This World," *Jerusalem Post*, Jan. 30, http://pqasb.pqarchiver.com/jpost/access/1643607781.html?dids=16436077 81:1643607781&FMT=ABS&FMTS=ABS:FT&date=Jan+30%2C+2009& author=Carl+Hoffman&pub=Jerusalem+Post&edition=&startpage=14&desc= Out+of+this+world.

Hofmann, Gustav; Kepplinger, Brigitte; Marckhgott, Gerhart; Reese, Hartmut. 2005. *Gutachten zur Frage des Amtes der Oö Landesregierung, "ob der Namensgeber der Landes-Nervenklinik Julius Wagner-Jauregg als historisch belastet angesehen werden muss*, Linz, Oct. 25, http://www.schloss-hartheim.at/redsyspix/download/Gutachten%20Wagner%20Jauregg.pdf.

Hommel, A.; Alexander, H. 1998. "Zu einigen Aspekten des Lebenswerkes von Ludwig Franekel (1870-1951) unter besonderer Berücksichtigung seiner sozialgynäkologischen und sexualwissenschaftllichen Arbeiten, " *Zentralblatt für Gynäkologie*, 120(10), 475-480.

Hughes, A.G. 1928. "Jews and Gentiles: Their Intellectual and Temperamental Differences, A Psychological Study Which Reveals the Innate Superiority of Jewish Children over Their Gentile School-Mates." *Eugenics Review*, vol. 20, January, 89-94.

Huisman, Biénne. 2007. "South African Jewry Rejects Theory as Racist Nonsense," *Sunday Times* (South Africa), April 15, http://www.jihadwatch.org/dhimmiwatch/archives/016083.php, accessed Dec. 1, 2007.

Human Betterment Foundation. 1934. Announcement.

Huxley, Julian. 1936. "Eugenics and Society." (The Galton Lecture given to the Eugenics Society), *Eugenics Review,* 28(1), Cold Spring Harbor Laboratory Archives, Image 1830.

Ilan, Shahar. 2006. "Fertility Rate among Israeli Women Steadily Declining," *Ha'aretz*, June 16, http://www.haaretz.com/hasen/pages/ShArt.jhtml?itemNo=727623, accessed February 2, 2009.

Ilan, Shahar. 2009. "Does a Clone Have a Soul," *Ha'aretz*, July 5, http://www.google.com/search?q=moshe+botschko&ie=utf-8&oe=utf-8&aq=t&rls=org.mozilla:en-US:official&client=firefox-a.

Ilani, Ofri. 2009. "How many Jews would there be if not for the Holocaust?" http://engforum.pravda.ru/showthread.php?t=248207, accessed June 4, 2009.

*Im deutschen Reich*. 1916. Jg 22, No. 7, 184.

"Improving the Breed," *Time Magazine*, http://www.time.com/time/magazine/article/0,9171,891157,00.html, accessed March 5, 2008.

Israel: Like this as if, "Loneliness, Religion and the Central Bureau of Statistics," http://israel-like-this-as-if.blogspot.com/2007/06/loneliness-religion-and-central-bureau.html, accessed July 25, 2008.

Israel Science and Technology Homepage. 1999-2005 (copyright). "Jewish Studies: Global Directory of Holocaust Museums," http://www.science.co.il/Holocaust-Museums.asp (accessed June 8, 2005).

Itzkoff, Seymour. 2006. *Fatal Gift: Jewish Intelligence and Western Civilization*. Ashfield, Massachusetts: Paideia Publishers.

Itzkoff, Seymour. 2008. *The World Energy Crisis and the Task of Retrenchment: Reaching the Peak of Oil Production,* foreword by Matthew R. Simmons. Lewiston/Queenston/Lampeter: The Edwin Mellen Press.

Jaber, Lutfi; Halpern, Gabrielle J. 2006. "Consanguinity among the Arab and Jewish populations in Israel," *Pediatric Endocrinology Review,* Aug., 3, Suppl 3: 437-46.

Jacobs, Joseph. 1886. "On the Racial Characteristics of Modern Jews," *The Journal of the Anthropological Institute of Great Britain and Ireland*, vol. 15, 23-62.

Jacobs, Joseph. 1899. "Are Jews Jews?" *Popular Science Monthly*, 502-511.

Jacobs, Joseph. 1910. "Works of Friedrich Nietzsche," *New York Times*, May 7, BR8.

Jacobsen, Kurt. 2007. "The Mystique of Genetic Correctness," *Logos,* 6(1-2), http://www.logosjournal.com/issue_6.1-2/jacobsen.htm.

Jacobson, Simon. 2004. "Pinchas: Elitism," July 8. http://www.meaningfullife.com/oped/2004/07.08.04$PinchasCOLON_Eliti sm.php.

Jaffe, A. J. 1942. Review, *American Journal of Sociology*, 48(1), June, 160.

Jakobovits, Immanuel. 1959. *Jewish Medical Ethics: A Comparative and Historical Study of the Jewish Religious Attitude to Medicine and Its Practice*. New York: Bloch Publishing Company.

Jenkins, Trevor. 2007. "Arthur G. Steinberg: 1912-2006," *The American Journal of Human Genetics*, vol. 80, June, 1009-1013.

*Jewish Chronicle*, "Too Few Marriages, Too Few People," Oct. 30, 8.

Jews and Race: A One-Day Conference, Temple University. 2008 Nov. 5, http://www.temple.edu/isrst/Events/documents/JewsandRaceConference11 _5_07.pdf, accessed July 9, 2008.

Jordan, David Starr. 1915. *War and the Breed: The Relation of War to the Downfall of Nations*. Boston: The Beacon Press.

Judt, J. M. 1903. *Die Juden als Rasse: Eine Analyse aus dem Gebiete der Anthropologie*. Berlin: Jüdischer Verlag.

Jumonville, Neil. 2002. "The Cultural Politics of the Sociobiology Debate," *Journal of the History of Biology,* vol. 35, 569-593.

Jungmann, Max. 1892. "Ist das jüdische Vok degeneriert?" *Die Welt*, No. 24, 3-4.

"Just Deserts," *Time Magazine*, Dec. 22, http://www.time.com/time/magazine/article/0,9171,793976,00.html, accessed March 4.

Kahn, Fritz. 1922. *Die Juden als Rasse nd Kulturvolk*. Berlin: Welt-Verlag.

Kahn, Susan Martha. 2000. *Reproducing Jews: A Cultural Account of Assisted Conception in Israel*. Durham, North Carolina: Duke University Press.

Kahn, Susan Martha. 2003. Response to Veronica Ouma, *Palestine Solidarity Review*, fall.

Kahn, Susan Martha. 2006. "Making Technology Familiar: Orthodox Jews and Infertility Support, Advice, and Inspiration," *Culture, Medicine and Psychiatry*, 30, 467-480.

Kaiser, Jochen-Christoph; Nowak, Kurt; Schwartz, Michael. 1992. *Eugenik, Sterilisation, 'Euthanasie': Politische Biologie in Deutschland 1895-1945*. Buchverlag Union, Halle.

Kaiser, Jo Ellen Green; Rifkin, Jeremy; Barglow, Raymond; Darnovsky, Marcy. 2002. "Stem Cell Wars," *Tikkun*, vol. 17, issue 4, July/August, 27-33.

Kallmann, Franz Josef. 1952. "Human Genetics as a Science, as a Profession, and as a Social-Minded Trend of Orientation," *American Journal of Human Genetics*, 4(4), 237-245.

Kallman, Franz Josef. 1955. "Review of Psychiatric Progress 1954: Heredity and Eugenics," *The American Journal of Psyhiatry*, Jan., 111, 502-505.

Kallmann, Franz Josef. 1960. "Discussion," *Psychosomatic Medicine*, xxii, No. 4, 258-259.

Kallmann, Franz J.; Baroff, George S. 1954. Review of *Höchstbegabung: Ihre Verhältnisse sowie ihre Beziehungen zu psychischen Anomalien*, Urban and Schwarzenber, Munich/Berlin.

Kamin, Leon J. 1974. *The Science and Politics of IQ.* Potomac, Maryland: Lawrence Erlbaum Associates.

Kandell, Jonathan. 1979. "Rightest Intellectual Groups Rise in France," *New York Times*, July 8, 3.

Kantner, John F. 1965. Book review of *The Concept of Race*, by Ashley Montagu, *Eugenics Quarterly*, Sept., 12(3), 173-174.

Kaplan, Arnold R. 1963. "Biology, Politics and Race," *Eugenics Quarterly*, Sept., 10(3), 188-190.

Kaplan, Arnold R. 1965. "On the Genetics of 'Schizophrenia,'" *Eugenics Quarterly*, Sept., 12(3), 132-135.

Kaplan, Steven. 1992. *The Beta Israel (Falasha) in Ethiopia: From Earliest Times to the Twentieth Century*. New York/London: New York University Press.

Kaplan, Steven. 2003. "If There Are No Races, How Can Jews Be a 'Race'?" *Journal of Modern Jewish Studies,* 2(1), 79-96.

Karpel, Dalia. 2006. "Culture Club," Haaretz.com, http://www.haaretz.com/hasen/spages/776995.html.

Kass, Leon. 2006. "A More Perfect Human: Part I," Human Life of Washington, Belleview, Washington, D.C. http://www.humanlife.net/view_ewpoera.hrm?rpid=14.

Kater, Michael H. 1987. "Hitler's Early Doctors: Nazi Physicians in Predepression Germany," *Journal of Modern History*, 59(1), March, 25-52.

Katz, Danny. 2000. "Between a Rock and a Master Race: This Life," *The Age*, Melbourne, Australia, "Today" section, 1.

Kaufman, Theodore N. 1941. *Germany Must Perish!* Newark, New Jersey: Argyle Press.

Kautsky, Karl. 1914. *Rasse und Judentum*. Stuttgart. Translated as *Are the Jews a Race?* Translator unknown, http://www.marxists.org/archive/kautsky/1914/jewsrace/ch01.htm, accessed Dec. 16, 2007.

Keith, Jim. 2993. "Eugenik (eugenics)," discussion in original English and in German of *Mass Control: Engineering Human Consciousness,* Illuminet Press, Litburg, 1999. http://www.smilenow.de/s00025.htm.

Kennedy, Hubert. 2003. "Institut für Sexualwissenschaft (1919–1933) – The Institute for Sexual Science – Instituto de Sexologia" (Internet exhibition), *Journal of the History of Sexuality,* 12(1), 122-126.

Kevles, Daniel. 1986. *In the Name of Eugenics: Genetics and the Uses of Human Heredity*, Berkeley: University of California Press.

Kevles, Daniel. 2003. "Here Comes the Master Race: Edwin Black Argues that the American Eugenics Movement Inspired Hitler," *New York Times Book Review*, Oct. 3, 8.

Khazaria.com. 2007. "Jewish Genetics: Abstracts and Summaries: A collection of abstracts and reviews of books, articles, and genetic studies," 1. Studies of Jewish Populations, http://www.khazaria.com/genetics/abstracts.html. 2. Studies of Cohens and Levites, http://khazaria.com/genetics/abstracts-cohen-levite.html.

Kirsh, Nurit. 2003. "Population Genetics in Israel in the 1950s: The Unconscious Internalization of Ideology," *Isis,* vol. 94, 631-655.

Kirsh, Nurit. 2003-2004. "Physicians in the Young State of Israel: Putting Jewish History into Its Historic Perspective," *Korot*, vol. 17, 71-97.

Kirsh, Nurit. 2004. "Geneticist Elisabeth Goldschmidt: A Two-Fold Pioneering Story," *Israel Studies*, 9.2, 71-105.

Klee, Ernest. 1983. *Euthanasie im NS-Staat: Die Vernichtung lebensunwerten Leben*, Frankfurt am Main: Max S. Fischer.

Klugman, Susan; Gross, Susan J. 2010. "Ashkenazi Jewish Screening in the Twenty-First Century," *Obstetrics and Gynecology Clinics of North America*, 37/1, March, 37-46.

Knafo, Ariel. 2006. "The Longitudinal Israeli Study of Twins (LIST): Children's Social Development as Influenced by Genetics, Abilities, and Socialization, *Twin Research and Human Genetics*, Dec., 9(6), 791-798.

Koenig, Robert. 1997. "Watson Urges 'Put Hitler Behind Us'," *Science*, vol. 276, May 9.

Koppitz, Ulrich; Labisch, Alfons. 1999. *Adolf Gottstein: Erlebnisse und Erkentnisse: Nachlass 1930/1940, Autobiographische und biographische Materialien.* Berlin: Springer Verlag.

Kossoff, Julian. 1993. "Radical Gay Group Targets Lord Jakobovits over Welsh Honour," *Jewish Chronicle*, Nov. 26, 1.

Kramer, Heide. 2003. "Hans Goslar: Ein jüdisches Politikerschicksal der Weimarer Zeit," haGalil.com, March, http://www.berlin-judentum.de/geschichte/goslar.htm.

Kramer, Martin. 2010. "Smear Intifada," Feb. 22, http://www.martinkramer.org/sandbox/2010/02/smear-intifada/, accessed Feb. 24, 2010.

Kraepelin, Emil. 1883. *Compendium der Psychiatrie. Zum Gebrauche für Studierende und Ärzte*, Leipig, A. Abel.

Kratz, Peter. 1980. "Das falsche Idol," July 8, http://www.trend.infopartisan.net/trd7800/t357800.htm, accessed Dec. 11, 2007.

"Kronfeld, Arthur," German *Wikipedia*. http://de.wikipedia.org/wiki/Arthur_Kronfeld, accessed Jan. 31, 2008.

Kühl, Stefan. 1994. *The Nazi Connection: Eugenics, American Racism, and German National Socialism*. New York/Oxford: Oxford University Press.

Kushner, Khilia Faivelovich. 1955. *Michurinskaia genetika i voprosy razvedeniia sel'skokhoziastvennykh zhivotnykh*. Moscow?

Landua, David. 1993. *Piety and Power: The World of Jewish Fundamentalism*. New York: Hill and Wang.

Landau, Ruth. 1996. "Assisted Reproduction in Israel and Sweden: Parenthood at Any Price?" *International Journal of Sociology and Social Policy,* 16(3), 29-46.

Landau, R. 1998. "The Management of Genetic Origins: Secrecy and Openness in Donor Assisted Conception in Israel and Elsewhere," *Human Reproduction*, Nov, 13(11), 3268-73.

Landau, Ruth. 2003. "Religiosity, Nationalism and Human Reproduction: The Case of Israel," *International Journal of Sociology and Social Policy,* 23(12), 64-80.

Lasker, Gabriel. 1959. "Recent Advances in Physical Anthropology," *Biennial Review of Anthropology*, vol. 1, 1-36.

Laski, Harold. 1910. "The Scope of Eugenics," *The Westminster Review*, July.

Lazin, Fred A. 2005. *The Struggle for Soviet Jewry in American Politics: Israel versus the American Jewish Establishment*. Lexington Books: Lanham, Boulder, New York, Toronto, Oxford.

Lederberg, Joshua; Stern, Harcey; Tendler, Moshe. 1996. "Genetic Screening for Breast Cancer," 7th International Conference on Judaism and Contemporary Medicine, April 28, New York, NY, http://www.nijm.org/oldnijm/NIJMBreastCancer.html, accessed Dec. 15, 2007.

Leff, Barry. 2008. "Competing Narratives," *The Persistence of Vision: Israel at Sixty*, http://www.israelatsixty.org.il/my_weblog/2008/03/competing-narra.html, accessed May 13, 2008.

Lengwiler, Martin. 2006. Wie nationalsozialisisch ist die Eugenik? Historisches Seminar, Universität Basel (Prof. Dr. Regina Wecker, Sabine Braunschweig, Gabriela Imboden, Hans-Jakob Ritter), Dr. Bernhard Küchenhoff (Psychiatrische Universitätsklinik Zürich), im Rahmen des Forschungsprojektes "Eugenik und Verwaltung im Kanton Basel-Stadt, 1880-1960" des Nationalen Forschungsprogramms 51 "Integration und Ausschluss," Basel, Feb. 17-18, http://hsozkult.geschichte.hu-berlin.de/tagungsberichte/id=1081, accessed Dec. 13, 2007.

Lenz, Fritz. 1928. Letter to Jüdischer Verlag, Nussbaum Archive, Leo Baeck Institute, New York, Dec. 5.

Leonard, Thomas. 2005. "Mistaking Eugenics for Social Darwinism," *History of Political Economy*, 37, Supplement, 200-233.

Letter to the editor of *Time Magazine, April 16, http://www.time.com/time/magazine/article/0,9171,775536,00.html, accessed March 4, 2008.*

Lewin, Gerson. 1934. *Ochrona zdrowia i eugenika w Biblii I Talmudzie*. Warsaw.

Levavi, Lea. 1990. "Genetic Marker Is Found for Fatal Illness," *Jerusalem Post*, Sept. 7.

Levin, Mark. 1999. "Screening Jews and Genes: A Consideration of the Ethics of Genetic Screening within the Jewish Community: Challenges and Responses." *Genetic Testing*, vol. 3(2), 207-213.

Levin, Samuel. 1964. "The Malthusian Heritage in Contemporary Life," *Eugenics Quarterly*, March, 11(1).

Levy, Gabby. 2008. "Israel Needs Palestinian State to Stay Jewish, Democratic," interview granted to Kerim Balci of *Turkish Weekly*, May 6, http://www.turkishweekly.net/news.php?id=55027, accessed July 9, 2008.

Lewit, Neta. 2003. "Psychiatry in Eretz Israel of the 1930s and its collaboration with the German doctrine of racial improvement," *International Association against Psychiatric Assault, no. 1*, May.

Lewontin, Richard. 1974. *The Genetic Basis of Evolutionary Change*. New York: Columbia University Press.

Lewontin, R. C.; Rose, Steven; Kamin, Leon J. 1984. *Not in Our Genes: Biology, Ideology, and Human Nature*. New York: Pantheon Books.

Lieber, David L.; Harlow, Jules. 2001. *Etz Havim: Torah and Commentary*. Philadelphia: Jewish Publication Society.

Lifesitenews.com. 2004. "Israel's Health Committee Approves Human Cloning in Principle," March 12. http://www.lifesitenews.com/ldn/2004/mar/04031206.html, accessed January 31, 2009.

Lilien, see: Brieger, 1922; Regener, 1905; Zweig, 1903.

Lippe, Karpel. 1887. *Symptome der antisemitischen Geisteskrankheit*. H. Goldner, Jassy.

Lipphardt, Veronika. 2006. "'Jewish Eugenics'? German Scientists with Jewish Context and their Notions of Eugenics, 1900-1935," (in German), Papier für die Tagung "Wie nationalsozialistisch ist die Eugenik?", Basel, Feb. 17-18.

Lipphardt, Veronika. 2008. *Biologie der Juden*. Göttingen: Vandenhoeck & Ruprecht.

Lipschutz, Joshua H. 1999. To Clone or Not to Clone – A Jewish Perspective," *Journal of Medical Ethics*, vol. 25, 105-107.

Lipsyte, Robert. 1996. "Coping: A Postcard from Morgan's Twilight World," *New York Times*, April 14, http://query.nytimes.com/gst/fullpage.html?res=9B03E1DB1139F937A257 57C0A960958260&sec=health&spon=&pagewanted=all.

Loeb, Jacques. 1911. *Das Leben*. Leipzig: A. Kröner.

Lombroso, Cesare. 1894. *Der Antisemitismus und die Juden im Lichte der modernen Wissenschaft*. Leipzig: Authorized German edition of Dr. H Kurella.

Lubarsch, Otto. 1931. *Ein bewegtes Gelehrtenleben: Erinnerung und Erlebnisse, Kämpfe und Gedanken*. Berlin: Verlag von Julius Springer.

Lucas, Alb. 1902. "Exasperation Expressed," *New York Times*, June 1, 30.

Ludmerer, Kenneth M. 1972. *Genetics and American Society: A Historical Appraisal*, Baltimore: Johns Hopkins Press.

Luschan, Felix von. 1892. "Die anthropologische Stellung der Juden: Eine Einführung in ihre Anthropologie," Deutsche Gesellschaft für Anthropologie, 23, 94.

Lutz, Heather. 2007. "Eugenics Roots Impact Health Practices Today," *Cleveland Jewish News*, Dec.7, vol. 111, issue 7, 10-11.

Lyman, Eric J. 2002. "Italy's Antinori Says He's Cloned Three People," United Press International, May 9, http://www.ericjlyman.com/upiclone.html, accessed February 1, 2009.

Lynn, Richard. 2001. *Eugenics: A Reassessment*. Westport, Connecticut/London: Praeger.

Lynn, Richard. 2006. *Race Differences in Intelligence: An Evolutionary Analysis*. Augusta, Georgia: Washington Summit Publishers.

Lynn, Richard. 2010. *The Chosen People: A Study of Jewish Intelligence and Achievement*, Washington Summit Publishers, manuscript in preparation.

Lynn, Richard; Kanazawa, Satoshi. 2008. "How to Explain High Jewish Achievement: The Role of Intelligence and Values," *Personality and Individual Differences*, 44, 801-808.

Maag, Christopher. 2009. "Spiritual Journey Leads to a Historic First," *New York Times*, June 6, A14.

MacDonald, Kevin 1994. *A People That Shall Dwell Alone: Judaism as a Group Evolutionary Strategy.* Westport, Connecticut: Praeger.

MacDonald, Kevin 1998a. *Separation and Its Discontents: Toward an Evolutionary Theory of Anti-Semitism.* Westport, Connecticut: Praeger.

MacDonald, Kevin 1998b. *The Culture of Critique: An Evolutionary Analysis of Jewish Involvement in Twentieth-Century Intellectual and Political Movements.* Westport, Connecticut: Praeger.

MacDonald, Victoria. 1997. "Abort babies with gay genes, says Nobel winner," *UK News: Electronic Telegraph,* Feb. 16, No. 632, http://www.telegraph.co.uk/htmlContent.jhtml?html=/archive/1997/02/16/n abort16.html, accessed July 1, 2008.

Maoz, B.; Levy, S.; Brand, N.; Halevi, H. S. 1966. "An Epidemiological Survey of Mental Disorders in a Community of Newcomers to Israel," *Journal of the College of General Practitioners,*" 11, 267-284.

Marchione, Marilynn. (Associated Press). 2010. Pre-Birth Genetic Tests Curb Inherited Diseases," *Newsday,* Feb. 17, A12.

Martin, Douglas. 2007. "Raul Hilberg, 81, Historian Who Wrote of the Holocaust as a Bureaucracy, Dies," *New York Times,* Aug. 7, C11.

Martin, Philip; Zürcher. 2008. *Managing Migration: The Global Challenge,* 63(1), March. Washington, D.C: Population Reference Bureau.

Matus, Ron; Winchester, Donna. 2008. "Foster links Darwin, Hitler," *St. Petersburg Times,* Jan. 12.

Marx, Alfred. 1928. "Kritische Bemerkungen zur Rassenhygiene," *Der Morgen,* Jg 4, No. 3, 255-264.

May, Todd. 2003. "Israel's new eugenics: Israel's new marriage law amounts to a genetic purification ritual," *Al-Ahram Weekly On-line,* August 28-September 3, no. 653.

Mayer, Harry H. 1926. *Eugenics: A Sermon for Mother's Day,* delivered May 9 at the Temple, Linwood Boulevard and Flora Ave., Kansas City, Missouri, at the special service for Mother's Day, conducted jointly by the Council of Jewish Women and the Temple Sisterhood.

McKigney, Erin. 2007. "Professor Battles Preconceived Notions about Jews and Race," *Forward,* Aug. 7, http://www.forward.com/article/11326/, accessed July 9, 2008.

McLean, Sheila A.M. 1998. "Interventions in the Human Genome," *Modern Law Review,* 61(5), 681-696.

McNeil, Donald G. 2008. "When Human Rights Extend to Nonhumans," *New York Times,* July 13, WK3.

Mead, Margaret. 1962. "The Social Responsibility of the Anthropologist," *The Journal of Higher Education,* 33(1), Jan., 1-12.

Meehan, Mary. 1997 (Nov-Dec.)/2001 (July). "Eugenics and the Power Elite," *Meehan Reports,* http://www.meehanreports.com/elite.html#N_9_, accessed April 10, 2008.

Mehler, Barry. 1994. "In Genes We Trust: When Science Bows to Racism," *Reform Judaism,* 23(2), cover, 11-12, 77-79; Gottesman's and Mehler's exchange of views in issue 3, 6.

Meisenberg, Gerhard. 2007. *In God's Image? The Natural History of Intelligence and Ethics.* Sussex: Book Guild Publishing.

Mendelsohn, Ezra. 1971. "From Assimilation in Zionism in Lvov: The Case of Alfred Nossig," *Slavic and East European Journal*, 14(117), 521-534.

"Mentor." 1912. "Eugenics: Judaism and Race Preservation," *Jewish Chronicle*, Aug. 2, 8.

Merkel, Howard. 1997. "Di goldene Medina (The Golden Land): Historical Perspectives of Eugenics and the East European (Ashkenazi) Jewish-American Community, 1880-1925," *Health Matric: Journal of Law Medicine*, 7(1), 49-65.

Meyer, Steven. Undated. "Will Israel Outlive Its Fascists? Jabotinsky: Mussolini's Favorite," Citizens' Electoral Council of Australia, http://www.cecaust.com.au/main.asp?sub=culture/jewish&id=p3/article3.htm, accessed Jan. 21, 2008.

Meyerson, A.1923. "Inheritance of Mental Disease," *Eugenics, Genetics and the Family: Second International Congress of Eugenics, 1921*, vol. 1, 218-225, Williams & Wilkins Company, Baltimore.

Michaelis, Curt. 1905. "Die jüdische Auserwählungsidee und ihre biologische Bedeutung," *Zeitschrift für Demographie und Statistik der Juden, Feb., No.2*, 1-4.

Midgley, Mary. 2000. "Biotechnology and Monstrosity: Why We Should Pay Attention to the 'Yuk Factor,'" *Hastings Center Report*, Sept., 30(5), 7-15.

Mildenberger, Florian. 2002. "Auf der Spur des 'scientific pursuit': Franz Josef Kallmann (1897–1965) und die Rassenhygienische Forschung," *Medizin Historisches Journal*, vol. 37, 183-200.

Mintz, Alan. 2001. *Popular Culture and the Shaping of Holocaust Memory in America*. Seattle/London: University of Washington Press.

Mironin, S. Undated. "Stalin i gosudarstvo: Stalin i genetika," http://stalinism.ru/index.php?option=com_content&task=view&id=878&Itemid=30, accessed Jan. 26, 2008.

Model, Alice. 1909. Interview granted to *Jewish Chronicle*, Mar. 19, 24.

"Moll, Albert," German *Wikipedia*, http://de.wikipedia.org/wiki/Albert_Moll, accessed Feb. 3, 2008.

Mono, Brian. 2001. "Questions and Answers: Mapping Medical Ethics [directed to Paul Root Wolpe, Center for Bioethics, University of Pennsylvania]," *Jewish Exponent*, 209(10), 14.

Montagu, Ashley. 1962, "The Concept of Race," *American Anthropologist*, 64(5), part 1, Oct., 919-928.

Moore, James. 2006. "R.A. Fisher: A faith in eugenics," *Studies in History and Philosophy of Science Part C: Studies in History and Philosophy of Biological and Biomedical Sciences*, 38(1), 110-135.

Morgan, Thomas B. 1964. "The Vanishing American Jew," *Look*, May 5, 42-26.

Morris-Reich, Amos. 2006a. "Arthur Ruppin's Concept of Race," *Israel Studies*, 11(3), 1- 30.

Morris-Reich, Amos. 2006b. "Project, Method, and the Racial Characteristics of Jews: A Comparison of Franz Boas and Hans F. K. Günther," *Jewish Social Studies*, 13(1), Fall, 136-169.

Morrison, Hyman. 1940. "A Biologic Interpretation of Jewish Survival," *Medical Leaves*, vol. 3, 97-103.

Moskowitz, Henry. 1917. "Palestine Not a Solution of Jewish Problem," *New York Times*, June 10, 66.

Motulsky, Arno. 1999. "If I had a gene test, what would I have and who would I tell?" *Lancet*, Supplement Molecular Med., 354(9176), pSI 35.

Mourant, Arthur Ernes; Kopeć, Ada C.; Domaniewska-Sobczak, Kazimiera. 1978. *The Genetics of the Jews*, Clarendon Press, Oxford.

MTs (authors's initials). 1936. "Protiv antinauchnykh vrazhdebnykh teorii" *Komsomol'skaia Pravda*, 1936.

Muhsam, Helmut Victor. 1965. "Differential Mortality in Israel by Socioeconomic Status," *Eugenics Quarterly*, 12(4), 227-232.

Muller, Herman J. 1936. Letter to Stalin, http://whatwemaybe.org/txt/other_materials.htm.

Muller, Herman J. 1961. "Human Evolution by Voluntary Choice of Germ Plasm," *Science*, vol. 134, No. 3480, Sept. 8, 643-649.

Müller-Hill, Benno. 2006. "Nazi Scientists," *Encyclopedia of Life Sciences,* John Wiley & Sons, Ltd., www.els.net.

Murray, Charles. 2007. "Jewish Genius," *Commentary,* April, 29-35, http://www.commentarymagazine.com/cm/main/viewArticle.html?id=10916&page=all.

Nathan, M.; Guttman, R. 1984. "Similarities in Test Scores and Profiles of Kibbutz Twins and Singletons," *ACTA Geneticae Medicae et Gemellologiae (Roma)*, 33(2), 213-218.

National Bioethics Advisory Commission.1997. "Report on Cloning by the U.S. Bioethics Advisory Commission: Ethical Considerations," Department of Health and Human Services, Washington, D.C.

*National Jewish Population Survey: 2000-1*. 2002. United Jewish Communities, http://www.ujc.org/local_includes/downloads/5086.pdf, accessed Jan. 7, 2008: Oct., http://www.jewishvirtuallibrary.org/jsource/US-Israel/ujcpop.html, accessed Jan. 7, 2008.

Nature. 1985. "Karl Illmansee Resigns," vol. 316, July, 98.

Nebel, A.; Filon, D.; Weiss, D.; Weale, M.; Faerman, M.; Oppenheim, A.; Thomas, M. 2000. "High resolution Y chromosome haplotypes of Israeli and Palestinian Arabs reveal geographic substructure and substantial overlap of haplotypes of Jews, " *American Journal of Human Genetics*, No. 107, 630-641.

Nebel, Almut; Filon, Dvora; Brinkmann, Bernd; Majumder, Partha P.; Faerman, Marina; Oppenheim, Ariella. 2001. "The Y Chromosome Pool of Jews as Part of the Genetic Landscape of the Middle East," *American Journal of Human Genetics*, No. 69, 1095-1112.

Nelkin, Dorothy; Michaels; Mark. 1998. "Biological Categories and Border Controls: The Revival of Eugenics in Anti-immigration Rhetoric," *International Journal of Sociology and Social Policy,* 18(5/6), 35-63.

N. Gt. 1971-1972. "New Left," *Encyclopedia Judaica*, 1031-1034.

Neugebauer, Wolfgang. 1998. "Rassenhygiene in Wien 1938," *Die Wiener klinische Wochenschrift*, Feb. 28, 110(4-5), 128-134.

Neuhaus, Richard John. 1998. *Commentary*, April, 15-26; reader's letters and Neuhaus's response Aug., 2-11.

Newman, Louis I.; Herskovits, Melville J.; Hankins, Frank H.; Goethe, C. M. 1930. "A Eugenic or Dysgenic Force?" *Eugenics: A Journal of Race Betterment*, vol. 3, 59-62.

Neumayr, George. 2005. "The New Eugenics: One of America's cures for disability is death." *American Spectator,* vol. 38(5), 22.

Neusner, Jacob. 2003. *World Religions in America: An Introduction* (3rd Edition). Westminster/John Knox Press, Louisville, Kentucky.

*New York Times.* 1867. "Southern Baptist Convention," May 12, 4; convention took place May 11.

*New York Times,* 1877. "Conversions to the Jewish Faith: The Induction of Christians to be Permitted in the Future – The Effect of Mixed Marriages," Mar. 25, 5.

*New York Times.* 1902. "East End Problems in London and New York," My 26, 6.

*New York Times.* 1910. "Schiff Would Check Jewish Immigrants: Urges East Siders Not to Tax the City's Resources by Bringing their Friends Here, Jan. 24, 16.

*New York Times.* 1911. "Topics of the Times," May 29, 8.

*New York Times.* 1913a. "Puts Beiliss Case to Reform Rabbis," Oct. 20, 4.

*New York Times,* 1913b. "Rabbi Attacks Eastern Council," May 19, 9.

*New York Times.* 1914. "Dedicate a Home for 500 Children," May 25, 10.

*New York Times.* 1916. "Dr. D. A. Gorton Dead in His 84th Year," Feb. 23, http://rds.yahoo.com/_ylt=A0geu70Q5z5KXuIA.V1XNyoA;_ylu=X3oDM TBybnZlZnRlBHNlYwNzcgRwb3MDMQRjb2xvA2FjMgR2dGlkAw-- /SIG=12lu4ur6i/EXP=1245722768/**http%3a//www.bklyn-genealogy-info.com/Newspaper/BSU/1916.Death.html.

*New York Times.* 1920. "F. Warburg Seeks to Check Exodus Here of Jews in Europe," Sept. 20, 16.

*New York Times.* 1930a. "To Pick Eugenics Sermon," Sept. 1, 12.

*New York Times.* 1930b. "Decries Mixed Marriages: Dr. Goldstein Preaches on Dangers of Inter-Racial Unions," Dec. 8, 24.

*New York Times.* 1934. "Relaxing Quotas for Exiles Fought," May 4, C4, http://www.eugenicsarchive.org/eugenics/image_header.pl?id=1111&print able=1&detailed=.

*New York Times.* 1936. "Science and Dictators," Dec. 17, 26.

*New York Times.* 1940a. "Asks Family Ideal of Four Children," July 27, 13.

*New York Times.* 1940b. "Bonus by Eugenics Puublishing," Dec. 12, 47.

*New York Times.* 1942. "Leases Entire Floor in 308 West 35th St.: Eugenics Publishing Company to Expand in Book Field," Sept. 19, 25.

*New York Times.* 1950. "Events Today," Oct. 6, 38.

*New York Times.* 1990. "Dr. Harry L. Shapiro, Anthropologist, Dies at 87," Jan. 9, D22.

*Nezavisimaya gazeta.* 2003. "'Sotvorim cheloveka po nashemu obrazu': Klonirovanie dopustimo, schitaet predesedatel' Kongressa evreiskikh religioznykh organizatsii I ob'edinenii v Rossii Zinoviii Kogan," Dec. 12, http://religion.ng.ru/people/2003-12-17/6_klon.html, accessed February 1, 2009.

Nordau, Max. 1968. *Degeneration* (translation of *Entartung*). Lincoln/London: University of Nebraska Press.

Nordau, Max (subject). 2004. "Max Nordau (1849-1923)," Jewish Virtual Library, copyright The American-Israeli Cooperative Enterprise, http://www.jewishvirtuallibrary.org/jsource/biography/nordau.

Nossig, Alfred.1905. "Die Auserwähltheit der Juden im Lichte der Biologie," *Zeitschrift für Demographie und Statistik der Juden,* vol. 3, 1-5.

Nossig, Alfred. 1921. *Polen und Juden: die polnisch-jüdische Verständigung zur Regelung der Judenfrage in Polen.* Vienna/New York: Renaissance.

Novick, Peter. 1999. *The Holocaust in American Life.* Boston/New York: Houghton Miffllin Company.

Nussbaum, Wilhelm. Untitled and undated. Nussbaum Archive, Leo Baeck Institute, Box 1, folder 18.

Nussbaum, Wilhelm. Undated. "Die Bedeutung der Vererbung für die Frau," Nussbaum Archive, Leo Baeck Institute, Box 1, folder 20.

Nussbaum, William. 1951, "Jewish Life," and submission response by Saul Bernstein, Nussbaum Archive, Leo Baeck Institute, New York, box 1, folder 29.

Oakley, Ann. 1992. "Social Medicine and the Career of Richard Titmuss in Britain 1935-1950," *British Journal of Sociology,* 42(2), June, 165-194.

Oh, Song. 2009. "Asians: The New Shiksas? *JewishJournal.com,* July 4, http://www.jewishjournal.com/singles/article/asians_the_new_shiksas_200 30418.

Ornstein, Leonard. 1967. "The Population Explosion, Conservative Eugenics, and Human Evolution," *BioScience,* 17/7, July, 461-464.

"ORTHOMOM." 2005. "Heroine of the Day," June 27, http://orthomom.blogspot.com/2005/06/heroine-of-day_27.html,accessed Jan. 6, 2007.

Ostrer, Harry. 2001. "A genetic profile of contemporary Jewish populations," *Nature Reviews,* vol. 2, Nov., 891-898.

Ouma, Veronica. 2003a. Review of Susan Kahn's *Reproducing Jews: A Cultural Account of Assisted Conception in Israel, Palestine Solidarity Review,* Summer.

Ouma, Veronika. 2003b. Letter in reply to Khan letter, *Palestine Solidarity Review,* Oct.

Pace, Eric. 1910, "Dr. Harry L. Shapiro, Anthropologist, Dies at 87," *New York Times,* January 9, D22.

Parfitt, Tudor; Egorova, Yulia. 2005. "Genetics, History, and Identity: The Case of the Bene Israel and the Lemba," *Culture, Medicine and Psychiatry,* June, 29(2), 193-224.

Pash, Barbara. 2006. Taking a Swipe at Genealogy," *Jewish Times,* Dec. 14, http://www.jewishtimes.com/3011.stm.

Passwell, J.; Adam, A.; Garfinkel, D.; Streiffler, M.; Cohen, B. E. 1977. "Heterogeneity of Wilson's disease in Israel," *Israel Journal of Medical Science,* Jan., 13(1), 15-19.

Patai, Raphael; Patai-Wing, Jennifer. 1975. *The Myth of the Jewish Race.* Scribner, New York.

Patai, Raphael. 1977. *The Jewish Mind.* Detroit: Wayne State University Press.

Patterson, Charles. May-June 2003. "The Great Divide: Animals and the Holocaust," *Tikkun,* vol. 18, Issue 3, 77-79.

Paul, Diane. 1984. "Eugenics and the Left," *Journal of the History of Ideas*, 45(4), Oct,- Dec., 567-590.

Paul, Diane B. 1998. *The Politics of Heredity*. Albany: State University of New York Press.

Paul, Kammerer Papers. America Philosophical Society, http://www.amphilsoc.org/library/mole/k/kammerer.xml, accessed Feb. 25, 2008.

Pchelov, E. V. 2006. "Evgenika i genealogiia v otechestvennoi nauke 1920-kh godov," *Gerboved*, no. 2, 76-146.

Pearson, Karl; Margaret Moul. 1925. "The Problem of Alien Immigration into Great Britain, Illustrated by an Examination of Russian and Polish Children," *Annals of Eugenics*, vol. 1, 5-55.

Penchaszadeh, V. B. 1994. "Genetics and Public Health," *Bulletin of the Pan American Health Organization*, March, 28(1), 62-72.

Perez, Shimon. 2010. "Israeli president addresses German Bundestag on Holocaust Remembrance Day," Federal News Service, http://www.lexisnexis.com.proxy-um.researchport.umd.edu/us/lnacademic/results/docview/docview.do?docLinkInd=true&risb=21_T8808701673&format=GNBFI&sort=RELEVANCE&startDocNo=1&resultsUrlKey=29_T8808701683&cisb=22_T8808701679&treeMax=true&treeWidth=0&csi=10962&docNo=6.

Perlin, Elliott. 1994. "Jewish Bioethics and Medical Genetics," *33(4), Winter, 333-340.*

*"Peter." 2003.* "Peter Singer and Eugenics," Institute for Social Ecology, Dec. 2, http://www.social-ecology.org/article.php?story=20031202122825648, accessed October 14, 2008.

Philippson, Ludwig. 1865. "Race oder Geschichte? und das 'Morgeblatt'," *Allgemeine Zeitung des Judenthums*, Nov. 14, vol. 46, 705-709.

Pickens, Donald K. 1968. *Eugenics and the Progressives*. Nashville, Tennessee: Vanderbilt University Press.

Pinker, Stephen. 2002. *The Blank Slate: The Modern Denial of Human Nature*. New York: Viking.

Pohlman, Edward. 1966. "Mobilizing Social Pressures toward Small Families," *Eugenics Quarterly*, June, 13(2), 122-127.

Popper, Nathaniel. 2005. "Study's Claim on Intelligence of Ashkenazim Spurs a Debate," June 10, *Forward*, http:/www.forward/articles/3623, accessed Mar. 26, 2008.

Post, R. H. 1965a. Summary of conference on genetics and demography, held in Princeton, New Jersey, Oct. 16-17, 1964, reported in *Eugenics Quarterly*, June, 1965, 12(2).

Post, R. H. 1965b. "Brief Report on 'Genetics, Jews and Disease,' by Louis B. Brinn," *Eugenics Quarterly*, Sept., 12(3), 162-164.

Post, R.H., editor in chief. 2003. Table of Contents, *Encyclopedia of Bioethics,* http://www.loc.gov/catdir/toc/ecip046/2003015694.html.

Prainsack, Barbara. 2006. "'Negotiating Life': The Regulation of Human Cloning and Embryo Stem Cell Research in Israel," *Social Studies of Science*, No. 36, 173-205.

Prainsack, Barbara; Firestine, Ofer. 2006. "'Science for Survival': Biotechnology Regulation in Israel," *Science and Public Policy,* 33(1), 33-46.

Prainsack, Barbara. 2009. E-mail to John Glad.

"Praise for Nazis," 1935. *Time Magazine*, Sept. 9, http://www.time.com/time/magazine/article/0,9171,748954-2,00.html, accessed March 4, 2008.

Presner, Todd Samuel. 2003. "'Clear Heads, Solid Stomachs, and Hard Muscles': Max Nordau and the Aesthetics of Jewish Regeneration, *Modernism*/Modernity, 10(2), 269-296.

Pretzel, Andreas. 1997. "Zur Geschichte der 'Ärtztlichen Gesellschaft für Sexualwissenschaft' (1913–1933) – Dokumentation und Forschungsbericht," *Mitteilungen der Magnus-Hirschfeld-Gesellschaft,* 24/25, October, 34-122, Forschungsstelle zur Geschichte der Sexualwissenschaft, Berlin.

*Pytheas Online.* 2005. "Quiet Eugenics Nearly Eliminates a Deadly Disease," http://pytheasonline.blogspot.com/2005_01_16_archive.html, accessed April 8, 2008.

Quirin, James Arthur. 1992. *The Evolution of the Ethiopian Jews: A History of the Beta Israel (Falasha) to 1920.* Philadelphia: University of Pennsylvania Press.

Qumsiyeh, Mazin. 2005. "Zionazi Racial Science," letter to Society of Histocompatibility and Immunology, July 3, http://eaazi.blogspot.com/search?q=eugenics; see also http://ambassadors.net/archives/issue11/opinions2.htm.

Rabinovich, Abraham. 1998. "A Jewish Geneticist's View on 'Jewish Genius'," Dec. 11, http:/www.mrcranky.com/movies/hurricane/42/145.html, accessed Jan. 17, 2008.

Rathenau, Walther. 1897. "Höre, Israel," *Die Zukunft*, No. 18, March 6, 454-462.

Ratner (Dr. med.). 1918. "Die Rassenhygiene, Familienforschung, Eugenik und einiges über die Vererbung geistiger Eigenschaften im altjüdischen Schriftum," *Hygienische Rundschau,* vol. 8, April 15, 249-253.

Ravitsky, Vardit. 2002. "Genetics and Education: The Ethics of Shaping Human Identity," *Mount Sinai Journal of Medicine,* 69(5), 312-316.

Raz, Aviad E. 2004. "'Important to test, important to support': Attitudes toward disability rights and prenatal diagnosis among leaders of support groups for genetic disorders in Israel," *Social Science and Medicine*, 59(9), November, 1857-1866.

Raz, Aviad E. 2005. "Disability Rights, Prenatal Diagnosis and Eugenics: A Cross-Cultural View," *Journal of Genetic Counseling*, vol. 14(3), 183-187.

Raz, Aviad. E. 2004, "Upright Generations of the Future: Tradition and Medicalization in Community Genetics," *Journal of Contemporary Ethnography*, 33(3), 296-322.

Raz, Aviad. 2009a. "Eugenic Utopias/Dystopias, Reprogenetics, and Community Genetics," *Sociology of Health and Illness*, 31(4), 602-616.

Raz, Aviad. E. 2009b. E-mail to John Glad, July 21.

Raz, Aviad E.; Glad, John. 2009c. e-mail, Sept. 22.

Raz, Aviad E.; Vizner, Yafa. 2008. "Carrier Matching and Collective Socialization in Community Genetics: Dor Yeshorim and the Reinforcement of Stigma," *Social Science and Medicine*, No. 67, 1361-1369.

Raz, Tahl; Rosner, Shmuel. 2007. "American Life Has Annihilated Jewish People-hood," *Jewcy*. http://www.jewcy.com/dialogue/2007-04-13/the_choosing_people.

Reaves, Jessica. 2001. "Human Cloning: Cause for Rejoicing or Despair?" *Time*, http://www.time.com/time/world/article/0,8599,101998,00.html , posted March 9, accessed June 14, 2005.

Regener, Edgar Alfred. 1905. *E. M. Lilien: Ein Beitrag der zeichnenden Künste*. Berlin/Leipzig: F. A. Lattmann in Goslar.

Reichler, Max.1941. "Eugenics," *The Universal Jewish Encyclopedia*, vol. 4, 191-192.

Reichler, Max. 1916. "Jewish Eugenics," reprinted in *Mankind Quarterly,* Spring 2001, http://whatwemaybe.org/txt/txt0001/Reichler.Max.1916.Jewish%20Eugenics.htm.

Reichsvertretung der deutschen Juden. 1934. Letter to the Arbeitsgemeinschaft für jüdische Erbforschung und Eugenik/Erbpflege, May 4, Nussbaum Archive, Leo Baeck Institute, New York.

Remennick, Larissa A. 2006. "The Quest for the Perfect Baby: Why Do Israeli Women Seek Prenatal Genetic Testing?" *Sociology of Health and Illness*, 28(1), 21-53.

Remennick, Larissa I.; Hetsroni, Amir. 2001. "Public Attitudes toward Abortion in Israel: A Research Note," *Social Science Quarterly*, 82(2), June, 420-431.

Renan, Ernest. 1883. *Le judaïsme comme race et comme religion: conference faite au cercle Saint Simon, le 27 Janvier*. Paris: C. Lévy.

Rense, Jeff. 2005. "Ever-Diminishing Official Numbers of Auschwitz Dead," http://www.rense.com/general69/dim.htm, accessed Mar. 15, 2005.

Report. 2010a. Harvard Fellow calls for genocidal measure to curb Palestinian births Report, *The Electronic Intifada,* 22 February, http://electronicintifada.net/v2/article11091.shtml, accessed Feb. 24, 2010.

Report. 2010b. "Harvard center condemns, then defends, fellow's pro-genocide statements," *The Electronic Intifada,* 23 Feb., http://electronicintifada.net/v2/article11097.shtml, accessed Feb. 24, 2010.

Revel, Michel. 1998. "Reproduction by Cloning: A New Ethical Challenge," *Assia: A Journal of Jewish Medical Ethics and Halacha*, vol. 3, No 2, September, http://www.daat.ac.il/daat/kitveyet/assia_english/revel.htm, accessed September 24, 2008.

Revel, Michel. 2008. "Ongoing Research on Mammalian Cloning and Embryo Stem Cell Technologies: Bioethics of their Potential Medical Applications," *Israel Medical Association Journal*, 8-14.

Rice, Thurman B. 1929. *Racial Hygiene: A Practical Discussion of Eugenics and Race Culture.* New York: Macmillan Company.

Rifkin, Jeremy. 1998. "Who Will Decide Between Defect and Perfect?" *Washington Post*, April 19, C4.

Robinson, Ira. 2007. "American Jewish Views of Evolution and Intelligent Design," *Modern Judaism*, May, vol. 27 (2), 174-192.

Robitscher, Jonas (ed.). 1973. *Eugenic Sterilization*, Springfield, Illinois: Charles C. Thomas.

Rodman, Hyman. 1965. "The Textbook World of Family Sociology," *Social Problems*, 12(4), Spring, 445-457.

Rogers, Lois; Follain, John. 2001. "Playing God – Focus," *Sunday Times*, March 11.

Romero, Simon. 2009. "Reborn Jews in Peru Make an Exodus to Israel," *New York Times*, June 21, http://www.nytimes.com/2009/06/22/world/americas/22peru.html?pagewanted=1&ref=world.

Rosanoff, Aaron J. 1923. "Inheritance of Mental Disorders," *Eugenics, Genetics and the Family: Second International Congress of Eugenics, 1921*, vol. 1, 226-230, Williams & Wilkins Company, Baltimore.

Rose, Gordon. 1958. "Trends in the Development of Criminology in Britain," *British Journal of Sociology*, 9(1), 53-65. (Originally delivered as paper at the British Sociological Association Conference, March, 1957.)

Rosen, Christine. 2003. "Eugenics – Sacred and Profane." *The New Atlantis,* Number 2, Summer, 79-89.

Rosen, Christine. 2004. *Preaching Eugenics: Religious Leaders and the American Eugenics Movement*, Oxford/New York: Oxford University Press.

Rosen, Gary. 2003. "Who's afraid of Leon Kass?" *Commentary*, Jan., 28-33.

Rosenthal, Erich. 1961. "Jewish Fertility in the United States," *American Jewish Yearbook*, vol. 62(3-27), 198-217.

Rosenthal, Erich. 1968. "Jewish Intermarriage in Indiana," *Eugenics Quarterly*, Dec., 15(4), 277-287.

Rosenzweig, Saul. 1946. "Clinical Psychology as a Psychodiagnostic Art," *Journal of Personality*, 15(2), Dec., 94-100.

Rosner, Fred. 1983. "Mass Screening for Tay-Sachs Disease," The *Hastings Center Report*, June, 44.

Rosner, Fred. 1998. "Judaism, Genetic Screening and Genetic Therapy," *Mount Sinai Journal of Medicine*, vol. 65, 406-413.

Rosner, Fred. 2000. "Genetic Screening, Genetic Therapy & Cloning in Judaism," *B'or Ha'Torah: Journal of Science, Art & Modern Life in the Light of Torah*, Weekly Q&A, http://www.borhatorah.org/home/question_archive/2000/week14.html.

Rosner, Fred; Bleich, J. David (ed.). 2000. *Jewish Bioethics*, KTAV Publishing House, Hoboken, New Jersey.

Rothman, Barbara Katz. 1998. "From the SWS President: A Sociological Skeptic in the Brave New World," *Gender and Society*, vol. 12, no. 5. Oct., 501-504.

Rothman, Stanley; Lichter, S. Robert. 1982. *Roots of Radicalism: Jews, Christians, and the New Left*, New York/Oxford: Oxford University Press.

Rothstein, Jane H. 2000. "'A new interest Jewish eugenics': Promoting the 'Purity of the Family' in the Early 20th Century United States," unpublished paper presented to a panel on "Constructing the Jewish Body," Association for Jewish Studies 2000, as part of author's dissertation.

Rubin, Herman H. 1946. *Eugenics and Sex Harmony: The Sexes, Their Relations and Problems*. New York: Pioneer Publications.

Rubin, Israel. 1934. "Gathering of Exiles from a Eugenic Perspective," *Moznayim*, vol. 1, 89-93 (in Hebrew).

Ruppin, Arthur. 1903. *Darwinismus und Sozialwissenschaft*, Jena: Gustav Fischer.

Ruppin, Arthur. 1930. *Soziologie der Juden*, 2 vols. Berlin: Jüdischer Verlag.

Ruppin, Arthur. 1931. *Der Kampf der Juden um ihre Zukunft* (vol. 2 of *Soziologie der Juden)*. Berlin: Jüdischer Verlag.

Ruppin, Arthur. 1940. *Milhemet ha-Yehudim le-kiyuman.* Tel Aviv: Hotsaat Mosad Byalik, al-yad Devir.

Michael Ruse. 1999. *Mystery of Mysteries: Is Evolution a Social Construction?* Cambridge: Harvard University Press.

S…, Lincoln. 2004. "My Letter to Presbyweb.com," http://aulula.blogspot.com/2004/08/my-letter-to-presbywebcom.html, accessed Jan.17, 2008.

Sachs, L.; Bat-Miriam, M. 1957. "The Genetics of Jewish Populations: Finger Print Patterns in Jewish Populations in Israel," *American Journal of Human Genetics*, June, 9(2), 117-126.

Saetz, Stephen B. 1985. "Eugenics and the Third Reich," *Eugenics Bulletin*, taken here from the *Future Generations* Web site (eugenics.net).

Sagi, Michal. 1998. "Ethical Aspects of Genetic Screening in Israel." *Science in Context* 11, (3-4), 419-429.

Sailer, Steve. 2003. "Cousin marriage conundrum: The ancient practice discourages democratic nation-building," *The American Conservative,* January 13, 20-22.

Sailer, Steve. 2005. "Jewish Telegraphy Agency on Ashkenazi Intelligence by Cochran and Harpending," iSteve.com Blog Archives, June 8, http://isteve.blogspot.com/2005/06/jewish-telegraph-agency-on-ashkenazi.html .

Salaman, Redcliffe, N. 1911. "Heredity and the Jew," *Journal of Genetics,* vol. 2, 273-292.

Salaman, Redcliffe N. 1921 (reprint 1985). "Some Notes on the Jewish Problem," *Eugenics in Race and State,* International Congress of Eugenics (2nd, 1921), American Museum of Natural History, Garland, New York, 134-153.

Saleeby, Caleb Williams. 1910. *Parenthood and Race Culture: An Outline of Eugenics.* New York: Moffat, Yard and Company.

Salisbury, Harrison E. 1949. "Genetics Is Linked to U.S. Imperialism," *Pravda*, April 6, 14.

Sand, Shlomo. 2009. *The Invention of the Jewish People.* Verso: London/New York.

Sandler, Aron. 1904. *Anthropologie und Zionismus: Ein populär-wissenschaftlicher Vortrag.* Brünn: Jüdischer Buch- und Kunstverlag.

Saperstein, Gilbert. 2004. "Jews and Eugenics," *Jewish Press*, April 6, http://www.jewishpress.com/displayContent_new.cfm?contentid=16349&mode=a&contentname=Letters_To_The_Editor&recnum=4&fromsect=1.

Sappenfeld, Burt R. 1942. "The Attitudes and Attitude Estimates of Catholic, Protestant, and Jewish Students," *Journal of Social Psychology*, 16, Nov. 2, 173-197.

Sartre, Jean-Paul. 1948. *Anti-Semite and Jew*, translated by George J. Becker. Grove Press: New York.

Schappacher, Norbert. 2005. "Felix Bernstein," *International Statistical Review*, 73(1), 3-7.

Scheinfeld, Amram. 1961. *The New YOU and HEREDITY.*" Philadelphia/New York: J. B. Lippincott.

Scheinfeld, Amram. 1965. *Your Heredity and Envronment.* Philadelphia/New York: J. B. Lippincott.

Schindler, Solomon. 1887. "Why Am I a Jew?" *Jewish Messenger*, January, 5-7.

Schmeck, Harold M. 1958. "Soviets Criticized on Genetics Issue," *New York Times*, Aug. 28, pg. 9.

Schmelz, Usiel Oscar; DellaPergola, Sergio. 2007. *Encyclopaedia Judaica,* ed. Michael Berenbaum and Fred Skolnik. Vol. 5. 2$^{nd}$ ed. Detroit: Macmillan Reference USA, 553-572. Reproduced in Gale Virtual Reference Library, http://go.galegroup.com.proxy-um.researchport.umd.edu/ps/retrieve.do?sgHitCountType=None&sort=RELEVANCE&inPS=true&prodId=GVRL&userGroupName=umd_um&tabID=T003&searchId=R4&resultListType=RESULT_LIST&contentSegment=&searchType=AdvancedSearchForm&currentPosition=1&contentSet=GALE%7CCX2587505093&&docId=GALE|CX2587505093&docType=GALE, accessed April 5, 2008.

Schmul, Hans-Walter. *The Kaiser Wilhelm Institute for Anthropology, Human Heredity, and Eugenics, 1927-1945: Crossing Boundaries.* Göttingen: Springer.

Schneider, Alison. 2000. "A California State Professor Is Attacked for His Defense of a Holocaust Denier," *Chronicle of Higher Education*, June 23, 46(42), http://web.ebscohost.com.proxy-um.researchport.umd.edu/ehost/detail?vid=4&hid=8&sid=62bf6173-b656-4db2-a226-c03f44733ec3%40sessionmgr3&bdata=JmxvZ2lucGFnZT1Mb2dpbi5hc3A mc2l0ZT1laG9zdC1saXZl#db=aph&AN=3225774.

Schüler, Alexander. 1912. *Der Rassenadel der Juden.* Berlin: Jüdischer Verlag.

Schult, Christoph. 2007. "Zuwanderer: Lockrufe aus Tel Aviv," *Der Spiegel*, vol. 40, 66.

"Science for the People." 2007. *Wikipedia*, http://en.wikipedia.org/wiki/Science_for_the_People, accessed Oct. 10.

Scripps Howard News Service. 2002. Quoted in Race and History, http://www.raceandhistory.com/cgi-bin/forum/webbbs_config.pl/noframes/read/1140, accessed on July 9, 2008.

Segal, Yossi. 1998. "The Human Genome Project." *Jewish Medical Ethics,* III(2), 20-29.

Segev, Tom; Weinstein, Arlen N. 1998. *1949: The First Israelis*, Owl Books, New York.

Seidelman, William E. 2001. "Science and Inhumanity: The Kaiser-Wilhelm/Max Planck Society," Feb. 18 revision, http://www.doew.at/thema/planck/planck1.html.

Selcuk, Deniz. 2006. "Are the Jews successful or are the successful ones called as Jewish?" July 17, Internet.

Seligman, Dan. 2002. "Good Breeding," *National Review*, 54(1), Jan 28, http://web.ebscohost.com.proxy-um.researchport.umd.edu/ehost/detail?vid=2&hid=107&sid=53331e84-c2d5-44a0-92e6-f0f6de887954%40sessionmgr103, accessed May 31, 2008.

Senior, Jennifer. 2005. "Are Jews Smarter? What Genetic Science Tells Us," *New York, http://nymag.com/nymetro/news/culture/features/1478.*

Senior, Jennifer. 2008. "Chronicle of a Death Foretold," *New York Times Book Review*, May 11, 2008, 28.

Shalev, Carmen. 2008. "Reflections on Human Dignity and the Israeli Cloning Debate," in *The Contingent Nature of Life*, Springer, 323-344.

Shapiro, Harry L. 1959. "Eugenics and Future Society," *Eugenics Quarterly*, 6(1), March, 3-7.

Shapiro, Harry L. 1960. *The Jewish People: A Biological History*. New York: UNESCO.

Shapiro, Kevin. 2004. "Good Breeding: *Preaching Eugenics: Religious Leaders and the American Eugenics Movement* by Christine Rosen," *Commentary,* July-Aug., 49-52.

Shapiro, N. I. 1966. "Pamiati A. S. Serebrovskogo," *Genetika*, vol. 9, 3-11.

Shapiro, Pauline C. 1967. "Large Families and Family Planning [1945-1967]," *Eugenics Review.* vol. 59, 257-262.

Shenhav, Yehouda; Yonah, Yossi (eds.). 2008. *Gizanut beyisrael* (Racism in Israel), Hakibbutz Hameucha / Jerusalem Van Leer Institute: Jerusalem.

Shenker, Israel. 1968. "Discussions Proliferate as Academics Hold Annual Scholarly Conferences," *New York Times*, Dec. 31, 16.

Sherwin, Byron L. 2000. "Designer Genes: Is Our Fashioning Fated?" *Bulletin of the Park Ridge Center*, Spertus Institute of Jewish Studies, Chicago.

Sherwin, Byron L. 2007, "Golems in the Biotech Century," *Zygon*, 42(1), March, 133-143.

Shiloh, S.; Reznik, H.; Bat-Miriam-Katznelson, M.; Goldman, B. 1995. "Pre-marital genetic counselling to consanguineous couples: attitudes, beliefs and decisions among counselled, noncounselled and unrelated couples in Israel," *Social Science and Medicine*, Nov., 41(9), 1301-1310.

Shulevitz, Judith. 2000. "Evolutionary Psychology's Anti-Semite," *Slate*, Jan. 24, http://www.slate.com/id/1004446/, accessed October 15, 2008. See also MacDonald's reply: "Slate: Police Thyself," reply to Judith Shulevitz, "Evolutionary Psychology's Anti-Semite," May 18, http://ww.csulb.edu/~kmacd/slate-SHULEVITZ.htm.

Slater, Eliot. 1947. "A Note on Jewish-Christian Intermarriage," *Eugenics Review*, vol. 39, 17-21.

Shvarts, Shifra; Davidovitch, Nadav; Seidelman, Rhona; Goldberg, Avishay. 2005. "Medical Selection and the Debate over Mass Immigration in the New State of Israel (1948–1951)," *Canadian Bulletin of Medical History,* 22(1), 5-34.

Siefer, Ted. 2005. "Study Linking Jewish IQ to Tay-Sachs Labeled 'Absurd' by Medical Ethics Expert," *The Jewish Advocate*, June 14-30, 196(25), A7.

Siegel, Bernard. 2006. "The Future of Stem Cell Research in Florida," B'nai B'rith International/Greater Florida Region, lecture, November 19, Deerfield Beach, Florida.

Siegel, Morris. 1939. *Population, Race, and Eugenics*. Ontario: Printed by Davic-Lisson.

Siegel-Itzkovich, Judy. 2001. "Israelis Agree on Terms for Embryonic Cell use," BMJ 2001; 323(7316): 771 (6 October), http://www.bmj.com/cgi/content/full/323/7316/771/b, accessed May 10, 2008.

Siegel-Itzkovich, Judy. 2009. "Sheba, NYU Researchers to Draw Genetic Map of Wandering Jew," *Jerusalem Post, July 20, http://www.jpost.com/servlet/Satellite?cid=1246443863406&pagename=J Post%2FJPArticle%2FPrinter.*

Siev, Izzy. 1978. *Social Eugenics*, Crucial Concepts, Ozone Park, NY.

Simon, Isidore. 1949. "La gynecologie, l'obstetrique, l'embryologie et la puericulture dans la Bible et le Talmud," *Revue d'histoire de la médecine hébraïque,* vol. 4, Sept.-Dec., 35-64.

Simon, Julian. 2001. *The Ultimate Resource*. Princeton, New Jersey: Princeton University Presses.

Simonson, Michael. 2006. *Guide to the Papers of William Nussbaum.* Leo Baeck Institute, +AR 10750/MF 740.

Simonstein, Frida F. 2004. "Germ-Line Engineering and Late-Onset Diseases: The Ethics of Self-Evolution," *Israel Medical Association Journal,* 6: November, 652-657.

Singer, Jefferson A. 2001. "Review of Kevin MacDonald's *Separation and Its Discontents*, Shofar 19(2), 166.

Singer, Sydney Ross. 2010. "Are Jews an Invasive Species?" *Jerusalem Post*, Jan. 14, 16.

Sinsheimer, Robert. 1990. "Whither the Genome Project?" *The Hastings Center Report*, 20(4), July-Aug., 5.

Smith, K.M. 1955. "Redcliffe Nathan Salaman, 1874-1955." *Biographical Memoirs of a Fellow of the Royal Society,* 239-245.

Snow, C.P. 1969. "Jewish Superiority Seen in Inbreeding," *Pittsburgh Post-Gazette,* pril 1, 26.

Snowman, J. 1913–1914. "Jewish Eugenics," *The Jewish Review,* vol. 4, 159-174.

Snyderman, Mark; Rothman, Stanley. 1986. "Science, Politics, and the IQ Controversy," *The Public Interest*, no. 83, Spring, 79-97.

The Society for the Amelioration of the Condition of the Jews. 1860. "The Condition of the Jews an Agency for the Spread of Christianity," *New York Times*, May 7, 8.

Solomon, Erwin S. 1956. "Social Characteristics and Fertility: A Study of Two Religious Groups [Catholic and Jewish] in Metropolitan New York." *Eugenics Quarterly,* 3(2), 100-103.

Sorsby, Arnold. 1952. "Eugenic Doctrines," *Jewish Chronicle*, June 6, 12.

Sparkes, Russell. 1999. "The Enemy of Eugenics" Chesterton Review, February-May, http://secondspring.co.uk/articles/sparkes.htm.

S.P.F. 1904. "Zangwill and Jewish Immigration," Oct. 23, 8.

Stadler, Friedrich. 1987. *Vertriebene Vernunft: Emigration und Exil Österreicher Wissenschaft*, Vienna: Jugend und Volk.

Stansky, Peter. 1996. "*Harold Laski: A Life on the Left* by Isaac Kramnick and Barry Sheerman," *Journal of Modern History*, 68(1), 184-186.

"State Chamber Assailed by Jews: Deutsch and Dr. S. S. Wise See Slur on Race in Eugenist's Immigration Report" 1937. *New York Times*, May 7, 1.

Stein, Ludwig 1905. "Die Rasse," *Die Zukunft*, Jan. 14, 85-92; Jan. 21, 131-143.

Steinberg, Avraham; Loike, John. 2004. "Human Cloning: Scientific, Ethical, and Jewish Perspectives," *Jewish Medical Ethics,* Dr. Falk Schlesinger Institute for Medical-Halachic Research, Book 1, 191-209; *Jewish Medical Ethics*, 3, 2, 11-19, http://daat.ac.il/daat///kitveyet/assia_english/steinberg.htm, accessed June 10, 2005.

Steinberg. 2005. "Israel's Stem Cell Pioneers," *Jewish Genetic Disorders*, Sept., 67.

Steiner, Hillel. 1995. "Persons of Lesser Value: Moral Argument and the 'Final Solution.'" *Journal of Applied Philosophy*, 12(2), 129-141.

Steinfels, Peter. 1989. "Auschwitz Revisionism: An Israeli Scholar's Case," *New York Times*, Nov. 12, Section 4, pg. 5.

Steinlight, Stephen. 2001. "The Jewish Stake in America's Changing Demography: Reconsidering a Misguided Immigration Policy," Center for Immigration Studies, Oct. http://www.cis.org/ChangingDemography-JewishInterestImmigrationPolicy, accessed July 5, 2010.

Steinlight, Stephen. 2010a. "Straight Talk about Jews and Immigration," Center for Immigration Studies, Feb. 2, http://www.cis.org/StraightTalk, accessed July 5, 2010.

Steinlight, Stephen. 2010b. "Jewish Establishment Censorship of Information on Immigration Policy, June 5, http://cis.org/steinlight/jewish-establishment-censorship, accessed July 5, 2010.

Stern, Andrew. 2007. "Hispanic women at risk of breast cancer gene," Reuters, Dec. 25.

Stern, Curt. 1949a. "Selection and Eugenics," *The Science Newsletter*, New Series, vol. 110, Aug. 26, 201-208.

Stern, Curt. 1949b. *Principles of Human Genetics*. San Francisco: W.H. Freeman.

Stern, Curt. 1957. "The Scope of Eugenics," *Proceedings of the National Academy of Sciences of the United States of America*, 43(8), Aug. 15, 744-749.

Stoler-Liss, Sachlav. 2003. "Mothers Birth the Nation": The Social Construction of Zionist Motherhood in Wartime in Israeli Parents' Manuals," *NASHIM: A Journal of Jewish Women's Studies and Gender Issues,* Project Muse, http://muse.jhu.edu, 110, 114.

Stone, Abraham. 1955. "Heredity Counseling: Eugenic Aspects of the Premarital Consultation," *Eugenics Quarterly*, March, 2(1), 51-52.

Stone, Naomi. 2000. "Erasing Tay-Sachs Disease," http://www.dartmouth.edu/cbbc/courses/bio4/bio4-2000/papers/NomiStone.html, accessed Dec. 25, 2007.

Tananbaum, Susan L. 2001. "Philanthropy and Identity: Gender and Ethnicity in London," *Journal of Social History,* Summer, 937-960.

Taylor-Allen, Ann. 1988. "German Radical Feminism and Eugenics, 1900-1908," *German Studies Review*, 17(1), Feb., 31-56.

Taylor-Allen, Ann. 2000. "Feminism and Eugenics in Germany and Britain, 1900-1940: A Comparative Perspective," *German Studies Review,* 23(3), Oct., 477-505.

Teehan, John. 2006. "The Evolutionary Basis of Religious Ethics," *Zygon*, 41(3), Sept., 747-773.

Tendler, Aron. 2001. "Netzavim – Passing Yichus," Rabbi's Notebook, http://www.Torah.org/learning/rabbis-notebook/5761/netzavim.html.

Tendler, Moshe David. 1988. *Pardes Rimonim: A Manual for the Jewish Family.* Hoboken, New Jersey: Ktav Publishing House.

Theilhaber, Felix A. 1911. *Der Untergang der deutschen Juden: Eine Volkswirtschaftliche Studie.* Munich: Ernst Reinhardt, Verlagsbuchhandlung.

Theilhaber, Felix. 1913a. *Das sterile Berlin: Eine volkwirtschaftcliche Studie,* Berlin.

Theilhaber, Felix. 1913b. "Zum Preisausschreiben: 'Bringt das materielle und soziale Aufsteigen den Familien Gefahren in rassenygienischer Beziehung? Dargelegt an der Entwicklung der Judenheit von Berlin,'" *Archiv für Rassen- und Gesellschafts-Biologie,* vol. 10, 67-92.

Tigay, Chanan. 2005. "Study on Ashkenazi Genes Sparks Intrigue, Debate – and Reflection," Jewish Telegraph Agency, June 7, posted by Steve Sailer, June 8, http://isteve.blogspot.com/2005/06/jewish-telegraph-agency-on-ashkenazi.html.

Tobin, Gary. 2008. "Stop Keeping out Non-Jews," The Jewish Telegraphic Agency (JTA, The Global News Service of the Jewish People), http://jta.org/news/article/2008/03/03/107292/tobinoped03022008, accessed June 11, 2009.

Tolts, Mark. 2003. "Mixed Marriage and Post-Soviet *Aliyah,"* www.brandeis.edu/hbi/pubs/ToltsTextFinal.doc

Traubmann, Tamara. 2004a. "Where Scientists Call the Shots: Advanced capabilities and lax regulation put Israel on the leading edge of cloning," *The Center for Public Integrity: Investigative Journalism in the Public Interest,* June 2, http://www.publicintegrity.org/genetics/report.aspx?aid=279 , accessed November 30, 2006.

Traubmann, Tamara. 2004b. "'Do not have children if they won't be healthy'," *Ha'aretz,* July 3, 5764, http://www.Ha'aretz.com/hasen/pages/ShArt.jhtml?itemNo=437879, accessed Jan. 18, 2008.

Traubmann, Tamara; Reznick, Dan. 2005. "A case of almost eugenics," *Ha'aretz,* Nov. 8, http://www.Ha'aretz.com/hasen/pages/ShArt.jhtml?itemNo=611189, accessed Jan. 18, 2008.

Treanor, Paul. Written 1997, revised Dec 2002. "Memory as ideology," http://web.inter.nl.net/users/Paul.Treanor/memory.tp.html.

Treanor, Paul. 2007?. "Why forget the 'Holocaust'?", September 15? http://web.inter.nl.net/users/Paul.Treanor/forget.html.

Troster, Lawrence. March 2002. "Cross-Generational Retribution and Genetic Engineering: Reflection on Chance and Free Will," *Conservative Judaism,* vol. 54, Issue 3, 33-41.

Trotsky, Lev. 1934. "If America Should Go Communist," *Trotsky Archive,* Aug. 17, http://www.marxists.org/archive/trotsky/1934/08/ame.htm, accessed April 8, 2008.

Turda, Marius; Weindling, Paul. 2007. *"Blood and Homeland"*: Eugenics and Racial Nationalism in Central and South-East Europe, 1900-1940. Budapest: Central Eastern European Press.

Turner, Terence. 1993. "Anthropology and Multiculturalism: What Is Anthropology that Multiculturalists Should Be Mindful of It?" *Cultural Anthropology,* 8(4), 411-429.

Tylor, Gus. 2007. "An Early Jewish-Black Marriage," *Forward*, July 26,
http://www.forward.com/blogs/tyler-too/11421/

Vergano, Dan. 2003. "Book Explores Eugenics' Origins," *USA Today*, posted Sept.
14, updated Sept. 16, http://www.usatoday.com/news/health/2003-09-14-
book-usat_x.htm, accessed May 14, 2008.

Vergasov, Fatekh. Undated. "Vladimir Pavlovich Èfroimson,"
http://www.pseudology.org/science/Ephroimson_VP.htm, accessed Dec.
28, 2007.

Vermel' (Vermelia?), S. W. 1923. "Prestupnost' evreev," report delivered to the Jewish Commission of the Russian Eugenics Society, Dec. 3.

Versweyveld., Leslie. 2002."World of genetically engineered people not so brave as
one might expect?" *Primeur: The monthly news service for the European
NPCN community*, June 20,
http://www.hoise.com/primeur/02/articles/live/LV-PL-06-02-1.html, accessed Dec. 24, 2007.

Vetter, Lara. 2007. "Theories of Spiritual Evolution, Christian Science, and the
'Cosmopolitan Jew': Mina Loy and American Identity," *Journal of Modern
Literature, Fall, 31(1), 47-63.*

Virchow, Rudolf. 1885. "Gesamtbericht über die Statistik der Farbe der Augen, der
Haare und der Haut der Schulkinder in Deutschland," *CBDAG* 16, 89-100.

Vogt, Annettte B. Undated. "Ursula Phillip," *Jewish Women's Archive*,
http://jwa.org/encyclopedia/article/philip-ursula.

Vogt, Carl. 1864. *Lectures on man: his place in creation, and in the history of the
earth.* London: Longman, Green, Longman, and Roberts.

Von Hartmann, Eduard. 1885. *Das Judenthum in Gegenwart und Zukunft.* Leipzig:
W. Friedrich.

Vronskaya, Jeanne; Chuguev, Vladimir. 1994. *Kto est' kto v Rossii i byvshem SSSR*,
Moscow: Terra.

Vsesojuznaia Akademiia Sel'skokhozjaistvennykh Nauk. 1953. "O polozhenii v
biologicheskoi nauke," conference transcript, Moscow, July 31-Aug. 7.

Vvedenskij, B. A. (editor). 1953. *Bol'shaia sovetskaia èntsiklopediia*, second edition,
vol. 15, Moscow, 372-373.

Wade, Nicholas. 2008. "Gene study shows Spain's Jewish and Muslim ancestry,"
*New York Times*, Dec. 5, A12.

Wahrman, Miryam Z. 1998. "Orthodox Scientists Pondering Genetics," *Forward*,
vol. CII, Issue 31,199; 25. August, accessed June 3, 2005.

Wahrman, Miryam Z. 2002. *Brave New Judaism: When Science and Scripture Col-
lide.* Hanover/London: Brandeis University Press.

Waldman, Amy. 1997. "For Childless Orthodox Jews, Fertility Treatment Is No Simple Solution," *New York Times*, August 10,
http://www.nytimes.com/1997/08/10/nyregion/for-childless-orthodox-jews-
fertility-treatment-is-no-simple-solution.html.

Waldman, Louis. 1998. "Jewish Gene Studies Pose No Threat," *New York Times*,
April 26, WK14.

Ward, Tom. Undated. "Conclusion," Diocese of Covington,
http://home.catholicweb.com/covingtonmessenger/index.cfm/NewsItem?ID
=202611&From=Home, accessed Dec. 21, 2007.

*Washington Post* (editorial). 2010. "The Quarrel with Israel," Mar. 16, A18.

Watson, James D. 1997. "Genes and Politics," *Journal of Molecular Medicine: Official Organ of the Gesellschaft Deutscher Naturforscher und Ärzte,* 75(9), 624-636.

Wattenberg, Ben. 2000. "The First Measured Century," *Washington Times,* Dec. 7, A16.

Wattenberg, Ben. 2003. "Why Control the Border?" *National Review,* Feb. 1, 45(2), sourced from *VDARE,* http://www.vdare.com/pb/control_borders.htm.

Wattenberg, Ben J.; Kadden, Jeremy. 2005. "Jewish Babies," *American Enterprise Institute for Public Policy Research,* http://www.aei.org/publications/filter.economic, accessed January, 1, 2006.

Weber, M. M. 1991. "A research institute for psychiatry..." The development of the German Research Institution for Psychiatry in Munich between 1917 and 1945," *Sudhoffs Arch Z Wissenschaftsgeschichte,* 75: 74-89.

Weikart, Richard. 2002. "Darwinism and Death: Devaluing Human Life in Germany 1859-1920," *Journal of the History of Ideas,* 63(2), 323-344.

Weikart, Richard. 2004. *From Darwin to Hitler: Evolutionary Ethics, Eugenics, and Racism in Germany.* New York: Palgrave MacMillan.

Weikart, Richard. 2008. "Was It Immoral for 'Expelled' to Connect Darwinism and Nazi Racism?" Discovery Institute, http://www.discovery.org/a/5069, accessed May 11, 2008.

Weingart, Peter. 1989. "German Eugenics between Science and Politics," *Osiris,* second series, Vol. 5, Science in Germany: The Intersection of Institutional and Intellectual Issues, 260-282.

Weingart, Peter. 2005. "Science and Political Culture: Eugenics in Comparative Perspective,"*The Scandinavian Journal of History,* 24(2), 163-177.

Weingart, Peter; Kroll, Jürgen; Bayertz, Kurt. 1992. *Rasse, Blut und Gene: Geschichte der Eugenik und Rassenhygiene in Deutschland.* Frankfurt am Main: Suhrkamp.

Weindling, Paul. 1989. *Health, Race and German Politics between National Unification and Nazism 1870-1945.* Cambridge/NewYork: Cambridge University Press.

Weininger, Otto. 1920. *Geschlecht und Character,* Wien/Leipzig William Braumüller, http://www.k-faktor.com/files/geschlecht-und-charakter.pdf, accessed Feb, 24, 2008.

Weiss, David W. 1989. "Science and Values," *Tradition,* Winter, 24(2), 150-160.

Weiss, Meira. 2002. *The Chosen Body: The Politics of the Body in Israel Society.* Standord, California: Stanford University Press.

Weiss, Sheila Faith. 1986. "William Schallmayer and the Logic of German Eugenics," *Isis,* 77(1), March, 33-46.

Weiss, Sheila Faith. 1987. "The Race Hygiene Movement in Germany," *Osiris,* Second Series, vol. 3, 193-236.

Weiss, Sheila Faith. 2005. "Essay Review: Racial Science and Genetics at the Kaiser Wilhelm Society," *Journal of the History of Biology,* 38(2), June, 367-369.

Weiss, Sheila Faith. 2006. "Human Genetics and Politics as Mutually Beneficial Resources: The Case of the Kaiser Wilhelm Institute for Anthropology, Human Heredity and Eugenics during the Third Reich," *Journal of the History of Biology,* vol. 39, 41-88.

Weiss, Volkmar. 1995. "The Advent of a Molecular Genetics of General Intelligence," Editorial, *Intelligence,* vol. 20, 115-124, http://www.volmar-weiss.de/index.html.

Weissenberg, S. 1905. "Das jüdische Rassenproblem," in *Z.D.S.J.*. vol. 1 (1905); M. Fishberg, *Beitrage zur phys. Anthropologie der nordafrikanischen Juden,* ditto; cited in http://forum.stirpes.net/60600-post2.html.

Wen, Patricia. 2000. "Jews Fear Stigma of Genetic Studies," *Boston Globe*, Aug. 15, F1.

West, William Lemore.1971. "The Moses Harman Story," *Kansas Collection: Kansas Historical Quarterlies,* Spring, 37(1), 41-63, http://www.kancoll.org/khq/1971/71_1_west.htm, accessed Dec.m26, 2008.

Weyl, Nathaniel. 1976. "Disease as a Eugenic Force," *Mankind Quarterly*, 16(4), April-June, 243-257.

Weyl, Nathaniel. 1989. *The Geography of American Achievement*, Washington, D.C: Scott-Townsend Publishers.

Whitisle, Harry. 1931. "Primitive Eugenics," Cold Spring Harbor Laboratory Archives, Image 1613, http://www.eugenicsarchive.org/html/eugenics/index2.html?tag=1613.

Whitney, Leon; Grossman, William. 1930. "Some Reasons for Jewish Excellence," *Eugenics: A Journal of Race Betterment,* 52-57.

Wilson, David Sloan; Wilson, Edward O. 2007. "Rethinking the Theoretical Foundation of Sociobiology," *The Quarterly Review of Biology*, Dec., 82(4), 327-348.

Wilson, E. O. 1995. "Science and Ideology," *Academic Questions*, No. 8, June 1, taken from Web site "Stalking the Wild Taboo," http://www.lrainc.com/swtaboo/taboos/wilson01.html, accessed July 18, 2008.

Wind, Rebecca. 2006. "A Tale of Two Americas for Women: Low-Income Women's Unplanned Pregnancy and Abortion Rates Are Increasing as Better-Off Women Continue Three Decades of Progress," Guttmacher Institute Media Center, May 4, http://www.guttmacher.org/media/nr/2006/05/05/index.html, accessed July 12, 2008.

Winkler, Daniel. 1998. "Eugenic Values," *Science in Context*, 11, vol. 3-4, 455-470.

Witte, Griff. 2008. "In Israel, A Clash Over Who Is a Jew: Ultra-Orthodox Conversions," *Washington Post*, Aug. 30, A1, A16-17.

Wolpe, Paul Root. 1997. "If I Am Only My Genes, What Am I? Genetic Essentialism and a Jewish Response," *Kennedy Institute of Ethics Journal*, 7.3, 213-230.

Wolpe, Paul Root. 2002. "Bioethics, the Genome, and the Jewish Body," *Conservative Judaism*, vol. 54(3S), 14-25.

Worldwide Religious News, 2002. "Pope, Religious Leaders Condemn Cloning Claim," Dec. 29, http://www.wwrn.org/article.php?idd=14751&sec=24&cont=5.

Wortis, Joseph. 1984. *Fragments of an Analysis with Freud*. New York: Simon and Schuster.

Wright, Lawrence. 1997. *Twins and What They Tell Us about Who We Are*. New York: John Wiley and Sons.

Wyman, David S. 1984. *The Abandonment of the Jews: American and the Holocaust*, 1941-1945. The New Press: New York/London.

Wyman, David S. and Medoff, Rafael. 2002. *A Race Against Death: Peter Bergson, America, and the Holocaust*. The New Press: New York.

Y.M.C.A. 1911. "Eugenics." Special circular offered to Evening Classes for Men at West Side Y.M.C.A., ad no. 14, *New York Times,* Sept. 16, 13.

SkyNews. 2009. "Eugenics-Cloned Humans on the Way," http://www.youtube.com/watch?v=Lu30yGvfEPY.

Zangwill. Israel. 1910. *Italian Fantasies.* New York: Macmillan.

Zangwill, Israel. 1912? *The Problem of the Jewish Race.* New York: Judaen Publishing Co.

Zavos, Panayiotis; Glad, John. 2009a. Telephone conversation, Lexington, Kentucky / Washington, D.C., Aug. 3.

Zavos, Panayiotis; Glad, John. 2009b. Telephone conversation, Lexington, Kentucky / Washington, D.C., Sept. 14.

Zezima, Katie. 2008. "More Women Than Ever Are Childless, Census Finds," *New York Times,* Aug. 19, A13.

Zhuravsky, D. 1993. "Terror," *Voprosy filosofii,* no. 7, 125-146, http://www.ihst.ru/projects/sohist/papers/jor93ph.htm, accessed Dec. 23, 2007.

Zimmer, Carl. 2008. "Engineering by Scientists on Embryo Stirs Criticism," *New York Times,* May 13, 2008, A14.

Zlotogora, Joël; van Baal, Sjozef; Patrinos, George P. 2007. "Documentation of inherited disorders and mutation frequencies in the different religious communities in Israel in the Israeli National Genetic Database," *Human Gene Mutation Database,* May 10.

Zohar, Noam J. 1991. "Prospects for 'Genetic Therapy' - Can a Person Benefit from Being Altered?" *Bioethics,* 5(4), 275-288.

Zohar, Noam J. 1997. *Alternatives in Jewish Ethics.* Albany: State University of New York Press.

Zohar, Noam J. 1998. "From Lineage to Sexual Mores: Examining 'Jewish Eugenics'," *Science in Context,* 11(3-4), 575-585.

Zohar, Noam J. (ed.). 2006. *Quality of Life in Jewish Bioethics,* Baltimore, Maryland: Lexington Books.

Zollmann, Paul. 2010. "Wrong Lesson on Euthanasia," *Gazette,* Mar. 5, A16.

Zoloth, Laurie. 2008. "Author Squares Jewish and Medical Ethics," *Ethics,* Aug. 21, http://www.forward.com/articles/14038.

Zollschan, Ignaz. 1920. *Das Rassenproblem unter besonderer Berücksichtigung der theoretischen Grundlager der jüdischen Rassenfrage.* Fourth edition, unaltered. W. Braumüller: Vienna.

Zweig, Stefan (preface). 1903. *E. M. Lilien: Sein Werk.* Berlin/Leipzig: Schuster & Loeffler.

Zweig, Stefan. 1964. *The World of Yesterday: An Autobiography.* Lincoln, Nebrasca: University of Nebrasca Press.

# Subject Index

*Micro-Chronology indexed according to year and item #, not page*

**Diabetes:** 2000/10

**Diagnosis:** 1911/8, 1935/10, 1946/6, 1982/3, 1983/4, 1992/4, 1993/5, 1994/3, 1994/10, 1996/5, 2003/9, 2005/8, 2006/22, 2007/16, 2008/10

**Diaspora** (see also individual countries): 12, 19, 28, 31, 33, 44, 46, 1960/2, 1960/3, 1964/3, 1970-2005/1, 1994/9, 2010/5

**Disability:** 81, 98, 99, 1993/5, 1999/6, 2000/16, 2004/14, 2005/8, 2005/10, 2008/10

**Dissidents, Jewish:** 91, 1903/2, 1928/6, 1967/3, 2000/6

**DNA (deoxyribonucleic acid):** 101, 102, 109, 2001/5, 2006/12, 2006/13, 2007/1

**Dor Yeshorim:** 64, 1973/2, 1985/3, 1995/1, 2000/9. 2003/11, 2008/19

**Down syndrome:** 1994/7, 2008/10

**Dysgenic phenomena** (not an exhaustive list; see also **Eugenics, negative**): 15, 16, 18, 19, 73, 87, 1904/6, 1976/4

**Egalitarianism / Environmentalism / Cultural relativism / Boasianisim:** 13, 19, 21, 26, 29, 47, 49, 50-51, 54-55, 59-61, 77, 78, 86, 89, 91, 106, 108, 109, 382, 1923/4, 1925/8, 1945/5, 1948/11, 1950/9, 1955/6, 1955/6, 1962/5, 1964/8, 1965/7, 1968/6, 1969/4, 1974/3, 1974/4, 1974/5, 1975/1, 1977/8, 1987/1, 1994/6, 1998/6, 2006/14, 2008/15

**Egg (ovum) donation:** 1998/3, 1998/16, 2001/8, 2007

**Egyptians:** 30, 43, 44, 46, 53, 62, 1908/4, 1909/4, 1999/1, 2009/17, 2010/2

**Embryo research** (see also **Cloning, Embryo transfer**, and **Embryo wastage**): 102, 1991/4, 1993/6, 1996/1, 1996/2, 1997/4, 1997/5, 1998/12, 1998/16, 2001/10, 2001/11, 2001/12, 2002/11, 2003/14, 2003/15, 2004/7, 2007/10, 2007/16, 2008/13, 2010/5, 2010/6

**Embryo wastage:** 2007/16

**Endogamy:** 30, 43, 44, 46, 382, 1902/3, 1905/5, 1906/1, 1911/7, 1911/11, 1874/1, 1881/2, 1884/2, 1892/2, 1892/3, 1904/2, 1904/5, 1905/5, 1906/2, 1910/6, 1911/7, 1912/5, 1921/6, 1924/3, 1924/4, 1929/1, 1930/10, 1930s/1, 1930/6, 1934/14, 1935/2, 1935/9, 1936/6, 1938/13, 1951/5, 1951/4, 1955/6, 1962/1, 1962/2, 1976/5, 1978/1, 1980/2, 1998/7, 1998/9, 1998/26, 1999/12, 2001/13, 2003/6, 2009/7

**England** (see also **Eugenics Society, Britain**, not an exhaustive list): 7, 8, 21, 23, 42, 45, 46, 55, 56, 57, 61, 70, 85, 102, 1861/1, 1875/1, 1885/1, 1885/2, 1891/1, 1891/2, 1892/2, 1902/4, 1905/7, 1906/4, 1907/4, 1909/3, 1910/4, 1910/6, 1919/9, 1910/10, 1011/3, 1911/7, 1911/10, 1911/6, 1912/1, 1912/3, 1912/8, 1912/11, 1913/4, 1913/6, 1913/7, 1914/3, 1915/4, 1915/5, 1916/6, 1917/3, 1921/6, 1922/7, 1922/8, 1923/3, 1923/7, 1925/3, 1925/6, 1926/3, 1930s/1, 1933/4, 1934/2, 1934/6, 1934/16, 1935/9, 1939/3, 1939/8, 1944/4, 1945-1947/1, 1946/5, 1948/10, 1949/7, 1951/8, 1951/9, 1957/6, 1960/2, 1963/4,

**Hybridity** (see also **Exogamy** and **Intermarriage**): 15, 1881/2, 1884/2, 1935/2, 2006/3

**Incest:** 30, 1995/6, 1998/4

**India:** 42, 44, 105, 111, 1891/1, 1922/2, 1954/2, 1962/1, 1964/4, 1966/2, 1980/2, 1994/4, 1997/1, 2000/3, 2005/2, 2005/3, 2007/8

**Individualized eugenics** (see also **Self-Evolution**): 18, 93, 1998/22, 2002/21, 2004/8

**Infanticide** (see also **Abortion**): 96-97, 104, 1939/8, 1965/6, 1967/6, 1968/4, 1968/5, 1975/5, 2007/10, 2009/18, 2009/19

**Infertility:** 30, 104, 1906/2, 1928/3, 1991/5, 2002/13, 2002/18, 2006/7, 2008/13

**Infiltration** (see also **Exogamy** and **Khazars**): 29-36, 1922/2, 1947/3, 1970/1, 1976/3, 2002/9, 2003/10, 2006/13, 2008/2, 2008/9, 2009/3, 2009/15

**Inquisition:** 2009/15

**Intelligence** (see also **I.Q.** and **Genius**): 11, 14, 15, 16, 20, 21, 24, 25, 29, 42, 46, 48, 74, 80, 85, 86, 87, 93, 102, 105, 110, 1884/2, 1886/2, 1896/4, 1894/1, 1905/4, 1908/4, 1910/6, 1917/3, 1917/4, 1923/4, 1925/8, 1926/3, 1931/5, 1933/7, 1961/3, 1964/6, 1967/4, 1969/2, 1975/1, 1977/2, 1994/6, 1994/13, 1998/13, 2000/8, 2002/12, 2005/4, 2005/9, 2006/10

**Intergenerational equity** (see also **Parental obligations**): 14, 15, 17, 48, 53, 89, 94, 106, 1930s/9, 1965/5, 1995/2, 1995/7, 1998/14, 1998/21, 2002/19, 2004/4, 2004/8

**Intermarriage:** see **Exogamy**

**Invasion biology:** 2010/5

**In vitro fertilization:** 15, 104, 1990/2, 1996/1, 1997/4, 2001/8, 2001/11, 2002/18, 2002/13, 2003/14, 2003/16

**IQ** (see also **Intelligence**): 13, 20, 29, 76, 84, 85, 93, 103, 1956/3, 1974/3, 1975/1, 1994/6, 203/5, 2005/9, 2006/6, 2007/2, 2007/3, 2008/4, 2010/9

**Iranian Jews:** 2008/2

**Iraqis:** 2010/11

**Islam:** 16, 18, 97, 102, 1990/7, 1992/1, 1999/4, 2009/17, 2010/2

**Isolation, genetic:** 15, 25, 42, 74, 1885/1, 1891/1, 1962/2, 1998/9, 2000/7, 2009/3

**Israel/Palestine** (not an exhaustive list): 10, 13, 17, 29, 30, 34, 35, 37, 39, 43, 44, 46, 54-58, 63, 65, 70, 72, 75, 76, 79, 80, 92, 93, 97, 98, 99, 100-105, 106, 109, 110, 111, 1899/3, 1902/3, 1908/2, 1908/3, 1917/2, 1919/4, 1920s/1, 1920-1960/1, 1921/1, 1921/3, 1922/5, 1922/6, 1923/7, 1926/6, 1927/5, 1930s/4, 1930/6, 1933/7, 1933/16, 1934/9, 1934/10, 1934/11, 1934/13, 1934/14, 1934/15. 1934/17, 1934/18, 1935/11, 1936/6, 1938/1, 1938/7, 1938/14, 1940/4, 1942/3, 1943/2, 1944/4, 1944/5, 1944/7, 1945/8, 1946/3, 1947/4, 1948/5, 1948/9, 1949/3, 1949/6, 1950s-Early 1960s/1, 1950/2, 1950/4,

1950/5, 1951/2, 1951/3, 1951/6, 1952/1, 1951/2, 1951/4, 1951/6, 1952/1, 1952/2, 1954/2, 1958/1, 1960/2, 1960/3, 1962/1, 1963/2, 1964/4, 1966/1, 1967/3, 1968/2, 1968/7, 1970-2005/1, 1973/2, 1975/6, 1975/8, 1977/3, 1977/5, 1979/2, 1979/5, 1980/2, 1983/2, 1984/1, 1985/4, 1986/2, 1989/2, 1990/2, 1990/3, 1991/1, 1991/2, 1991/3, 1991/4, 1991/5, 1991/6, 1992/1, 1992/2, 1993/1, 1993/5, 1994/1, 1994/2, 1994/4, 1994/10, 1996/1, 1997/1, 1997/4, 1997/5, 1997/7, 1997/8, 1998/1, 1998/3, 1998/9, 1998/12, 1998/15, 1998/17, 1998/18, 1998/19, 1998/22, 1999/1, 1999/2, 1999/4, 1999/10, 2000/3, 2000/7, 2000/11, 2001/2, 2001/3, 2001/4, 2001/10, 2001/11, 2003/19, 2003/21, 2004/4, 2004/6, 2004/7, 2004/8, 2004/9, 2004/12, 2005/2, 2005/3, 2005/6, 2005/8, 2006/1, 2006/3, 2006/6, 2006/9, 2006/10, 2006/13, 2006/16, 2006/17, 2007/3, 2007/8, 2007/9, 2007/12, 2007/16, 2007/17, 2008/1, 2008/3, 2008/11, 2008/18, 2009/3, 2009/6, 2009/10, 2009/15, 2009/16, 2009/17, 2009/20, 2010/1, 2010/2, 2010/7

**Israeli National Committee of Science:** 2001/1

**Italians/Romans:** 43, 2001/11

**Jewish Central Committee (Germany, Zentralrat der Juden):** 2005/5

**Jewish studies:** 35, 54-55, 91, 92, 111, 382, 1946/5, 1992/1, 1999/4, 2000/13, 2001/6, 2003/3, 2003/12, 2007/7, 2007/10, 2008/16

**Jewishness, definition of:** 8, 9, 30, 54, 73, 108, 39-46, 382, 1880/1, 1881/2, 1928/1, 1970/1, 1983/3, 1998/5, 2002/5, 2003/3, 2003/8, 2003/10, 2003/19, 2004/15, 2008/1, 2008/11, 2009/3, 2010/11

**Judaism** (not an exhaustive list; see also **Theology**): 8, 10, 12, 16, 18, 30, 40-41, 43, 45, 46, 52-53, 57, 63, 64, 93, 96- 99, 97, 99, 104, 105, 108, 1846/1, 1865/1, 1870/1, 1873/1, 1879/2, 1880/1, 1884/2, 1885/1, 1887/1, 1893/1, 1896/3, 1905/6, 1906/2, 1907/3, 1907/4, 1911/3, 1911/5, 1911/6, 1912/4, 1912/9, 1912/1, 1913/1, 1913/5, 1914/3, 1914/4, 1917/2, 1917/4, 1917/6, 1917/8, 1917/9, 1918/1, 1919/3, 1921/6, 1922/5, 1925/7, 1926/4, 1927/4, 1927/7, 1929/1, 1929/4, 1929/5, 1930/1, 1930/2, 1930/4, 1930/7, 1930/8, 1930/10, 1930/11, 1931/5, 1931/6, 1933/13, 1934/4, 1935/6, 1935/7, 1937/4, 1938/7, 1938/9, 1939/6, 1939/8, 1941/6, 1947/3, 1949/4, 1954/2, 1956/2, 1962/1, 1964/4, 1965/6, 1966/2, 1967/6, 1968/3, 1968/4, 1968/5, 1969/5, 1965/6, 1970/1, 1973/2, 1975/4, 1975/6, 1975/7, 1995/6, 1977/6, 1978/3, 1979/4, 1981/4, 1982/2, 1982/4, 1983/3, 1983/4, 1985/3, 1985/6, 1988/2, 1989/5, 1990/2, 1990/5, 1990/8, 1991/3, 1993/4, 1995/6, 1996/1, 1996/3, 1996/4, 1997/1, 1997/3, 1997/6, 1998/3, 1998/4, 1998/13, 1998/16, 1998/22, 1998/23, 1998/24, 1999/4, 1999/10, 2000/5, 2000/12, 2000/13, 2000/14, 2001/5, 2001/14, 2002/8, 2002/11, 2002/15, 2002/18, 2002/19, 2002/20, 2002/21, 2003/18, 2003/20, 2004/5, 2005/2, 2005/3, 2006/4, 2006/5, 2007/4, 2008/1, 2008/2, 2008/11, 2009/9, 2009/12, 2009/14, 2009/15, 2009/18, 2009/19

**Kant/Kantianism:** 89, 2000/11

**Khazars:** 36, 43, 110, 1906/2, 1936/7, 1976/3, 2003/10, 2008/14, 2009/15

**Kin selection:** 52, 2006/9

**Knesset:** 102, 1949/3, 1950/5, 1951/3, 1970/1, 1985/1, 1986/6, 1999/10, 2001/15, 2003/2, 2004/4, 2009/1

**Kohanim (Cohanim):** 111, 1906/2, 1998/1, 2000/1, 2003/10

**Lamarck/Lamarckianism** (see also **Lysenko**): 6, 47, 58, 59-61, 108, 382, 1896/5, 1904/5, 1907/5, 1908/2, 1911/9, 1911/15, 1919/2, 1923/2, 1926/8, 1927/8, 1929/3, 1929/6, 1932/1, 1934/15, 1936/1, 1936/9, 1936/11, 1936/12, 1937/2, 1938/19, 1939/5, 1944/1, 1947/1, 1953/3, 1955/3, 1958/1

**Law, Jewish,** see **Judaism** and **Halakha**

**Leakage:** 2000/7

**Left, political:** 3, 17, 19, 46, 50, 77, 60, 74-92, 100, 106, 108, 1862/1, 1900/2, 1900-1930/2, 1907/5, 1908/5, 1914/1, 1916/13, 1923/4, 1923/6, 1926/5, 1928/2, 1928/4, 1929/7, 1900-1930/4, 1934/1, 1936/13, 1948/9, 1968/2, 1969/4, 1974/4, 1979/1, 1980/4, 2005/10, 2005/11

**Legislation, Israeli** (see also **Knesset**): 35, 99, 1950/5, 1951/6, 1952/5, 1992/2, 1994/4, 1996/1, 1996/6, 1997/5, 1999/10, 2001/4, 2003/2, 2005/3, 2006/13, 2007/16

**Lemba:** 1999/4, 2005/2

**Levites:** 2003/10

**Leprosy:** 1930/11, 1978/3, 1998/21

**Libya:** 6

**Liberal eugenics:** 2009/20

**Manic-depressive syndrome,** see **Bipolar disorder**

**Matchmaking:** 30, 1912/9, 1917/3

**Maternity leave,** see **Pronatalist policies**

**Matrilineal descent:** 40-41, 2006/7

**Media:** 10, 21, 38, 62, 71, 79, 80-82, 90, 94, 102, 106, 110, 1973/2, 1985/6, 1992/3, 1994/8, 1995/7, 2001/10, 2003/4, 2004/10, 2005/11, 2009/7

**Mental illness, among Jews:** see **Psychiatry**

**Mesopotamia:** 6, 42, 46,

**Migration:** 16, 19, 30, 31, 39, 45, 58, 73, 74, 77, 108, 109, 110, 111, 180/2, 1882/1, 1902/4, 1910/3, 1917/1, 1917/3, 1920/2, 1920/3, 1923/3, 1924/1, 1925/1, 1925/3, 1926/6, 1926/9, 1933/8, 1934/7, 1936/6, 1937/4, 1939/14, 1941/5, 1942/4, 1943/3, 1944/6, 1946/3, 1949/6, 1950/4, 1950/5, 1951/3, 1951/6, 1962/1, 1964/4, 1979/2, 1994/4, 1995/7, 1997/1, 2000/8, 2000/15, 2001/3,

2002/1, 2003/7, 2004/6, 2005/3, 2005/5, 2005/6, 2006/15, 2007/1, 2009/10, 2009/12

**Mishnah** (or **Mishna**, the first major written redaction of the Jewish oral traditions called the 'Oral Torah') 1995/6, 1996/3

**Mitochondrial DNA (mtDNA):** 2007/1

**Mixed offspring:** 9, 31-34, 1881/2, 1943/4, 1954/2, 2007/16, 2010/9

**Mizrachi (Mizrahi, Misrahi):** 1952/1, 2001/14, 2007/1, 2008/3, 2009/6

**Monogamy:** 18, 107, 1946/7

**Moratorium on germ line genetic manipulation in humans:** 2004/2

**Moratoriums on cloning:** 103, 1999/10, 2004/4, 20009/1

**Mosaic Law:** 1934/5, 2006/9

**Mucolipdosis:** 1974/2, 1985/3, 2007/13

**Multi-method studies:** 2006/10

**Multi-trait studies:** 2006/10

**Muscular (Muscle) Jew:** 1898/1, 1900/4

**Mutations, genetic:** 39, 87, 98, 1927/2, 1948/6, 1998/24, 1998/26

**Mysticism:** 1930s/1, 2007/10

**Nanomedicine:** 2000/13

**National Bioethics Advisory Committee on Cloning:** 2001/5

**National Jewish Population Survey:** 20, 40, 1982/2, 2002/5, 2007/6

**National Socialism:** 21, 44, 67, 69, 71, 79, 94, 103, 108, 1931/3, 1933/20, 1934/5, 1934/8, 1934/10, 1934/12, 1934/14, 1934/17, 1935/1, 1935/10, 1936/5, 1936/9, 1937/4, 1938/6, 1938/13, 1940/1, 1940/4, 1940/8, 1943/1, 1944/6, 1965/8, 1976/5, 1980/4, 1983/1, 1986/3, 1989/3, 1989/6, 1992/3, 1993/3, 1998/2, 1998/7, 1998/10, 2001/5, 2001/10, 2002/22, 2003/5, 2004/10, 2004/12, 2004/13, 2004/14, 2004/15, 2004/16, 2005/7, 2006/1, 2006/21, 2007/14, 2008/5, 2008/21, 2009/7, 2010/7

**Nature/Nurture:** 52, 59, 1861/1, 1865/1, 1902/1, 1905/5, 1930/5, 1998/13, 1999/12

**Neandertals:** 25, 1974/4

**New Freedom Initiative:** 2005/11

**Niemann-Pick disease:** 1985/3, 2007/13

**Noachidic Laws:** 1966/2

**Nordic:** 68, 73, 1899/2, 1909/1, 1923/1, 1924/1, 1925/8, 1927/1, 1928/6, 1933/5, 1923/21, 1946/7

# Names Index

*Micro-Chronology indexed according to year and item #, not page*

# Learn More about Eugenics

*I am with you, you men and women of a generation,*
*or ever so many generations hence.*

Walt Whitman
"Crossing Brooklyn Ferry"

1. John Glad's *Future Human Evolution* is a brief, basic primer on modern eugenics. It may be downloaded free of charge in thirteen languages at www.whatwemaybe.org, with still other languages in preparation.

2. If you are a native speaker of a language other than English and wish to volunteer to translate either *Future Human Evolution* or *Jewish Eugenics* into your native tongue, please contact John Glad at WoodenShore@gmail.com or jglad@umd.edu. If his e-mail address changes, the new address may be learned from the website www.whatwemaybe.org.

3. Assign these books to your students if you are a teacher dealing with any of the following areas: academic freedom, anthropology, bioethics, biology, biopolitics, cloning, crime, demographics, ecology, egalitarianism, environmentalism, ethics, eugenics, euthanasia, evolution, fertility, futurology, intergenerational equity, genetics, history, the holocaust, human rights, migration / emigration / immigration, philosophy, political science, population studies, religion, sociobiology, sociology, testing, welfare.

4. Write a book review.

5. Ask your librarian to order both books.

6. Tell your friends.

7. Support Wooden Shore financially.